T0335606

AN INTRODUCTION TO MACROSCOPIC QUANTUM PHENOMENA AND QUANTUM DISSIPATION

Reviewing macroscopic quantum phenomena and quantum dissipation, from the phenomenology of magnetism and superconductivity to the presentation of alternative models for quantum dissipation, this book develops the basic material necessary to understand the quantum dynamics of macroscopic variables.

Macroscopic quantum phenomena are presented through several examples in magnetism and superconductivity, developed from general phenomenological approaches to each area. Dissipation naturally plays an important role in these phenomena, and therefore semi-empirical models for quantum dissipation are introduced and applied to the study of a few important quantum mechanical effects. The book also discusses the relevance of macroscopic quantum phenomena to the control of meso- or nanoscopic devices, particularly those with potential applications in quantum computation or quantum information. It is ideal for graduate students and researchers.

A. O. CALDEIRA is a Professor at the Instituto de Física Gleb Wataghin, Universidade Estadual de Campinas (UNICAMP), Brazil. His main research interests are in condensed matter systems at low temperatures, in particular, quantum statistical dynamics of non-isolated systems and strongly correlated systems in low dimensionality.

AN INTRODUCTION TO MACROSCOPIC QUANTUM PHENOMENA AND QUANTUM DISSIPATION

AMIR O. CALDEIRA

Instituto de Física Gleb Wataghin
Universidade Estadual de Campinas

CAMBRIDGE
UNIVERSITY PRESS

CAMBRIDGE
UNIVERSITY PRESS

University Printing House, Cambridge CB2 8BS, United Kingdom

One Liberty Plaza, 20th Floor, New York, NY 10006, USA

477 Williamstown Road, Port Melbourne, VIC 3207, Australia

314-321, 3rd Floor, Plot 3, Splendor Forum, Jasola District Centre, New Delhi - 110025, India

79 Anson Road, #06-04/06, Singapore 079906

Cambridge University Press is part of the University of Cambridge.

It furthers the University's mission by disseminating knowledge in the pursuit of
education, learning and research at the highest international levels of excellence.

www.cambridge.org
Information on this title: www.cambridge.org/9780521113755

First published 2014

A catalogue record for this publication is available from the British Library

ISBN 978-0-521-11375-5 Hardback

to Sueli, Rodrigo, and Natalia

Contents

Preface

On deciding to write this book, I had two main worries: firstly, what audience it would reach and secondly, to avoid as far as possible overlaps with other excellent texts already existing in the literature.

Regarding the first issue I have noticed, when discussing with colleagues, supervising students, or teaching courses on the subject, that there is a gap between the standard knowledge on the conventional areas of physics and the way macroscopic quantum phenomena and quantum dissipation are presented to the reader. Usually, they are introduced through phenomenological equations of motion for the appropriate dynamical variables involved in the problem which, if we neglect dissipative effects, are quantized by canonical methods. The resulting physics is then interpreted by borrowing concepts of the basic areas involved in the problem – which are not necessarily familiar to a general readership – and adapted to the particular situation being dealt with. The so-called macroscopic quantum effects arise when the dynamical variable of interest, which is to be treated as a genuine quantum variable, refers to the collective behavior of an enormous number of microscopic (atomic or molecular) constituents. Therefore, if we want it to be appreciated even by more experienced researchers, some general background on the basic physics involved in the problem must be provided.

In order to fill this gap, I decided to start the presentation of the book by introducing some very general background on subjects which are emblematic of macroscopic quantum phenomena: magnetism and superconductivity. Although I wanted to avoid the presentation of the basic phenomenological equations of motion as a starting point to treat the problem, I did not want to waste time developing long sections on the microsocopic theory of those areas since it would inevitably offset the attention of the reader. Therefore, my choice was to develop this required basic knowledge through the phenomenological theories of magnetism and superconductivity already accessible to a senior undergraduate student. Microscopic details have been avoided most of the time, and only in a few situations are these concepts (for example, exchange interaction or Cooper pairing) employed in order to ease the understanding of the introduction of some phenomenological terms when necessary. In so doing, I hope to have given the general reader the tools required to perceive the physical reasoning behind the so-called macroscopic quantum phenomena.

Once this has been done, the next goal is the treatment of these quantum mechanical systems (or the effects which appear therein) in the presence of dissipation. Here too, a semi-empirical method is used through the adoption of the now quite popular "system-plus-reservoir" approach, where the reservoir is composed of a set of non-interacting oscillators distributed in accordance with a given spectral function. Once again this is done with the aim of avoiding encumbering the study of dissipation through sophisticated many-body methods applied to a specific situation for which we may know (at least in principle) the basic interactions between the variable of interest and the remaining degrees of freedom considered as the environment. The quantization of the composite system is now possible, and the harmonic variables of the environment can be properly traced out of the full dynamics, resulting in the time evolution of the reduced density operator of the system of interest only.

At this point there are alternative ways to implement this procedure. The community working with superconducting devices, in particular, tends to use path integral methods, whereas researchers from quantum optics and stochastic processes usually employ master equations. As the former can be applied to underdamped as well as overdamped systems and its extension to imaginary times has been tested successfully in tunneling problems, I have chosen to adopt it. However, since this subject is not so widely taught in compulsory physics courses, not many researchers are familiar with it and, therefore, I decided to include some basic material as guidance for those who know little or nothing about path integrals.

Now let me elaborate a little on the issue of the overlap with other books. By the very nature of a book itself, it is almost impossible to write anything without overlaps with previously written material. Nevertheless, in this particular case, I think I have partially succeeded in doing so because of the introductory material already mentioned above and also the subjects I have chosen to cover. Although at least half of them are by now standard problems of quantum dissipation and also covered in other books (for example, in the excellent books of U. Weiss or H. P. Breuer and F. Petruccione), those texts are mostly focused on a given number of topics and some equally important material is left out. This is the case, for instance, with alternative models for the reservoir and its coupling to the variable of interest, and the full treatment of quantum tunneling in field theories (with and without dissipation), which is useful for dealing with quantum nucleation and related problems in macroscopic systems at very low temperatures. These issues have been addressed in the present book.

The inclusion of the introductory material on magnetism, superconductivity, path integrals, and more clearly gives this book a certain degree of self-containment. However, the reader should be warned that, as usual, many subjects have been left out as in any other book. I have tried to call the attention of the reader whenever it is not my intention to pursue a given topic or extension thereof any further. In such cases, a list of pertinent references on the subject is provided, and as a general policy I have tried to keep as close as possible to the notation used in the complementary material of other authors in order to save the reader extra work.

In conclusion, I think that this book can easily be followed by senior undergraduate students, graduate students, and researchers in physics, chemistry, mathematics, and engineering who are familiar with quantum mechanics, electromagnetism, and statistical physics at the undergraduate level.

Acknowledgments

There are a few people and institutions whose direct or indirect support for the elaboration of this book should be acknowledged.

Initially, I must thank F. B. de Brito and A. J. R. Madureira for helping me with the typesetting of an early version of some lecture notes which have become the seed for this book. I thank the former also for helpful discussions and insistence that I should write a book on this subject.

The original lecture notes were enlarged slightly when I taught a course entitled "Macroscopic Quantum Phenomena and Quantum Dissipation" as a Kramers Professor at the Institute for Theoretical Physics of the University of Utrecht, the Netherlands, whose kind hospitality is greatly acknowledged.

The project of transforming those lecture notes into a book actually started in 2010, and two particular periods were very important in this respect. For those, I thank Dionys Baeriswyl and A. H. Castro Neto for the delightful time I had at the University of Fribourg, Switzerland and Boston University, USA, respectively, where parts of this book were written. I must also mention very fruitful discussions with Matt LaHaye, Britton Plourde, and M. A. M. Aguiar during the late stages of elaboration of the project, and the invaluable help of Hosana C. Oliveira in designing the cover of the book.

Finally, I wish to acknowledge support from CNPq (Conselho Nacional de Desenvolvimento Científico e Tecnológico) and FAPESP (Fundação de Amparo à Pesquisa no Estado de São Paulo) through the "Instituto Nacional de Ciência e Tecnologia em Informação Quântica" from 2010 onwards.

1

Introduction

Since its very beginning, quantum mechanics has been developed to deal with systems on the atomic or sub-atomic scale. For many decades, there has been no reason to think about its application to macroscopic systems. Actually, macroscopic objects have even been used to show how bizarre quantum effects would appear if quantum mechanics were applied beyond its original realm. This is, for example, the essence of the so-called *Schrödinger cat paradox* (Schrödinger, 1935). However, due to recent advances in the design of systems on the meso- and nanoscopic scales, as well as in cryogenic techniques, this situation has changed drastically. It is now quite natural to ask whether a specific quantum effect, collectively involving a macroscopic number of particles, could occur in these systems.

In this book it is our intention to address the quantum mechanical effects that take place in properly chosen or built "macroscopic" systems. Starting from a very naïve point of view, we could always ask what happens to systems whose classical dynamics can be described by equations of motion equivalent to those of particles (or fields) in a given potential (or potential energy density). These can be represented by a generalized "coordinate" $\varphi(\mathbf{r}, t)$ which could either describe a field variable or a "point particle" if it is not position dependent, $\varphi(\mathbf{r}, t) = \varphi(t)$. If not for the presence of dissipative terms which, as we will discuss later, cause problems in the canonical quantization procedure, these equations of motion can, in general, be derived from a Hamiltonian which allows us, through canonical methods, to immediately write down their quantum mechanical versions. Now, depending on the structure of the potential energy of the problem, we can boldly explore any effect there would be if we were dealing with an ordinary quantum mechanical particle.

Everything we have said so far is perfectly acceptable, at least operationally, if two specific requirements are met. Firstly, the parameters present in our problem must be such that any combination thereof, leading to a quantity with dimensions of angular momentum (or action), is of the order of $\hbar$. Secondly, we must be sure that the inclusion of dissipative terms, no matter how it is done at this stage, does not interfere greatly with the quantum effects resulting from the former requirement.

Needless to say, following this prescription, we are implicitly assuming that there is no natural limitation for the application of quantum mechanics to any physical system. Although it does pose several conceptual questions, we would have to propose alternative theories if we wanted to think otherwise. So, we have chosen to rely on this hypothesis and explore it as far as we can, not bothering, at least openly, about questions concerning the foundations of quantum mechanics, a promise that is obviously hard to keep fully in such a subtle application of the quantum theory.

In a sense, we will be looking for situations where the microscopic parameters involved in the description of a given phenomenon only appear in particular combinations which rule the dynamics of a few collective macroscopic variables. Moreover, these resulting combinations are such that their numerical values are comparable to those compatible with the application of quantum mechanics to these systems. The remaining degrees of freedom constitute what we shall call the environment, and the signature of their existence is the presence of dissipative terms in the classical equations of motion of the variables of interest. Therefore, it is mandatory to learn how to include dissipative effects in quantum mechanics if we want to understand its application to our target systems. In other words, quantum dissipation is a natural consequence of the study of macroscopic quantum phenomena.

There are many systems that fulfill our requirements to display quantum effects at the macroscopic level. Unfortunately, they are not tiny marbles tunneling across a wall but rather a somewhat more subtle variation thereof. Usually they are magnetic or superconducting samples of reduced dimensions subject to very low temperatures. Although these are not the only examples we could mention, they will be the ones elected as our favorite throughout this book. Magnetic systems provide us with quantum effects that, on top of being experimentally realizable, are easier to interpret whereas superconductors (in particular, superconducting devices) are those systems where the search for quantum effects at the macroscopic level really started and also present the most reliable tests of their existence.

Regarding superconducting devices, we have chosen to deal with current-biased Josephson junctions (CBJJs), superconducting quantum interference devices (SQUIDs), and Cooper pair boxes (CPBs), and investigate the quantum mechanics of the corresponding dynamical variables of interest. We realize that their behavior may also be viewed as a manifestation of genuine quantum mechanical effects as applied to macroscopic bodies. Accordingly, the difficulty of perfect isolation of our tiny (but still macroscopic) devices brings dissipative effects into play which forces us to search for a systematic treatment of the influence of damping on many different quantum mechanical phenomena. It will be shown that this environmental influence tends to inhibit the occurrence of quantum effects on the macroscopic level in the great majority of cases.

However, it is not only this fundamental aspect of the subject that attracts the attention of the scientific community to these systems. Since under specific conditions the behavior of some of these devices is well mimicked by a two-state system dynamics, they can be regarded as *qubits*. Therefore, the hope of controlling the destructive influence of dissipation on complex networks using our devices as their basic elements raises expectations toward the fabrication of quantum computers where the new qubits could be accessed by ordinary electronic circuitry.

We have organized the chapters in the following way. We start by introducing some basic concepts on the phenomenology of magnetism and superconductivity in Chapters 2 and 3, respectively, in order to give the reader some background material to understand the specific physical situations where macroscopic quantum phenomena can take place in each of these areas. Nucleation problems, vortex and wall dynamics, and device physics are all analyzed within the quantum mechanical context.

In Chapter 4 we review the classical theory of Brownian motion in order to show the reader how the physics of those systems can be understood if they obey classical mechanics. Then, we develop the general approach for the quantum mechanics of non-isolated systems, the *system-plus-reservoir approach*, and establish the general program to be followed from then onwards.

Once this has been done, we argue, in Chapters 5 and 6, in favor of semi-empirical approaches for treating the so-called dissipative quantum systems and introduce a few models for the reservoir coupled to the system of interest. In particular, we introduce what we call the *minimal model*, where the system of interest is coupled to a set of non-interacting harmonic oscillators through a coordinate–coordinate interaction Hamiltonian. We impose the conditions under which the specific coupling we have chosen allows us to reproduce the expected dynamics of the system in the classical limit and study the quantum mechanics of this composite system. From this study we are able to describe the influence of the environment on the quantum dynamics of the system of interest solely in terms of the phenomenological constants present in the classical dynamics of that system. The way to deal with the same effect with regard to the equilibrium state of the system of interest is also addressed in Chapter 6 for the specific case of the minimal model. We should stress here that although the treatment is broad enough to cope with many different kinds of dissipative systems, we will be focusing on the so-called *ohmic dissipation*, since it is ubiquitous in most systems of interest to us.

Chapters 7, 8, and 9 are devoted to the application of our methods to the dynamics of wave packets in the classically accessible region of the phase space of the system (where a thorough analysis of decoherence is also presented), the decay of "massive" particles and field configurations from metastable states by quantum tunneling and coherent tunneling between bistable states, respectively.

Finally, in Chapter 10 we apply some of these results to the superconducting devices presented in this book, aiming at the possibility of using them as reliable qubits. Further applications and experimental results are briefly analyzed in this section.

Functional methods – path integrals in particular – are at the heart of the mathematical techniques employed throughout most of the chapters of this book, and a review of some of them is presented in the appendices.

2

Elements of magnetism

Magnetic phenomena have been observed for a very long time, and by many ancient civilizations. The very fact that a piece of *magnetite*, the so-called *lodestone*, has the ability to attract some particular materials has been reported many times over the centuries. Even the use of this sort of material for building instruments – such as compasses – to orient navigators has been the subject of about three millennia of uncertainty (Mattis, 1988).

However, only recently have more profound and systematic studies been performed on materials which present that kind of property, and a very rich collection of phenomena has appeared related to them. Paramagnetism, diamagnetism, and the ordered ferromagnetic or anti-ferromagnetic phases of some materials are examples we could mention (Mattis, 1988; White, 2007). The lodestone itself is an example of a ferromagnetic substance which provides us with a permanent magnetic field at room temperature.

It is an experimental fact, first observed by Pierre Curie, that many of those materials attracted by the lodestone, which we call *magnetic materials*, do not present any magnetic property if isolated from the external influence of the latter. Actually, their static magnetic susceptibilities, which are a measure of the response of the material to the external stimulus (see below), behave, at sufficiently high temperatures, in the following way:

$$\chi_M \equiv \lim_{H \to 0} \frac{M}{H} = \frac{C}{T}. \tag{2.1}$$

In this expression, C is a positive constant that depends on the material under investigation, M is the magnetization, and H is the external magnetic field, whose definitions will be given shortly. Materials presenting this sort of behavior are called *paramagnetic*.

On the contrary, there are materials whose behavior contrasts with that presented above. Instead of being attracted by the external magnetic field, they are repelled

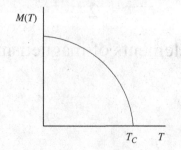

Figure 2.1 Magnetization as a function of temperature

by its presence. In this case, the magnetic susceptibility is negative and varies very slowly with temperature. This is the so-called *diamagnetism*.

Finally, there is a third kind of magnetic material; those materials which present a spontaneous magnetization even in the absence of an externally applied field, the *ferromagnetic materials*. Although there can be paramagnetic substances which keep their magnetic properties as given by (2.1) down to zero temperature, it may happen that the magnetic susceptibility of some materials diverges at a critical temperature, T_C, the so-called Curie temperature. Below that, the material develops a spontaneous magnetization, $M(T)$, which has temperature dependence as shown in Fig. 2.1.

2.1 Macroscopic Maxwell equations: The magnetic moment

In order to explain these three basic phenomena, the natural starting point would be to analyze the macroscopic Maxwell equations (White, 2007) in a given material medium. They read:

$$\nabla \cdot \mathbf{D} = 4\pi\rho,$$
$$\nabla \cdot \mathbf{B} = 0,$$
$$\nabla \times \mathbf{E} = -\frac{1}{c}\frac{\partial \mathbf{B}}{\partial t},$$
$$\nabla \times \mathbf{H} = \frac{4\pi}{c}\mathbf{J} + \frac{1}{c}\frac{\partial \mathbf{D}}{\partial t}, \tag{2.2}$$

where $\mathbf{E} \equiv \mathbf{E}(\mathbf{r}, t)$ and $\mathbf{B} \equiv \mathbf{B}(\mathbf{r}, t)$ refer, respectively, to averages of the microscopic fields $\langle \mathbf{e}(\mathbf{r}', t) \rangle_\mathbf{r}$ and $\langle \mathbf{h}(\mathbf{r}', t) \rangle_\mathbf{r}$ over a macroscopic volume ΔV about the position $\mathbf{r}$. $\rho \equiv \rho(\mathbf{r}, t)$ and $\mathbf{J} \equiv \mathbf{J}(\mathbf{r}, t)$ represent, respectively, the same sort of average of the free charge and free current densities and the fields $\mathbf{D} \equiv \mathbf{D}(\mathbf{r}, t)$ and $\mathbf{H} \equiv \mathbf{H}(\mathbf{r}, t)$ are defined as usual:

$$\mathbf{D} = \mathbf{E} + 4\pi\mathbf{P},$$
$$\mathbf{H} = \mathbf{B} - 4\pi\mathbf{M}, \tag{2.3}$$

where $\mathbf{P} \equiv \mathbf{P}(\mathbf{r}, t)$ and $\mathbf{M} \equiv \mathbf{M}(\mathbf{r}, t)$ are, respectively, the polarization and magnetization fields of the material being considered. As we are interested in magnetic phenomena we shall mostly be discussing the role played by $\mathbf{H}$ and, particularly, $\mathbf{M}$.

From the above equations (White, 2007) we conclude that the magnetization is actually due to the existence of the microscopic current density $\mathbf{J}_{\mathrm{mol}}(\mathbf{r}, t)$, which ultimately results from the stationary atomic motion of the electrons. Attributing a local current density $\mathbf{J}^{(i)}_{\mathrm{mol}}(\mathbf{r}, t)$ to the electronic motion about a given molecular or ionic position $\mathbf{r}_i$, we can associate a magnetic moment

$$\boldsymbol{\mu}_i(t) \equiv \boldsymbol{\mu}(\mathbf{r}_i, t) = \frac{1}{2c} \int d\mathbf{r}'(\mathbf{r}' - \mathbf{r}_i) \times \mathbf{J}^{(i)}_{\mathrm{mol}}(\mathbf{r}', t) \tag{2.4}$$

with that position. From this expression and a general representation of $\mathbf{J}^{(i)}_{\mathrm{mol}}(\mathbf{r}', t)$ in terms of $\boldsymbol{\mu}_i(t)$ itself it can be shown (White, 2007) that the magnetization is written

$$\mathbf{M}(\mathbf{r}, t) = \left\langle \sum_i \Delta(\mathbf{r}' - \mathbf{r}_i)\boldsymbol{\mu}_i(t) \right\rangle_{\mathbf{r}}, \tag{2.5}$$

where $\Delta(\mathbf{r}' - \mathbf{r}_i)$ is a function normalized to unity and strongly peaked about $\mathbf{r}_i$. Integrating the latter expression over the whole volume of the sample, we easily see that $\mathbf{M}(\mathbf{r}, t)$ is the total magnetic moment per unit volume.

If we consider the presence of N_e point electrons per ion (or molecule) at positions $\mathbf{r}^{(i)}_k$ relative to $\mathbf{r}_i$ with velocities $\mathbf{v}^{(i)}_k$, the local current density reads

$$\mathbf{J}^{(i)}_{\mathrm{mol}}(\mathbf{r}) = \sum_{k=1}^{N_e} e\mathbf{v}^{(i)}_k \delta(\mathbf{r} - \mathbf{r}_i - \mathbf{r}^{(i)}_k), \tag{2.6}$$

which we can use in (2.4) to show that

$$\boldsymbol{\mu}_i = \frac{e}{2mc} \sum_{k=1}^{N_e} \mathbf{r}^{(i)}_k \times \mathbf{p}^{(i)}_k = \frac{e}{2mc}\mathbf{L}_i, \tag{2.7}$$

where $\mathbf{p}^{(i)}_k = m\mathbf{v}^{(i)}_k$ is, in the absence of an external field, the canonical momentum of the kth electron at $\mathbf{r}_i$ and $\mathbf{L}_i \equiv \sum_{k=1}^{N_e} \mathbf{r}^{(i)}_k \times \mathbf{p}^{(i)}_k$ is the total electronic angular momentum at the same site.

Now, it is a standard example of statistical mechanics textbooks to compute the total magnetization of a set of non-interacting magnetic moments at a finite temperature (see, for example, Reif (1965)) using as a starting point its energy when acted on by an external field **H**, which reads

$$E = -\sum_i \boldsymbol{\mu}_i \cdot \mathbf{H}. \tag{2.8}$$

As a result we find that the magnetization so obtained agrees with the empirical form suggested by Langevin, which stated that

$$M = f\left(\frac{H}{T}\right), \tag{2.9}$$

where f is an odd function of its argument.[1] We see that for sufficiently high temperatures, the argument of the above function is small, allowing us to replace it by the lowest-order term of its Taylor expansion. Then, evaluating the paramagnetic static susceptibility with this expression, we see that it obeys the Curie expression (2.1) for a material in its paramagnetic phase and, therefore, classical physics can explain paramagnetism without any problem.

The same does not hold if we try to explain, for example, diamagnetism by the same token. For this, suppose we apply a constant external field **H** to our sample. In this case, the expression for the magnetic moment in terms of the angular momentum must be modified by replacing

$$\mathbf{p}_k \rightarrow \mathbf{p}_k - \frac{e}{c}\mathbf{A}(\mathbf{r}_k) \tag{2.10}$$

in (2.7), where $\mathbf{A}(\mathbf{r})$ is the vector potential which, in the symmetric gauge, reads

$$\mathbf{A}(\mathbf{r}) = \frac{1}{2}\mathbf{H} \times \mathbf{r}. \tag{2.11}$$

Accordingly, the Hamiltonian of the electronic system must also be replaced by

$$\mathcal{H}(\ldots \mathbf{p}_k, \mathbf{r}_k \ldots) = \sum_k \frac{1}{2m}\left(\mathbf{p}_k - \frac{e}{c}\mathbf{A}(\mathbf{r}_k)\right)^2 + U(\ldots \mathbf{r}_k \ldots), \tag{2.12}$$

where $U(\ldots \mathbf{r}_k \ldots)$ represents any possible interaction regarding the electronic system.

Now we can use (2.12) in the classical Boltzmann factor to compute the magnetization of the system at finite temperatures, as we have done for paramagnetic materials. However, in this case, the Bohr–van Leeuwen theorem (Ashcroft and Mermin, 1976) states that the phase space integral which determines this quantity must vanish. In other words, there is no classical explanation for diamagnetism and this is the first situation where we do need quantum mechanics to deal with magnetic materials.

---

[1] For a classical magnetic moment this function is $n\mu\left(\coth\theta - \frac{1}{\theta}\right)$, where $\theta = \frac{\mu H}{k_B T}$ and n is the density of magnetic moments.

2.2 Quantum effects and the order parameter

Let us start the analysis of quantum mechanics applied to magnetic systems by writing the quantum mechanical form of (2.7). In order to do that we must remember the results of the quantization of angular momentum (Merzbacher, 1998), which tell us that the orbital angular momentum eigenstates of the electronic motion are given by $|\ell, m_\ell\rangle$ such that

$$L^2|\ell, m_\ell\rangle = \ell(\ell+1)\,\hbar^2|\ell, m_\ell\rangle; \qquad \ell = 0, 1, 2 \ldots$$
$$L_z|\ell, m_\ell\rangle = m_\ell\,\hbar|\ell, m_\ell\rangle; \qquad m_\ell = 0, \pm 1, \pm 2, \ldots, \pm\ell \qquad (2.13)$$

where the conventional notation $\mathbf{L} = \mathbf{r} \times \mathbf{p}$ has been used and all the dynamical variables are now regarded as operators. Then, if we are interested in one of the components of the magnetic moment operator, say μ_z, we can rewrite the z component of (2.7) applied on $|\ell, m_\ell\rangle$ as

$$\mu_z|\ell, m_\ell\rangle = -m_\ell\,\mu_B|\ell, m_\ell\rangle, \qquad (2.14)$$

where we have used the fact that $e = -|e|$ and

$$\mu_B \equiv \frac{\hbar|e|}{2mc} = 9.27 \times 10^{-21}\ \text{erg} \cdot \text{G}^{-1}\ (\text{or} \times 10^{-24}\ \text{J} \cdot \text{T}^{-1}) \qquad (2.15)$$

is the *Bohr magneton*.

However, this is not the whole story. The Zeeman effect and the Stern–Gerlach experiment (Merzbacher, 1998) indicate that the electron itself must carry an intrinsic magnetic moment which is related to the generator of rotations in a two-dimensional Hilbert space spanned by the states $|s, m_s\rangle$. The eigenvalue problem associated with these generators is

$$S^2|s, m_s\rangle = s(s+1)\,\hbar^2|s, m_s\rangle; \qquad s = \frac{1}{2}$$
$$S_z|s, m_s\rangle = m_s\,\hbar|s, m_s\rangle; \qquad m_s = \pm\frac{1}{2} \qquad (2.16)$$

and the relation equivalent to (2.7) now reads

$$\boldsymbol{\mu} = -g_s \frac{\mu_B}{\hbar}\mathbf{S} \equiv \gamma_g\mathbf{S}, \qquad (2.17)$$

where $\mathbf{S} = \hbar\boldsymbol{\sigma}/2$, $\boldsymbol{\sigma}$ is a vector whose components are the well-known 2×2 Pauli matrices and $g_s \approx 2$ and γ_g are the electron gyromagnetic factor and ratio, respectively. Notice that γ_g has the same sign as the particle's charge.

Consequently, assuming that the electron is subject to a localized potential and an external field along the **z** direction, strong enough so that the energy eigenstates can be considered as $|\ell, s, m_\ell, m_s\rangle = |\ell, m_\ell\rangle \otimes |s, m_s\rangle$, its energy eigenvalues now read

$$E_{m_\ell m_s} = (m_\ell + 2m_s)\,\mu_B H. \tag{2.18}$$

For weak fields, it is more appropriate to characterize the magnetic state by the eigenstates of the total angular momentum $\mathbf{J} = \mathbf{L} + \mathbf{S}$, $|j, \ell, s, m\rangle$, from which we can write (Ashcroft and Mermin, 1976; Merzbacher, 1998)

$$\boldsymbol{\mu} = -g(j, l, s)\frac{\mu_B}{\hbar}\mathbf{J} \equiv \gamma_j \mathbf{J}, \tag{2.19}$$

where $g(j, l, s)$ is the well-known Landé g-factor.

Now we can compute the statistical Boltzmann factor corresponding to the interaction $-\boldsymbol{\mu} \cdot \mathbf{H}$ and reproduce again expression (2.1) for the paramagnetic susceptibility (see, for example, Ashcroft and Mermin (1976)) where the constant C is determined in terms of the total angular momentum quantum number j and the Bohr magneton.

2.2.1 Diamagnetism

Let us turn now to the diamagnetic phenomenon and suppose we start by taking the quantum mechanical version of (2.12). So, if we consider the system under the influence of an external field $\mathbf{H}$ we can write

$$\mathcal{H}(\dots \mathbf{p}_k, \mathbf{r}_k \dots) = \sum_k \frac{p_k^2}{2m} + U(\dots \mathbf{r}_k \dots) + \Delta\mathcal{H}, \tag{2.20}$$

where the full magnetic correction $\Delta\mathcal{H}$ in the symmetric gauge (2.11) reads

$$\Delta\mathcal{H} = \mu_B(\mathbf{L} + 2\mathbf{S}) \cdot \mathbf{H} + \frac{e^2}{8mc^2} H^2 \sum_k \left(x_k^2 + y_k^2\right), \tag{2.21}$$

with $\mathbf{L} + 2\mathbf{S} = \sum_k \mathbf{L}_k + 2\mathbf{S}_k$. Notice that the total angular momentum is not given by this expression but by $\mathbf{J} = \mathbf{L} + \mathbf{S}$ instead. It should be stressed that in (2.21) we are neglecting spin–orbit terms (see below).

If the full Hamiltonian (2.20) has eigenstates $|n\rangle$ and the external field can be treated perturbatively, we find corrections of the order of H^2 to the energy eigenvalue E_n (Ashcroft and Mermin, 1976):

$$\Delta E_n = \mu_B\langle n\,|\,(\mathbf{L} + 2\mathbf{S}) \cdot \mathbf{H}\,|\,n\rangle + \sum_{n' \neq n} \frac{|\mu_B\langle n\,|\,(\mathbf{L} + 2\mathbf{S}) \cdot \mathbf{H}\,|\,n'\rangle|^2}{E_n - E_{n'}}$$

$$+ \frac{e^2}{8mc^2} H^2 \sum_k \langle n\,|\,x_k^2 + y_k^2\,|\,n\rangle, \tag{2.22}$$

from which we see that the paramagnetic effects are dominant for small fields whenever $\langle \mathbf{L}\rangle$ or $\langle \mathbf{S}\rangle \neq 0$. However, for a system with filled atomic shells, its ground state $|0\rangle$ is such that (Ashcroft and Mermin, 1976) $\mathbf{L}|0\rangle = \mathbf{S}|0\rangle = \mathbf{J}|0\rangle = 0$, and only the last term survives in (2.22). Therefore, using the fact that

$$\mu = -\frac{\partial E}{\partial \mathbf{H}}, \qquad (2.23)$$

we can compute the magnetic susceptibility as (Ashcroft and Mermin, 1976)

$$\chi_M = -\frac{N}{V}\frac{\partial^2 \Delta E_0}{\partial H^2} = -\frac{Ne^2}{6mc^2V} \langle 0|\sum_k r_k^2|0\rangle, \qquad (2.24)$$

which explains the formation of a magnetization contrary to the applied field and consequently why the diamagnetic material is repelled by the source of that field.

2.2.2 Curie–Weiss theory: Ferromagnetism

Let us now digress a little to introduce the phenomenological theory of Curie and Weiss for ferromagnetism.

As we have anticipated before, there are materials which present spontaneous magnetization even in the absence of an externally applied magnetic field. Without knowing what sort of microscopic mechanism would be responsible for the electric and magnetic properties of these materials, Pierre-Ernest Weiss established that each atom, ion or molecule of the system would feel, in addition to an external field, the so-called *molecular field*, which would be affected by the presence of all the other component particles of the material and proportional to the magnetization at that point. In this way he proposed a modification of the Langevin expression (2.9), which reads

$$M = f\left(\frac{H + NM}{T}\right). \qquad (2.25)$$

Measurements of the factor N showed that for ferromagnetic materials it was extremely high and could not be explained by any classical mechanism. For example, if we attributed the existence of N to the demagnetizing effect due to the presence of fictitious magnetic poles on the surface of the ferromagnetic sample (see below), its value would be negative and not large.

The expansion of (2.25) for high temperatures would then give us

$$M = C\frac{H + NM}{T} \qquad \Rightarrow \qquad M = \frac{CH}{T - CN}, \qquad (2.26)$$

from which we get the magnetic susceptibility

$$\chi_M = \frac{C}{T - T_C}; \qquad (2.27)$$

where the Curie temperature, $T_C \equiv CN$, is the temperature at which the susceptibility diverges. For $T \gg T_C$ we clearly recover the Curie paramagnetic susceptibility (2.1).

For low temperatures (2.26) is no longer valid and we would need to have the specific form of f to solve (2.25) for M. In particular, for $H = 0$, we would find a monotonically decreasing function of temperature, $M(T)$, which would vanish at $T = T_C$ in order to agree with the experimental results (see Fig. 2.1). Moreover, due to the abnormally high value of N, finite external field effects would be very small for fields $H \ll H_C(T)$, where $H_C(T) = -NM(T)$ is the field which destroys the spontaneous magnetization of the sample. The question to be answered here is then, why is the value of N so high?

This is again an effect that can only be understood within the scope of quantum mechanics. The main mechanism responsible for the appearance of such a term is the so-called *direct exchange* (Mattis, 1988; Merzbacher, 1998; White, 2007). This is basically due to the fact that, because the states of a many-electron system are Slater determinants of the spatial wave functions multiplied by the spin states of single electrons at given occupied orbitals, the average Coulomb interaction energy between any two of them presents a term without classical analogue, which is known in the literature as the *exchange integral*. In contrast to the *direct* Coulomb interaction term

$$K_{ij} \equiv \langle ij \mid V \mid ij \rangle = \int d\mathbf{r} \int d\mathbf{r}' \mid \phi_i(\mathbf{r}) \mid^2 \frac{e^2}{\mid \mathbf{r} - \mathbf{r}' \mid} \mid \phi_j(\mathbf{r}') \mid^2, \qquad (2.28)$$

which is always present and has a clear classical analogue, the exchange term reads

$$\tilde{J}_{ij} \equiv \langle ij \mid V \mid ji \rangle = \int d\mathbf{r} \int d\mathbf{r}' \, \phi_i^*(\mathbf{r}) \, \phi_j^*(\mathbf{r}') \frac{e^2}{\mid \mathbf{r} - \mathbf{r}' \mid} \phi_i(\mathbf{r}') \phi_j(\mathbf{r}), \quad (2.29)$$

and results from the indistinguishability of the particles we are considering.

The particular case of only two electrons occupying the lowest localized energy levels described by the orbitals $\phi_i(\mathbf{r}) \equiv \phi(\mathbf{r} - \mathbf{r}_i)$ and $\phi_j(\mathbf{r}) \equiv \phi(\mathbf{r} - \mathbf{r}_j)$, about positions $\mathbf{r}_i$ and $\mathbf{r}_j$, respectively, can be studied in the basis generated by the Slater determinants (White, 2007):

$$\begin{vmatrix} \phi_i(\mathbf{r}_1)\alpha(1) & \phi_i(\mathbf{r}_2)\beta(2) \\ \phi_j(\mathbf{r}_1)\alpha(1) & \phi_j(\mathbf{r}_2)\beta(2) \end{vmatrix}, \qquad (2.30)$$

where $\alpha(n)$ and $\beta(n)$ represent the components of the spin of particle $n = 1, 2$ along the two-dimensional states

$$\mid \uparrow \rangle \equiv \begin{pmatrix} 1 \\ 0 \end{pmatrix} \qquad \text{or} \qquad \mid \downarrow \rangle \equiv \begin{pmatrix} 0 \\ 1 \end{pmatrix}. \qquad (2.31)$$

This basis has four states, $\{|\psi_1\rangle, |\psi_2\rangle, |\psi_3\rangle, |\psi_4\rangle\}$ corresponding to the possibilities $|\alpha\,\beta\rangle = |\uparrow\uparrow\rangle, |\uparrow\downarrow\rangle, |\downarrow\uparrow\rangle, |\downarrow\downarrow\rangle$, in which our model Hamiltonian can be represented as

$$\mathcal{H} = \begin{pmatrix} E_{ij} - \tilde{J}_{ij} & 0 & 0 & 0 \\ 0 & E_{ij} & -\tilde{J}_{ij} & 0 \\ 0 & -\tilde{J}_{ij} & E_{ij} & 0 \\ 0 & 0 & 0 & E_{ij} - \tilde{J}_{ij} \end{pmatrix}, \tag{2.32}$$

with $E_{ij} \equiv E_i + E_j + K_{ij}$. Rewriting the above equation as an explicit representation of the direct product of the two-dimensional spin subspaces of particles 1 and 2 (White, 2007), we have

$$\mathcal{H} = \frac{1}{4}(E_s + E_t) - \frac{1}{4}(E_s - E_t)\sigma_i \cdot \sigma_j, \tag{2.33}$$

where $E_{s/t} = E_i + E_j + K_{ij} \pm \tilde{J}_{ij}$ are the eigenenergies corresponding, respectively, to the singlet

$$|0, 0\rangle = \frac{|\uparrow\downarrow\rangle - |\downarrow\uparrow\rangle}{\sqrt{2}} \tag{2.34}$$

and triplet

$$|1, 1\rangle = |\uparrow\uparrow\rangle, \quad |1, 0\rangle = \frac{|\uparrow\downarrow\rangle + |\downarrow\uparrow\rangle}{\sqrt{2}} \quad \text{and} \quad |1, -1\rangle = |\downarrow\downarrow\rangle \tag{2.35}$$

eigenstates of the problem. The states on the l.h.s. of (2.34), (2.35) are the eigenstates $|S, S_z\rangle$ of the total angular momentum operator, $\mathbf{S} = \mathbf{S}_1 + \mathbf{S}_2$, of the system (Merzbacher, 1998). Using the explicit forms of $E_{s/t}$ defined below (2.33) we can finally rewrite it, apart from a constant value, as

$$\mathcal{H} = -\tilde{J}\sigma_i \cdot \sigma_j, \tag{2.36}$$

where $\tilde{J}$ is proportional to the exchange integral (2.29).

Therefore, we see that it is possible to write the exchange energy, in appropriate units, as the product of spin operators $\mathbf{S}_i \cdot \mathbf{S}_j$ of each particle and, depending on the sign of the exchange integral, they can be either parallel or anti-parallel to one another. In other words, they can have a ferromagnetic or anti-ferromagnetic interaction.

Actually, the reasoning presented here can also be applied, for example, to the case of N_e electrons which occupy the same number of orthogonal localized orbitals $\phi(\mathbf{r} - \mathbf{r}_n)$. It can be shown that, within first-order perturbation theory, the effective Hamiltonian of this problem (Merzbacher, 1998; White, 2007) can also be written as

$$\mathcal{H} = -\sum_{ij} J_{ij} \, \mathbf{S}_i \cdot \mathbf{S}_j, \tag{2.37}$$

where i and j refer to the different sites and we have defined $J_{ij} \equiv \tilde{J}_{ij}/\hbar^2$. Moreover, since the exchange integrals are strongly dependent on the overlap of the localized orbitals, the main contribution to (2.37) comes from neighboring sites, which means that the summation must be performed over i and j referring to nearest neighbors only. This is the famous *Heisenberg Hamiltonian*.

It is the Hamiltonian (2.37) which allows us to obtain a spontaneous magnetization for a ferromagnetic sample. If we have a positive exchange integral, the ground state of the system will tend to have all spins aligned in the same direction. We can also study, at least approximately, the statistical mechanics of this system and, if the temperature is not high enough, conclude that there will not be enough thermal fluctuations to completely destroy this state. In the anti-ferromagnetic case (negative exchange integral), the tendency is to have the neighboring spins pointing in opposite directions, which results in a vanishing total magnetization but in a finite so-called *staggered magnetization*. In this case the system can be viewed as two distinct interpenetrating sub-lattices, A and B, each with finite but opposite magnetizations. The staggered magnetization turns out to be twice the value of the magnetization of one of these sub-lattices and is given by

$$\mathbf{M}_S = \mathbf{M}_A - \mathbf{M}_B, \tag{2.38}$$

where $\mathbf{M}_A = -\mathbf{M}_B$.

In any case, the average value of the spin at a given site will depend on the neighboring spins and, moreover, since the exchange integral has a purely electrostatic origin, its value is by no means smaller than the direct Coulomb interaction between the electrons. These results clearly corroborate the phenomenological approach proposed by Curie and Weiss for ferromagnetism.

Before extending our discussion toward the construction of a general phenomenological theory for the magnetization of a given magnetic system, let us briefly mention a couple of other equally important microscopic interactions which justify the inclusion of more terms in our future theory.

In treating the relativistic electron theory, we have to deal with the Dirac equation (see, for example, Bjorken and Drell (1964)) which, when expanded in powers of E/mc^2, gives rise to separate equations for positive and negative energy spinors which contain terms like the non-relativistic kinetic and potential energy terms, the Zeeman interaction

$$\mathcal{H}_Z = \mu_B(\ell + \sigma) \cdot \mathbf{H}, \tag{2.39}$$

the spin–orbit term

$$\mathcal{H}_{so} = \zeta \boldsymbol{\ell} \cdot \boldsymbol{\sigma} \qquad \text{with} \qquad \zeta = \frac{e\hbar^2}{4m^2c^2}\frac{1}{r}\frac{\partial V}{\partial r} \qquad \text{and} \qquad \mathbf{L} \equiv \hbar\boldsymbol{\ell}$$

$$(2.40)$$

and also a diamagnetic correction like the last term in (2.22) which we have neglected here.

Now, considering the presence of many electrons in the problem, it is a simple matter to convince ourselves that the magnetic field which acts in one of the electrons as in (2.39) will be partly due to the presence of all the other electrons in the system and this establishes a magnetic dipole interaction between them. The overall effect of this term is to create a demagnetizing field (Akhiezer *et al.*, 1968; White, 2007) which depends on the geometry of the specimen under consideration and, in some cases, will only be a perturbation of the exchange and external field effects. This same term happens to be obtainable purely by magnetostatic arguments and is found in the literature as the *magnetostatic energy* (Bertotti, 1998).

The spin–orbit term may also present important corrections to the whole magnetic Hamiltonian. Since this term describes the interaction of the electronic spin with the orbital angular momentum, even for a single electron, there may be effects of anisotropy on the spin operator because the orbital angular momentum **L** ultimately carries information about the symmetry of the lattice where the electrons are immersed because of the *crystal field effect*. This is known as the *single ion anisotropy*. However, this is not the only possible source of anisotropy in the problem. When more electrons are present, exchange terms together with the spin–orbit interaction may cause what is called the *exchange anisotropy*. A very accessible presentation of the general mechanisms of anisotropy in magnetic systems can be found in Bertotti (1998).

The above brief analysis of the different magnetically active microscopic terms of the electronic Hamiltonian led us basically to four important observed effects: exchange, anisotropy, demagnetization, and external field energies. In what follows we will approach the physics of the magnetic system from a slightly different point of view, which will be more appropriate for our future needs. However, since the microscopic knowledge is so tempting to evoke for magnetic systems, we shall keep it not far from mind remembering always that it is the phenomenological approach which is our main goal here.

2.2.3 Magnetization: The order parameter

As has already been pointed out by the phenomenological Curie–Weiss theory, below a certain critical temperature, T_C, some magnetic materials develop a finite magnetization even in the absence of an externally applied field. Moreover, the static magnetic susceptibility diverges at that temperature, which is a sign of a continuous phase transition (Huang, 1987).

In contrast, we have also seen that this sort of behavior is only possible because the exchange interaction between electrons, when positive, favors the alignment of their spins and, therefore, the spontaneous magnetization results. Although the initial Hamiltonian of the system is rotationally invariant, which can easily be seen from (2.37) – a scalar in spin space, the magnetization chooses one specific direction along which it will point. This is what is known in the literature of phase transitions and field theory as a *spontaneous symmetry breaking* and the magnetization is the so-called *order parameter* of the system. It is said that the ground state, or the thermodynamic equilibrium state in the case of finite temperatures, breaks the underlying symmetry of the system. In practice, there are always finite perturbations which will tell the magnetization where to point (see below), but it by no means invalidates our conclusions. More formally, in order to obtain the symmetry-broken state, or any consequence thereof, we have to compute thermal averages using the partition function of the system at a finite temperature $T = 1/\beta k_B$ and subject to an external field, which reads

$$\mathcal{Z} = \text{tr}\left[\exp{-\beta(\mathcal{H} - \mathbf{H} \cdot \mathbf{M})}\right], \tag{2.41}$$

and then take the limit when $H \rightarrow 0$ of this expression. However, this must be done carefully. Actually, we must first take the thermodynamic limit ($N \rightarrow \infty$, $V \rightarrow \infty$ but $N/V = n$ (finite)) and then $H \rightarrow 0$. If this is performed the other way round, we get a symmetric result, or, in other words, zero magnetization. This is really a sign that only in the thermodynamic limit does the system naturally show this tendency for developing a finite-order parameter or symmetry-breaking term.

The phenomenon of spontaneous symmetry breaking is also related to another important effect, which is the development of *long-range order*. As a matter of fact, the tendency for alignment was already present in (2.36), which means that the system always has *short-range order*. Nevertheless, as we approach T_C, the magnetic susceptibility starts to diverge. This means that any small disturbance in a given spin will be felt by other very distant ones. In other words, the typical length which determines the behavior of the spin–spin correlation function, the magnetic coherence length, $\xi_m(T)$, diverges as $T \rightarrow T_C$.

Let us now propose a phenomenological Hamiltonian to the magnetic system written as a functional of the magnetization. The microscopic origin of all its terms can be explained if we use (2.5) and (2.17) to create a *magnetization operator*

$$\mathcal{M}(\mathbf{r}, t) = \gamma_g \sum_i \mathbf{S}_i \delta(\mathbf{r} - \mathbf{r}_i), \tag{2.42}$$

where we have assumed a point-like charge distribution and replaced $\Delta(\mathbf{r} - \mathbf{r}_i)$ in (2.5) by $\delta(\mathbf{r} - \mathbf{r}_i)$. Here we should notice that although (2.42) is still a quantum mechanical expression, we could, at least for high spin values ($s \gg 1$), replace $\mathcal{M}(\mathbf{r})$ by its average value, namely, the classical magnetization $\mathbf{M}(\mathbf{r})$.

Bearing this in mind we write the following Hamiltonian functional:

$$
\mathcal{H}[\mathbf{M}(\mathbf{r})] = \int d\mathbf{r} \left[\sum_{i=x,y,z} \frac{1}{2} \alpha_i \nabla M_i \cdot \nabla M_i - \frac{1}{2} \beta_a (\mathbf{M} \cdot \hat{\mathbf{n}})^2 - \frac{1}{2} \mathbf{M} \cdot \mathbf{H}_M(\mathbf{r}) \right.
$$

$$
\left. - \mathbf{M} \cdot \mathbf{H}(\mathbf{r}) \right],
\tag{2.43}
$$

from which a simple dimensional analysis tells us that $\alpha_i = [L]^2$ and β_a is dimensionless.

The first term in (2.43) is the continuum version of the Heisenberg Hamiltonian (2.37) and measures basically the energy to distort the magnetization of the system. The minimum energy configuration is clearly the one which is uniform like the ferromagnetic phase. Since $[\alpha_i] = L^2$ we can assume it is a dimensionless factor (a combination of the microscopic parameters of the problem) times the typical distance between the orbitals where the electrons are localized. Therefore, the distortion we are talking about is measured in units of this microscopic parameter. The remaining dimensionless factors reflect the fact that there can be, as we mentioned in our previous analysis, some exchange anisotropy in the system due to the spin–orbit coupling combined with the exchange terms. This term can actually involve more complicated objects instead of the factors α_i (Akhiezer *et al.*, 1968), but we consider here only the situation where they reduce to the present form.

The second term in (2.43) is the on-site uniaxial anisotropy and originates from the spin–orbit coupling (2.40). Although in general this term can also be more complicated (Akhiezer *et al.*, 1968), we assume that the symmetry of the local orbital is such that it creates, for $\beta_a > 0$, only one preferred axis of orientation for the magnetization which is $\hat{\mathbf{n}}$, the *easy axis*. In what follows we shall assume, without loss of generality, that $\hat{\mathbf{n}} = \hat{\mathbf{z}}$, unless otherwise stated. This term tells us that once the magnetization is created, it will tend to point along $\hat{\mathbf{z}}$ either in the positive or negative direction. These two configurations are completely degenerate. When $\beta_a < 0$ the magnetization will tend to align perpendicular to this axis, lying in the so-called *easy plane*.

The third term in (2.43) is the demagnetization or magnetostatic contribution and results, as mentioned above, from the magnetic dipole interaction between the electrons. The magnetic field $\mathbf{H}_M$ in (2.43) is solely due to the finite magnetization present in the system which generates the magnetization current $\mathbf{J}_M = c\nabla \times \mathbf{M}$.

Employing the electrostatic analogy (Bertotti, 1998; White, 2007), this field is given by the solution of

$$\nabla \cdot \mathbf{H}_M = 4\pi \rho_M,$$
$$\nabla \times \mathbf{H}_M = 0, \tag{2.44}$$

where $\rho_M = -\nabla \cdot \mathbf{M}$. Consequently, we can write $\mathbf{H}_M = -\nabla \phi_M$ where

$$\phi_M(\mathbf{r}) = -\int_V d\mathbf{r}' \, \frac{\nabla \cdot \mathbf{M}(\mathbf{r}')}{|\mathbf{r} - \mathbf{r}'|} + \oint_S ds' \, \frac{\hat{\mathbf{n}}' \cdot \mathbf{M}(\mathbf{r}')}{|\mathbf{r} - \mathbf{r}'|}, \tag{2.45}$$

with V and S being, respectively, the volume and surface of the sample and $\hat{\mathbf{n}}'ds'$ the element of area about $\mathbf{r}'$.

For a specimen with constant magnetization, the volume integral in (2.45) vanishes and we are left with a surface term which gives

$$\mathbf{H}_M = -\widetilde{\mathbf{N}} \cdot \mathbf{M}, \tag{2.46}$$

where $\widetilde{\mathbf{N}}$ is the so-called demagnetization tensor. It can be shown (Mattis, 1988; White, 2007) that for ellipsoidal samples with principal axis coinciding with $\hat{\mathbf{x}}$, $\hat{\mathbf{y}}$ and $\hat{\mathbf{z}}$, the tensor elements N_{ij} reduce to a diagonal form with elements N_x, N_y and N_z such that

$$\sum_{i=x,y,z} N_i = 4\pi. \tag{2.47}$$

These elements depend on the geometry of the sample one uses. For a sphere, for example, $N_x = N_y = N_z = 4\pi/3$. This term tends to disorder the system by favoring a situation where the components of the magnetization are null.

Finally, the last term of that expression is self-explanatory and represents the energy of the system due to the presence of an external field. It tends to align the magnetization along the direction of the field.

In order to see how the macroscopic explanation of these terms does follow from the microscopic parts of the Hamiltonian, we only need to use (2.42) in (2.43) to recover the discrete form of the latter in terms of the spin operator $\mathbf{S}_i$ and then find the appropriate approximations in the microscopic terms to reach the desirable form (2.43).

Another important fact about the macroscopic form (2.43) as a functional of the classical magnetization $\mathbf{M}(\mathbf{r}, t)$ is that it is an expression to which one can apply the canonical quantization procedure to study many different quantum mechanical effects involving the system magnetization, as we will see in the future. Moreover, it also allows us to analyze the competition between the different energy contributions and consequently guides us to build systems with some desired features.

For instance, if the magnetic material we are studying is such that the anisotropy parameter $\beta_a \gg 1$, we can safely neglect the magnetostatic term. This is what is known as a *hard material*. In the opposite limit, $\beta_a \ll 1$, corresponding to a *soft material*, it is the magnetostatic contribution that dominates the energetic balance. On top of that, we can also define characteristic lengths in this problem (see, for example, Bertotti (1998)), which allow us to classify the system as *large* or *small* as we will see in the specific examples given below.

2.2.4 Walls and domains

Let us assume we want to study the minimum energy configuration of the magnetization of a thin ferromagnetic slab of a hard magnetic material ($\beta_a \gg 1$), with easy axis $\hat{z}$, placed on the xy plane. We also assume that the exchange energy is isotropic ($\alpha_i = \alpha$) and there is no external field. Moreover, suppose we are looking for static solutions which do not depend on y or z, which means $\mathbf{M}(\mathbf{r}, t) = \mathbf{M}(x)$. Therefore, if we write the magnetization vector in terms of the usual spherical angles θ and ϕ, we obtain from (2.43) an energy functional

$$\mathcal{H}[\theta(x), \phi(x)] = S \int_{-\infty}^{+\infty} dx \left[\frac{\alpha M^2}{2} \left(\frac{d\theta}{dx} \right)^2 + \frac{\alpha M^2 \sin^2 \theta}{2} \left(\frac{d\phi}{dx} \right)^2 \right.$$
$$\left. + \frac{\beta_a M^2}{2} \sin^2 \theta \right], \tag{2.48}$$

where S is the total area of the slab perpendicular to $\hat{x}$ and M is the saturation magnetization.

In order to find the minimum energy configurations of $\mathcal{H}$ one has to appeal to variational calculus and compute its first functional derivatives with respect to θ and ϕ, which must vanish at the desired configuration. These read

$$\frac{\delta \mathcal{H}[\theta(x), \phi(x)]}{\delta \theta} = -\alpha S M^2 \frac{d^2\theta}{dx^2} + \frac{\alpha S M^2}{2} \sin 2\theta \left(\frac{d\phi}{dx} \right)^2 + \frac{\beta_a}{2} S M^2 \sin 2\theta = 0,$$
$$\frac{\delta \mathcal{H}[\theta(x), \phi(x)]}{\delta \phi} = -\alpha S M^2 \sin^2 \theta \frac{d^2\phi}{dx^2} = 0, \tag{2.49}$$

which must be solved with appropriate boundary conditions.

The second equation of (2.49) is trivially solved for $\phi = $ constant since, for any other solution of the form $\phi(x) = ax + b$, the energy of the configuration would always increase. Let us take, for example, $\phi = \pi/2$.

Now, defining $\beta_a/\alpha \equiv 1/\zeta^2$ ($\ell_W = \zeta$ is generally known as the *wall length*), we can rewrite the first of the above equations as

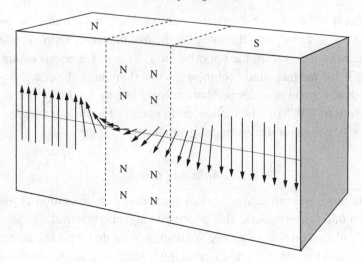

Figure 2.2 Bloch wall

$$\frac{d^2\theta}{dx^2} - \frac{1}{2\zeta^2}\sin 2\theta = 0, \tag{2.50}$$

which we now discuss in more detail.

The differential equation (2.50) clearly admits the trivial solution $\theta = 0$ which is compatible with the fact that the system is a ferromagnet and all the spins could be pointing along $\hat{z}$. However, since there is also an anisotropy term in (2.48) which makes both configurations $\theta = 0$ and $\theta = \pi$ completely degenerate, one could try to find a solution which interpolates between $\theta = \pi$ when $x \to \mp\infty$ and $\theta = 0$ when $x \to \pm\infty$. This soliton-like solution (Rajaraman, 1987) is the so-called *Bloch wall* and has the form shown in Fig. 2.2:

$$\theta(x) = 2\arctan\left(\exp\mp\frac{x}{\zeta}\right), \tag{2.51}$$

where ζ has been defined just above (2.50) and represents the width of the wall. From that definition we see that this width is proportional to $\sqrt{\alpha} \equiv \ell_{EX}$, the *exchange length*, which is itself proportional to the lattice spacing a. Therefore, our description in terms of a continuous magnetization only makes sense when the width of the wall comprises many lattice spacings.

We can also compute the energy stored in the system due to the presence of the wall. For that one only needs to insert (2.51) in (2.48) to evaluate the energy per unit area of the wall as

$$\frac{E}{S} = 2M^2\sqrt{\alpha\beta_a}. \tag{2.52}$$

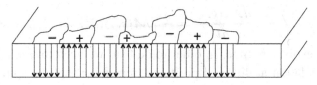

Figure 2.3 Magnetic domains

Notice that this energy is positive and proportional to the width of the wall through a new length scale $\ell_D \equiv \sqrt{\alpha \beta_a}$, which is a direct consequence of the transverse distortion of the magnetization over a finite length ζ along $\hat{x}$. Therefore, this configuration is only a local minimum of the energy functional which has higher energy than the uniformly ordered configuration. Nevertheless, for an infinite slab, the former is separated from the latter by an infinite anisotropy energy barrier and, consequently, is stable. This means that in order to go from one configuration to the other, one has to flip an infinite number of spins over the anisotropy barrier to which each one of them is subject. When this happens we say that the two configurations carry different topological charges (Rajaraman, 1987). However, by the same token, we see that in order to create two walls – a clockwise followed by a counter-clockwise distortion – the spins must overcome a finite energy barrier which depends roughly on the distance between the centers of the two walls. Therefore, depending on how strong the anisotropy is, thermal fluctuations can create a large number of pairs of walls (solitons and anti-solitons (Rajaraman, 1987)) which separate *domains* of reversed magnetization as in Fig. 2.3.

Although the energy barriers we have just mentioned depend on the number of spins to be flipped, the energy of the final multi-domain configuration is the sum of the energy of each wall. This is true at least for well-separated walls. Therefore, within the approximation we have been using of a very strong anisotropy, this situation is unlikely to happen. Nevertheless, if we relax this condition and take into account the whole energy stored in the magnetic field originated by the magnetization configuration of the sample (Kittel, 2004), one sees that the presence of many domains indeed reduces the total magnetic energy compared with that resulting from the fully magnetically oriented configuration. That is the reason why some substances, even below their Curie temperature, do not show any spontaneous magnetization. In order for this to happen we need to apply a small external field to reorient all these small domains.

2.3 Dynamics of the magnetization

As the energy functional (2.43) can be transformed into a quantum mechanical operator if one replaces the classical magnetization $\mathbf{M}(\mathbf{r}, t)$ by $\mathcal{M}(\mathbf{r}, t)$, we can use it directly in the Heisenberg equation of motion

$$\frac{d\mathcal{M}(\mathbf{r}, t)}{dt} = \frac{1}{i\hbar}[\mathcal{M}(\mathbf{r}, t), \mathcal{H}] \tag{2.53}$$

to find how it evolves in time. In order to do that, we need to generalize the spin commutation relation at different positions (remember that the spin components are operators)

$$\mathbf{S}_i \times \mathbf{S}_j = i\hbar\delta_{ij}\mathbf{S}_i \tag{2.54}$$

to the continuous magnetization operator $\mathcal{M}(\mathbf{r}, t)$ which can be achieved using (2.17) and (2.42), leading us to

$$\mathcal{M}(\mathbf{r}, t) \times \mathcal{M}(\mathbf{r}', t) = -ig_s\mu_B\mathcal{M}(\mathbf{r}, t)\delta(\mathbf{r} - \mathbf{r}'), \tag{2.55}$$

where we have explicitly considered the expression in the electronic case. However, from now on, we shall always use γ_g in our expressions and adopt our sign convention.

Then, proceeding as before and replacing $\mathcal{M}(\mathbf{r}, t) \rightarrow \mathbf{M}(\mathbf{r}, t)$ to return to the classical description, we reach the equation of motion for the magnetization which reads (Akhiezer *et al.*, 1968)

$$\frac{d\mathbf{M}(\mathbf{r}, t)}{dt} = \gamma_g\mathbf{M}(\mathbf{r}, t) \times \mathbf{H}^{(eff)} \tag{2.56}$$

where

$$\mathbf{H}^{(eff)} = -\frac{\delta\mathcal{H}}{\delta\mathbf{M}(\mathbf{r}, t)} = \mathbf{H} + \beta_a\hat{\mathbf{n}}(\hat{\mathbf{n}} \cdot \mathbf{M}) + \mathbf{H}_M + \tilde{\boldsymbol{\alpha}} \cdot \nabla^2\mathbf{M}, \tag{2.57}$$

with $\tilde{\boldsymbol{\alpha}}$ being a diagonal matrix whose elements are $\alpha_x, \alpha_y, \alpha_z$. So, we see that the magnetization precesses about an effective field provided by the external field plus various other terms that can be cast in a form which depends on the magnetization itself. It is clearly a very intricate motion but there are some very general conclusions we can draw from that just by analyzing its contents qualitatively. However, before doing that for three specific situations presented below, let us introduce another very important term in that equation of motion.

In dealing with the dynamics of the magnetization in a ferromagnet, one can never forget that there is an enormous number of effects which we have neglected but are by no means unimportant. As in any other macroscopic motion, we always have to account for the losses that inevitably take place here too. However, in realistic situations, the sources of damping are generally very hard to treat and usually a phenomenological term is introduced in the equation of motion. In this particular case, this has been done by Landau and Lifshitz (1935) and the proposed equation is

$$\frac{d\mathbf{M}}{dt} = \gamma_g\mathbf{M} \times \mathbf{H}^{(eff)} - \frac{\lambda}{M^2}\left(\mathbf{M} \times \left(\mathbf{M} \times \mathbf{H}^{(eff)}\right)\right). \tag{2.58}$$

It is easy to see that the second term on the r.h.s. of (2.58) indeed causes damping since it is a torque perpendicular to the magnetization vector and directed to the effective field axis. In other words, it tends to align the magnetization with the effective field. Later on, Gilbert (2004) deduced from a variational principle another phenomenological expression, which reads

$$\frac{d\mathbf{M}}{dt} = \gamma_g \mathbf{M} \times \left[\mathbf{H}^{(eff)} - \eta \frac{d\mathbf{M}}{dt} \right]$$

$$= \gamma_g \mathbf{M} \times \mathbf{H}^{(eff)} - \frac{\alpha_d}{M} \mathbf{M} \times \frac{d\mathbf{M}}{dt}, \quad (2.59)$$

where $\alpha_d = \eta \gamma_g M$ is the dimensionless damping constant. He also showed that (2.58) could be written in this same form with a redefined gyromagnetic ratio as

$$\frac{d\mathbf{M}}{dt} = \bar{\gamma}_g \mathbf{M} \times \mathbf{H}^{(eff)} - \frac{\alpha_d}{M} \mathbf{M} \times \frac{d\mathbf{M}}{dt}, \quad (2.60)$$

where $\alpha_d = \lambda/\gamma_g M$ is the same as above and $\bar{\gamma}_g = \gamma_g(1 + \alpha_d^2)$, the redefined gyromagnetic ratio, implies a faster precession of the magnetization as the damping is increased. Therefore, for $\alpha_d \ll 1$ we see that (2.59) and (2.58) are approximately the same equation.

Before turning our attention to the analysis of some particular problems related to the magnetization dynamics, let us digress a little from the case of a ferromagnetic material to discuss the magnetization dynamics in paramagnetic materials just for the sake of completeness.

In this case, the equation of motion for the magnetization does not obey any of the two forms (2.58) or (2.60) but rather is written as the famous *Bloch equations* (Slichter, 1996; Kittel, 2004)

$$\frac{dM_z}{dt} = \gamma_g (\mathbf{M} \times \mathbf{H})_z + \frac{M_0 - M_z}{T_1},$$

$$\frac{dM_x}{dt} = \gamma_g (\mathbf{M} \times \mathbf{H})_x - \frac{M_x}{T_2},$$

$$\frac{dM_y}{dt} = \gamma_g (\mathbf{M} \times \mathbf{H})_y - \frac{M_y}{T_2}, \quad (2.61)$$

where $M_0 = n\mu \tanh(\mu B/k_B T)$ is the equilibrium value of the magnetization. μ is the magnetic moment of the particles we are treating, n the number of magnetic moments per unit volume and $\mathbf{H}$ is an external field with a static component along $\hat{z}$.

Equation (2.61) is very useful for studying magnetic resonance problems where the magnetic field $\mathbf{H}$ is always decomposed as $\mathbf{H}_0 + \mathbf{H}_1(t)$ with $\mathbf{H}_0$ being the static field component along $\hat{z}$ and $H_0 \gg H_1$. The damping terms are now completely different from those proposed for ferromagnetic materials and depend on the *longitudinal relaxation time*, T_1, and the *transverse relaxation time*, T_2, whose origins we describe in what follows.

The longitudinal relaxation time originates from the spin–lattice interaction (Slichter, 1996; Kittel, 2004), which is ultimately the main mechanism responsible for the approach to equilibrium of a paramagnetic sample. For instance, in the absence of $\mathbf{H}_1(t)$, we can easily show (Kittel, 2004) that

$$M_z(t) = M_0 \left(1 - \exp -\frac{t}{T_1} \right). \tag{2.62}$$

In contrast, the transverse, or *dephasing time*, is due to the local fluctuation of the magnetic field felt by a single magnetic moment which makes it precess at a random Larmor frequency. These fluctuations are caused by the neighboring magnetic moments, mainly via the magnetic dipole interaction and, as a result, both average values $\langle M_x \rangle = \langle M_y \rangle = 0$ in thermal equilibrium. The relation between T_1 and T_2 depends on the specific system with which one is dealing, and may vary from $T_1 \approx T_2$ to $T_1 >> T_2$.

A final remark about the dynamics of the magnetization is now in order. In either case treated above we have always dealt with the magnetization vector, which indeed represents an average either of the classical (2.5) or quantal (2.42) magnetization expressions. In reality, to all of the above-mentioned damped equations of motion there must be added noise terms (fluctuating torques) (Coffey *et al.*, 1996), which would ultimately be responsible for the description of the equilibrium properties of our systems at high or low temperatures. Although they do not influence the relaxation of the average value of the magnetization to equilibrium, they are crucial to account for the fluctuations shown by the dynamical variable about its equilibrium configuration.

2.3.1 Magnetic particles

The first problem we want to treat now is that of a small number of magnetic moments which interact ferromagnetically. The number of constituents, although small compared with Avogadro's number, is considered large enough to allow the system to become ordered below a certain critical temperature. So, this whole set of particles would behave as a single huge spin with $S >> 1$. This is what we call a *magnetic particle* and a possible example thereof is a small ferromagnetic region with linear nanoscopic dimensions.

In such a case, the dynamics of the particle can be described by equation (2.60), where the effective field (2.57) contains no exchange factor. In other words, we are assuming that the linear dimensions of the particle are all much shorter than the

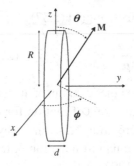

Figure 2.4 Magnetic disk

characteristic length ζ of possible distortions of the magnetization vector studied above.

In order to analyze the motion of this megaspin, we consider the particular case for which the magnetic energy (2.43) has $\alpha_i = 0$, $\hat{\mathbf{n}} = \hat{\mathbf{z}}$, $\mathbf{H}_M$ given by (2.46) with $N_y \approx 4\pi$, $N_x = N_z \approx 0$ and external field $\mathbf{H} = H\hat{\mathbf{z}}$. The uniform energy density of the particle is then

$$\mathcal{E}(\mathbf{M}) = \frac{E(\mathbf{M})}{\mathcal{V}} = -\frac{1}{2} \beta_a M_z^2 + \frac{1}{2} N_y M_y^2 - \mathbf{M} \cdot \mathbf{H}, \tag{2.63}$$

where $E(\mathbf{M})$ is the total magnetic energy of the particle and $\mathcal{V}$ its volume. This choice means that the magnetic particle has, for example, a geometric form close to that of a very thin disk (radius $R \gg$ thickness d) (see Fig. 2.4) placed on the xz plane, which has a uniaxial anisotropy along the $\hat{\mathbf{z}}$ direction and is subject to an external field in this same direction. However, since the structure of (2.63) can apply to other geometrical forms with uniaxial and in-plane anisotropies, we shall rewrite it more generically as

$$E(\boldsymbol{\mu}) = -k_1 \mu_z^2 + k_2 \mu_y^2 - \mu_z H, \tag{2.64}$$

where $\boldsymbol{\mu} = \mathcal{V}\mathbf{M}$ is the total magnetic moment of the particle and k_1 and k_2 have dimensions of inverse volume. For simplicity we will neglect the damping term in (2.60) and write the equations of motion of the components of the magnetic moment as

$$\frac{d\mu_x}{dt} = 2\gamma_g (k_1 + k_2)\mu_y \mu_z + \gamma_g H \mu_y,$$

$$\frac{d\mu_y}{dt} = -\gamma_g (2k_1 \mu_z + H)\mu_x,$$

$$\frac{d\mu_z}{dt} = -2\gamma_g k_2 \mu_y \mu_x, \tag{2.65}$$

where we have used for the effective field (2.57)

$$\mathbf{H}^{(eff)} = -\frac{\partial E(\boldsymbol{\mu})}{\partial \boldsymbol{\mu}} = -2k_2\mu_y\hat{\mathbf{y}} + (2k_1\mu_z + H)\hat{\mathbf{z}}. \qquad (2.66)$$

We can now express the magnetic moment in terms of its polar representation and solve the resulting equations of motion for the spherical angles θ and ϕ. However, there is a much more straightforward way to achieve these equations for θ and ϕ through the Hamiltonian formulation of the problem (Chudnovsky and Gunther, 1988b; Stamp *et al.*, 1992).

As we know, the spin component S_z is the momentum canonically conjugate to the azimuthal angle ϕ. Consequently, remembering that $S_z = (\mu/\gamma_g)\cos\theta$, with $\mu = |\boldsymbol{\mu}|$, the Hamilton equations read

$$\dot{\theta}\sin\theta = \frac{\gamma_g}{\mu}\frac{\partial E(\theta,\phi)}{\partial\phi},$$

$$\dot{\phi}\sin\theta = -\frac{\gamma_g}{\mu}\frac{\partial E(\theta,\phi)}{\partial\theta} \qquad (2.67)$$

where, from (2.64), $E(\theta,\phi)$ is, up to a constant, given by

$$E(\theta,\phi) = (K_1 + K_2\sin^2\phi)\sin^2\theta + \mu H(1 - \cos\theta), \qquad (2.68)$$

where $K_i \equiv k_i\mu^2$. From the above expression for the energy of the magnetic particle we can easily deduce that this function has minima at $\theta = 0$ and $\theta = \pi$, and $\phi = 0$. This is a physically plausible result, indicating that the in-plane anisotropy forces the magnetic moment to lie in the xz plane whereas the uniaxial anisotropy directs it along $\pm\hat{\mathbf{z}}$. Between these two minimal energy configurations, $\boldsymbol{\mu} = \pm\mu\hat{\mathbf{z}}$, there is also an energy barrier whose maximum value is located at $\theta = \theta_m$, where $\cos\theta_m = -H/H_c$ and $H_c = 2K_1/\mu$ is the so-called *coercivity field*. For $H \gg H_c > 0$, expression (2.68) is dominated by the external field contribution, which means that the magnetic moment precesses about the $\hat{\mathbf{z}}$ axis like a spinning top. If the particle is in its magnetic ground-state configuration one has $\boldsymbol{\mu} = +\mu\hat{\mathbf{z}}$. As the modulus of the external field is reduced, the magnetic moment still points along $\hat{\mathbf{z}}$ until $H < H_c$ when the structure of maxima and minima of (2.68) develops, as we have described above. In this case the motion of $\boldsymbol{\mu}$ is about the xz plane, which allows us to perform some approximations in (2.67). Taking the derivative of the first of those equations with respect to time, replacing the resulting $\dot{\phi}$-dependent term by the second equation and using the fact that $\phi \approx 0$, we are led to an equation of motion solely parametrized by the polar angle θ, which reads

$$\ddot{\theta} = -\frac{dU}{d\theta}, \qquad (2.69)$$

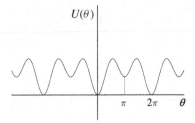

Figure 2.5 Magnetic energy for $H_c > H > 0$

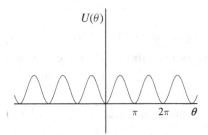

Figure 2.6 Magnetic energy for $H = 0$

where

$$U(\theta) = \frac{\omega_0^2}{2} \sin^2 \theta + \omega_0^2 \frac{H}{H_c}(1 - \cos \theta), \tag{2.70}$$

with $\omega_0 \equiv 2\gamma_g \sqrt{K_1 K_2}/\mu$. It should be stressed here that the approximations we have just mentioned, and consequently their resulting equation (2.69), are only valid for fields not so close to H_c. For $H_c > H > 0$, the potential energy $U(\theta)$ develops local minima $U(0) = 0$ and $U(\pi) = 2\omega_0^2 H/H_c$ as shown in Fig. 2.5. The former is the absolute minimum of the function, representing the stable configuration of the magnetic moment, whereas the latter, a relative minimum of the energy, is a metastable configuration of that quantity. Further analyzing (2.70) it can be shown that the magnetic moment can oscillate with frequencies $\omega_0(1 \pm H/H_c)$ about each minimum of the potential, where the plus sign applies to $\theta = 0$ and the minus sign to $\theta = \pi$.

When $H = 0$, the potential $U(\theta)$ becomes symmetric about $\theta = \pi/2$ and consequently $U(0) = U(\pi) = 0$, which means that the configurations $\boldsymbol{\mu} = \pm\mu\hat{z}$ are now degenerate (see Fig. 2.6).

Now, if we invert the direction of the external field, which means that $H < 0$ or $H = -|H|$, the role played by the two minima of the potential function is exchanged. The previous absolute minimum $U(0) = 0$ is now a relative minimum of the function, whereas $U(\pi) = -2\omega_0^2|H|/H_c$ has become the absolute

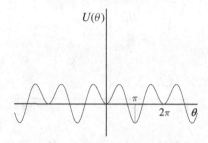

Figure 2.7 Magnetic energy for $H < 0$

minimum of the potential energy (see Fig. 2.7). Therefore, the former metastable configuration $\boldsymbol{\mu} = -\mu\hat{\mathbf{z}}$ is now the most stable magnetic configuration of the particle, whereas $\boldsymbol{\mu} = +\mu\hat{\mathbf{z}}$ has become the new metastable configuration of the system.

As we increase the modulus of the external field further in the direction opposite to the initial configuration of the magnetic moment, $\boldsymbol{\mu} = +\mu\hat{\mathbf{z}}$, the local minimum at $\theta = 0$ becomes shallower and the magnetic moment is only kept along this direction by a potential barrier $U_0 = \epsilon\omega_0^2/2$, where $\epsilon \equiv 1 - (|H|/H_c)$. When $H = H_c$, the coercivity field, the minimum at $\theta = 0$ becomes a maximum of the potential energy and the magnetic moment rotates about $\hat{\mathbf{y}}$ and ends up at $\boldsymbol{\mu} = -\mu\hat{\mathbf{z}}$. Actually, if it were not for dissipative effects, the magnetic moment would be indefinitely oscillating in the xz plane about the latter configuration due to energy conservation. Therefore, in order to fully account for the realistic dynamics of the magnetic moment of this particle, we need to introduce a phenomenological dissipative term in (2.69) which would originate from the damping term in the Landau–Lifshitz–Gilbert equation (2.60). Moreover, as we have mentioned before, we also need to consider the presence of a noise term in that equation to account for the equilibrium properties of the magnetic particle.

The behavior of the magnetic moment we have just described is the well-known example of *hysteresis* and is illustrated in Fig. 2.8. As we move along the segment *abcd* in Fig. 2.8 by changing the external field, the magnetization (remember $\boldsymbol{\mu} = \mathcal{V}\mathbf{M}$) changes from the stable configuration between *ac* to a bistable configuration at *c* and then becomes metastable along *cd*. When we reach point *d* the configuration becomes unstable and makes a transition to the more stable configuration at *d'*. If we now trace this contour backwards by changing the direction of the external field, the magnetization remains stable along *d'c'*, bistable at *c'*, and metastable along *c'b'*, beyond which it returns to the original magnetization $\mathbf{M} = +M\hat{\mathbf{z}}$. However, this description does not reflect the more realistic situation, which is depicted in Fig. 2.9 and explained below.

If the temperature of the particle is $T < T_c$ (the ferromagnetic transition temperature), but high enough so the total magnetic moment can still be described

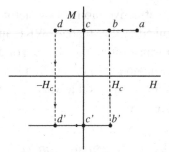

Figure 2.8 Hysteresis: Ideal case

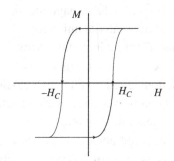

Figure 2.9 Hysteresis: Realistic case

classically, thermal fluctuations can make the magnetization jump over the potential barrier U_0 introduced above. Actually, the closer the external field is to the coercivity value $-H_c$ ($\epsilon \ll 1$) the smaller this barrier gets and, consequently, it is easier for the magnetization to make a transition to the more stable configuration $\mathbf{M} = -M\hat{\mathbf{z}}$ before that extremal value of the external field is reached. This effect clearly rounds off the corners at b, d, d', b' in Fig. 2.8, which results in Fig. 2.9.

As the temperature is lowered it becomes harder for the magnetization to be driven out of the well by thermal fluctuations and it might become frozen at the metastable configuration until the coercivity field is reached. Nevertheless, if we allow for the possibility of quantum mechanical effects taking place, the megaspin about which we have been talking will behave as a quantum rotor and the above-mentioned transitions will occur by the rotation of the magnetization vector (or magnetic moment) under the potential barrier created by $U(\theta)$; in other words, it might take place by quantum mechanical tunneling of a fictitious particle whose classical dynamics is described by the equation of motion (2.69).

Once again neglecting dissipation, we easily see that the quantum mechanical behavior of the system classically described by (2.69) is governed by the Schrödinger equation

$$i\hbar\frac{\partial \psi(\theta)}{\partial t} = -\frac{\hbar^2}{2I}\frac{\partial^2 \psi(\theta)}{\partial \theta^2} + IU(\theta)\psi(\theta), \tag{2.71}$$

where $I \equiv \mu^2/2\gamma_g^2 K_2$ is the effective moment of inertia associated with the magnetic particle. This can be solved by the standard WKB approximation and provides us with the quantum rate of transition in the absence of dissipation. The combined effect of the tunneling phenomenon with dissipation is what we expect the mechanism to be for the magnetization transition at very low temperatures, and its study will be one of our aims in this book.

2.3.2 Homogeneous nucleation

Now let us turn our attention to a more subtle situation. Suppose we return to our previous example of a very thin ferromagnetic slab placed on the xy plane which this time has, on top of the uniaxial anisotropy along the $\hat{z}$ direction, an in-plane anisotropy of the same form as in (2.64). Notice, however, that this is not a demagnetization effect as before, and the reason is twofold. Firstly, we want to allow for a position-dependent magnetization which would invalidate the particular form described in (2.46) and secondly, even if the magnetization were uniform, the present geometry would not generate a demagnetization term of that sort. Moreover, let us assume that the uniaxial anisotropy along $\hat{z}$ is much stronger than the demagnetization along this same direction. Therefore, its Hamiltonian functional in terms of the polar angles $\theta(\mathbf{r})$ and $\phi(\mathbf{r})$ must be a combination of (2.48) and (2.68), which reads (Chudnovsky *et al.*, 1992; Stamp *et al.*, 1992)

$$\mathcal{H}[\theta(\mathbf{r}), \phi(\mathbf{r})] = b \int\limits_{-\infty}^{+\infty} \int\limits_{-\infty}^{+\infty} dx dy \left[\frac{\alpha M^2}{2} (\nabla_2 \theta)^2 + \frac{\alpha M^2 \sin^2 \theta}{2} (\nabla_2 \phi)^2 \right.$$

$$\left. + (K_1 + K_2 \sin^2 \phi) \sin^2 \theta + M H (1 - \cos \theta) \right], \quad (2.72)$$

where b is the thickness of the slab and we have implicitly assumed that we are looking for solutions which are uniform along z. Consequently, ∇_2 contains partial derivatives along x and y only. Notice that now, K_1 and K_2 are anisotropy energy densities.

As we have seen before, the dynamics of the magnetization of the system can be obtained within the Hamiltonian formulation from the variation of the magnetic action; $\delta S[\theta, \phi] = 0$, where

$$S[\theta(\mathbf{r}), \phi(\mathbf{r})] = b \int dt \int\limits_{-\infty}^{+\infty} \int\limits_{-\infty}^{+\infty} dx dy \frac{M}{\gamma_g} \cos \theta(\mathbf{r}) \dot{\phi}(\mathbf{r})$$

$$- \int dt \mathcal{H}[\theta(\mathbf{r}), \phi(\mathbf{r})], \quad (2.73)$$

and now $\dot{f}(\mathbf{r})$ stands for $\partial f(\mathbf{r}, t)/\partial t$.

The equations of motion then read

$$\frac{\delta S[\theta(\mathbf{r}), \phi(\mathbf{r})]}{\delta \theta} = 0 \Rightarrow + \frac{M}{\gamma_g} \sin\theta \, \dot{\phi} - \alpha \, M^2 \, \nabla_2^2 \theta + \frac{\alpha M^2}{2} \sin 2\theta \, \nabla_2^2 \phi$$
$$+ \left(K_1 + K_2 \sin^2 \phi\right) \sin 2\theta + M H \sin\theta = 0$$

(2.74)

and

$$\frac{\delta S[\theta(\mathbf{r}), \phi(\mathbf{r})]}{\delta \phi} = 0 \Rightarrow + \frac{M}{\gamma_g} \sin\theta \, \dot{\theta} + \alpha M^2 \sin^2 \theta \, \nabla_2^2 \phi$$
$$+ \alpha M^2 \sin 2\theta \, \nabla_2 \phi \cdot \nabla_2 \theta - K_2 \sin^2 \theta \, \sin 2\phi = 0.$$

(2.75)

This pair of equations would be exactly the same as (2.67) if not for the gradient terms we have just introduced to account for possible distortions of the magnetization of the system. The whole discussion following (2.67) can indeed be transposed here without any modification, and even the approximations which led us to (2.69) can also be applied here. An equation equivalent to the latter now reads

$$\ddot{\theta} = c_s^2 \, \nabla_2^2 \theta - \frac{dU}{d\theta},$$

(2.76)

where $c_s \equiv \sqrt{2 \alpha K_2 \gamma_g^2}$ is the characteristic spin-wave velocity in this system and $U(\theta)$ is exactly the potential defined in (2.70).

This equation admits several solutions depending on the symmetries of the configurations for which one is looking. For instance, in the absence of an external field, we can also have static solutions which are uniform along y and z. However, contrary to what we have seen in (2.49), where we had the freedom to choose $\phi = \pi/2$, we now have solutions with $\phi = 0$ and θ still given by (2.50). This is an example of a *Néel wall*. Another important solution can be obtained for $-H_c < H < 0$. Keeping $\phi = 0$ and defining the new variable $\rho \equiv \sqrt{r^2 - c_s^2 t^2}$ in terms of which (2.76) can be rewritten as

$$\frac{d^2\theta}{d\rho^2} + \frac{2}{\rho} \frac{d\theta}{d\rho} - \frac{dU}{d\theta} = 0,$$

(2.77)

we can show that it admits a solution $\theta(\rho)$ of the form displayed in Fig. 2.10, which has a very interesting interpretation.

If we start with a positive external field, the magnetization of the sample will point along the $\hat{z}$ direction and it can be kept as such if we carefully change the magnetic field and allow it to point in the opposite direction as long as $-H_c < H < 0$. As we have seen before, in the example of a magnetic particle, this situation is metastable and thermal fluctuations, for example, could drive the magnetization vector out of this configuration even before the coercivity field is

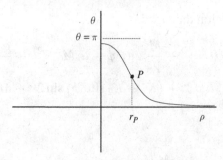

Figure 2.10 Magnetic droplet

reached. The difference here is that we now have an extended object and need to analyze how this sort of transition can take place in this new situation.

Suppose now a droplet of spins is formed with a magnetization pointing in the direction of the external field. It is easy to convince ourselves that in this case there is a reduction in the energy of the system proportional to the volume of this droplet of reversed magnetization. However, this is not the only effect we should take into account. The droplet formation is followed by a distortion of the magnetization, which is energetically costly and competes with the previous effect we have just described. Consequently, depending on the size of the region formed with reverse magnetization, the positive surface energy due to the distortion can impede the droplet expansion. This is the case for droplets with a volume smaller than a critical volume. The latter is defined exactly as the volume above which the volumetric energy reduction achieved by inverting a given number of spins wins over the surface energy caused by distortion. Therefore, droplets can appear and shrink at any point of the sample until one of them is formed with volume above the critical size, which expands and converts the whole sample into the more stable configuration. This is the phenomenon of nucleation, which is very well known in statistical mechanics (Landau and Lifshitz, 1974).

The solution whose behavior is presented in Fig. 2.10 represents exactly the physics discussed above. At $t = 0$ the variable ρ is nothing but the radial coordinate r relative to a given origin. This solution interpolates between $\theta \lesssim \pi$ at the origin and $\theta = 0$ at $r \to \infty$, and this excursion represents the distortion (wall) we have just mentioned. If we follow the time evolution of the point P represented in Fig. 2.10 we see that if its initial distance to the center of the droplet is $r = r_p$ it will evolve in time as $r^2(t) = r_p^2 + c_s^2 t^2$, as expected from the argument above.

When $-H_c \lesssim H < 0$, $\theta(0)$ can be much smaller than π, meaning that close to the instability, a very small distortion of a few spins will be enough to make the whole system undergo a transition to the more stable configuration about $\theta = \pi$. The excess energy stored in the droplet disappears as relaxation effects are taken

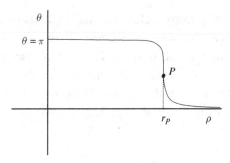

Figure 2.11 Magnetic droplet in the thin-wall approximation

into account and the magnetization stabilizes at $\theta = \pi$. In contrast, if $H \lesssim 0$ the size of the critical droplet must be huge in order to store a volumetric energy greater than the surface distortion energy, as shown in Fig. 2.11. This is the so-called *thin-wall* approximation. In the extreme limit $H \to 0$ the size of the droplet becomes infinite, which means that only if all the spins in the system revert their direction would the transition be possible. This clearly represents an extremely rare event if the transition is driven by any kind of fluctuation.

The same sort of analysis carried out for fluctuations of magnetic particles can also be done here. For high temperatures we expect that thermal fluctuations are responsible for the transition from the metastable to the more stable configuration. Nevertheless, as the temperature is lowered, that is no longer the main mechanism for the transition. Instead, we should expect quantum mechanical fluctuations (decay by quantum tunneling) to play a major role in the new situation. Later on we will show the relevance of the imaginary time version of the kind of solution we have been discussing here to this sort of matter. Notice that once again we have neglected completely the effect of dissipation on this phenomenon, which will be addressed at the proper time.

Before leaving the subject of nucleation a few words should be said about the existence of stable lattices of the so-called *magnetic bubbles* (Eschenfelder, 1980; Bertotti, 1998) in almost two-dimensional magnetic samples. Everything we have said so far about the magnetic slab used as the example in this section assumes that we are dealing with a very hard magnetic material, which means that demagnetization effects can be entirely neglected. If we relax this requirement and take into account the presence of a demagnetization field along the $\hat{z}$ direction, the energetic analysis of our problem changes somewhat.

Now, even for external fields still pointing along the spontaneous magnetization, $0 < H < H_c$, the demagnetization field contribution resulting from the surface term in (2.45) favors the formation of a domain of reversed magnetization whose size depends on the competition among the strengths of the external, demagnetization, and anisotropy fields. As the modulus of the external field is reduced, more and larger domains are formed, giving rise to a magnetic bubble lattice. If the field is further reduced, these domains coalesce (Eschenfelder, 1980) and a *striped phase* results. Once again this energetic analysis can be enriched with fluctuation arguments to explain the dynamical formation of these objects in the same fashion as we have applied in the previous situation.

2.3.3 Wall dynamics

Let us now return to the case where there is only one Néel wall present in the example given above for $H = 0$. As (2.76) is invariant under a Lorentz transformation where the speed of light is replaced by the spin-wave velocity c_s, it admits solutions, as in (2.51), of the form

$$\theta(x, t) = 2 \arctan\left(\exp \mp \frac{x - vt}{\sqrt{1 - (v/c_s)^2}\zeta}\right), \tag{2.78}$$

for $v < c_s$, which means that walls with constant speed are also allowed in the system with no energy cost.

In contrast, if the external field is now turned on, part of the system will have its magnetization pointing opposite to the field direction and therefore its size has to be reduced in order to lower the total energy of the system. This clearly implies the acceleration of the wall in a way similar to that presented for the nucleation problem.

In order to deduce the equation of motion for the displacement of the wall, let us assume that its center can be described by $x_0(t)$, independently of the coordinates y or z, and its profile is chosen based on (2.78) as

$$\theta(x, t) = 2 \arctan\left(\exp -\frac{x - x_0(t)}{\zeta}\right). \tag{2.79}$$

Therefore, substituting (2.79) into (2.76) we can show that, at least for speeds $\dot{x}_0(t) \ll c_s$, the equation obeyed by $x_0(t)$ is

$$\ddot{x}_0 = -\left(\frac{H}{H_c}\right)\frac{c_s^2}{\zeta}. \tag{2.80}$$

Notice that since our choice for the sign of the exponent in (2.78) was negative, one has $\theta(-\infty) = \pi$ and $\theta(\infty) = 0$ and consequently, when subject to an external field along $\hat{z}$, the wall will move along the negative $\hat{x}$ direction, justifying the minus sign in (2.80). Here it should be stressed that the appearance of H_c in (2.80) does not give it special physical meaning in this problem and we should regard it only as a natural scale for the magnetic field in dealing with the wall motion in this specific model.

To analyze the motion of a magnetic wall in a general medium we must remember that usually there are many imperfections in the sample, such as point-like or extended defects, which generally act as *pinning* centers and can either slow down or efficiently trap the magnetic wall (Chudnovsky and Gunther, 1988a; Stamp *et al.*, 1992; Braun *et al.*, 1997; Brazovskii and Nattermann, 2004). These create an effective potential, $\tilde{V}_{pin}(u(y, t))$, to the generalized coordinate, $u(y, t)$, along the $\hat{x}$ direction which now describes the displacement of an elastic line representing the position of the wall and whose equation of motion reads

$$\frac{1}{c_s^2}\frac{\partial^2 u(y, t)}{\partial t^2} - \frac{\partial^2 u(y, t)}{\partial y^2} + \frac{\partial \tilde{V}_{pin}(u)}{\partial u} = -\left(\frac{H}{H_c}\right)\frac{1}{\zeta}. \tag{2.81}$$

A simple dimensional analysis shows that $\tilde{V}_{pin}(u(y, t))$ above is a dimensionless potential.

Notice that what we have just done is introduce a dependence of the wall center $x_0(t)$ on y since the wall no longer needs to be uniform along that direction. We could also have allowed for a y-dependent width of the wall, $\zeta(y)$, to make it as general as possible, but under very general and reasonable hypotheses we can still show that the wall will preserve both its profile and width and the relevant dynamical variable will be only its displacement $u(y, t)$. This also means that we are assuming the presence of the potential generated by the pinning centers, $\tilde{V}_{pin}(u(y, t))$, will not affect the internal structure of the wall. Another important remark we should make about the wall motion is that there should be an extra term in (2.81), for example of the form $\eta \, \partial u/\partial t$, to account for the viscous motion it presents in realistic situations. Although damping has been phenomenologically introduced in our equation for the magnetization dynamics by expressions like (2.58), (2.59), or (2.60), the wall has its damped motion due to other mechanisms (for example, scattering of spin waves) that we shall address in future chapters of this book. For the time being, we shall keep on neglecting dissipative terms in our forthcoming analysis of the problem.

Now let us analyze the pinning potential in two special circumstances. The first is the so-called *strong pinning* case, where the impurities are either strong, point-like but sparsely spaced, or of extended nature. In particular, assume that it creates a potential $\tilde{V}_{pin}(u(y, t))$ characterized by a well of dimensionless energy depth $\tilde{V}_0$ and centered along a line at $x = 0$ as shown in Fig. 2.12 if the external field is zero. This is due, for example, to the presence of a line of ions which interact with the neighboring atoms of the host lattice through a different exchange coupling. The wall can then be accommodated along this line and only suffer thermal fluctuations about this equilibrium position. However, if the external field is turned on, a

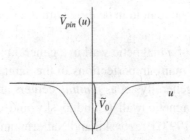

Figure 2.12 Pinning potential

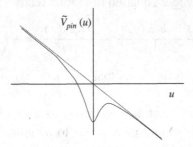

Figure 2.13 Pinning potential for $H \neq 0$

constant force like the one present in (2.81) appears and provides us with a linear potential which tilts the previously introduced pinning potential; this now acquires the form shown in Fig. 2.13, rendering the wall configuration metastable.

As the external field is increased it will eventually reach the value H_d, the depinning field, at which the maximum and minimum of Fig. 2.13 coalesce and become an inflection point of that function. When this happens the wall is entirely free to move downhill. However, this depinning transition can take place even before the depinning field H_d is reached, by the same mechanism we have introduced to explain homogeneous nucleation in the previous section. If thermal fluctuations activate a given length of the wall over the tilted pinning barrier there is, as before, competition between the energy lost by this line segment on the other side of the barrier and the distortion energy that must be paid to create it. This competition generates a critical length L_c above which the energy lost by the depinned region on the wall increases if it drags the rest of the wall downhill. Operationally, this depinning phenomenon can be treated exactly as the homogeneous nucleation problem if one makes the following replacements in (2.76): $\theta \rightarrow u$, $\nabla_2 \rightarrow \partial/\partial y$ and $U \rightarrow V_{pin} + Hu/(H_c\zeta)$ according to the sign of (2.80). Once again, although H_c naturally appears in our latest replacement, the meaningful field scale that should be used here is the depinning field H_d whose computation depends on the specific details of $\tilde{V}_{pin}$ and the natural field scale H_c for the particular model employed.

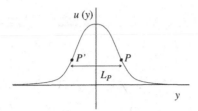

Figure 2.14 Droplet wall

The critical configuration of the line can then be obtained by the equivalent form of (2.77), which now reads

$$\frac{d^2u}{d\rho^2} + \frac{2}{\rho}\frac{du}{d\rho} - \frac{d\tilde{V}_{pin}}{du} = -\left(\frac{H}{H_c}\right)\frac{1}{\zeta}, \tag{2.82}$$

where $\rho \equiv \sqrt{y^2 - c_s^2 t^2}$. The same reasoning followed below (2.77) to obtain the critical droplet profile can be repeated here with the appropriate replacements mentioned above. Note that the cylindrical symmetry of the nucleation problem is now lost, because $y \in (-\infty, +\infty)$ contrary to $r \in (0, +\infty)$. Nevertheless, this does not change any of our previous conclusions and the time evolution of the critical wall profile now follows $y^2(t) = (L_p/2)^2 + c_s^2 t^2$, which means that at $t = 0$ instead of a droplet of radius r_p we have a configuration characterized by the length L_p measuring the distance between the two symmetrical points P and P' (see Fig. 2.14) located at the steepest slopes of the distortions of the line $u(y, 0)$. Once this critical distortion is formed, the wall is dragged downhill.

Another important difference between the two analyzed situations is that, contrary to the nucleation case, there is no other potential minimum which can accommodate the line at a later time, unless we consider the presence of other defect lines composing the pinning potential. For example, if only two closely spaced, very strong pinning lines at $x = 0$ and $x = a$ are present, there is a perfect analogy between the two problems as the wall depinning can be regarded as a nucleation from its metastable configuration at $x = 0$ (for $-H_d < H < 0$) to the stable one at $x = a$. This takes place by the formation of a pair of distortions (kink–anti-kink; see Rajaraman (1987)) of the elastic line $u(y, t)$, since its previous unstable growth ceases now at $u(y, t) = a$. It is always important to warn the reader that there must be relaxation effects in order to drive an excited state of the line about $u = 0$ to its final stable configuration exactly at $u = a$. The hysteresis phenomenon is also present here if we now invert the direction of the external field to $H_d > H > 0$.

We could also consider the pinning potential as formed by the presence of slightly weaker defect lines parallel to y and periodically distributed along x (see

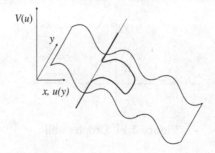

Figure 2.15 Periodic potential energy density for $H \neq 0$

Fig. 2.15). In this case there are many potential barriers to be overcome by the wall which we assume to be possible, for example, by thermal fluctuations. This hopping-like motion of the wall is an example of what is known in the literature as *wall creep*.

The second important circumstance in which we must analyze the depinning problem is the *weak pinning* case. Now we have a random distribution of weak pinning centers and the wall becomes trapped, not because of a localized and well-defined pinning potential as above, but due to the interplay between its elastic properties and the randomness of the defects distribution giving rise to the so-called *collective pinning theory* (Larkin, 1970; Brazovskii and Nattermann, 2004). Although the development of this theory is much more subtle and involving than that previously presented here for strong pinning, a simple dimensional analysis might help us to gain general insight into the physical phenomenon in this case. For example, let us now return to dimensionful quantities and replace the pinning potential $\tilde{V}_{pin}(u)$ by a potential energy per unit area $V_R(y, u)$ given by

$$V_R(y, u) = \int_{-\infty}^{+\infty} dx \, v_R(x, y) \, \rho(x, y, u), \qquad (2.83)$$

where $\rho(x, y, u) = \delta(x - u(y))$ is the wall density, which can be smeared out over the wall width ζ, and $v_R(x, y)$ is such that

$$\langle v_R(x, y) \rangle_R = 0 \quad \text{and} \quad \langle v_R(x_1, y_1) \, v_R(x_2, y_2) \rangle_R = v_R^2 \delta(x_1 - x_2) \delta(y_1 - y_2),$$
$$(2.84)$$

where $\langle \ldots \rangle_R$ stands for average over disorder. In other words, we are saying that $v_R(x, y)$ is Gaussian distributed and short-range correlated with correlation length l (which means that the delta function arguments are strongly peaked at 0 and smeared out over l). Armed with these hypotheses one can show that

$$\langle V_R(y, u)\rangle_R = 0 \quad \text{and} \quad \langle V_R(y_1, u_1)\, V_R(y_2, u_2)\rangle_R = R(u_1 - u_2)\delta(y_1 - y_2).$$

$$(2.85)$$

If we assume, for simplicity, that the wall profile $\rho(x, y, u)$ is a Gaussian centered at $u(y)$ with width ζ, we have $R(u) = v_R^2 k(u)$ where $k(u)$ is also a Gaussian but now spread over $\sqrt{2}\zeta$. In order to check for the veracity of our forthcoming expressions through dimensional analysis, we should notice that v_R^2 has dimensions of energy squared per area.

Let us consider next that a slightly distorted form of the wall can be represented by

$$\theta(x, t) = 2\arctan\left(\exp -\frac{x - u(y)}{\zeta}\right), \tag{2.86}$$

which means that the displacement $u(y)$ is very small on a scale we will determine in what follows.

Now, if we replace (2.86) in (2.72) (for $\phi = 0$), evaluate the integrals on x and introduce the above-defined potential $V_R(y, u)$ in order to recover the elastic part of (2.81), we have

$$\mathcal{H}[u(y)] = b \int_{-\infty}^{+\infty} dy \left[\frac{k_w}{2}\left(\frac{\partial u}{\partial y}\right)^2 + V_R(y, u) - f_w\, u\right], \tag{2.87}$$

where the wall line tension $k_w = 2\alpha M^2/\zeta$ and the external surface tension $f_w = -2HM$. Here it should be said that the linear term in u has been obtained from a variation of the external field energy up to first order in this displacement.

Expression (2.87) allows us to obtain very important information about the wall energetics for the weak pinning situation. Let us start by considering the energy $E_w(L)$ of a stiff wall of length L. In this case it is easy to conclude that due to the first relation in (2.85) one has $\langle E_w(L)\rangle_R = 0$, which means that a stiff wall has average pinning energy equal to zero. However, one can still use the same expression and the second relation of (2.85) to evaluate

$$\langle E_w^2(L)\rangle_R = b^2 \int_0^L \int_0^L dy\, dy'\langle V_R(y, 0)V_R(y', 0)\rangle_R = b^2 v_R^2 \frac{L}{\zeta}, \tag{2.88}$$

which implies that $\sqrt{\langle E_w^2(L)\rangle_R} \propto \sqrt{L}$. This sublinear growth with increasing L is due to the competition between different pinning centers and makes us conclude that a stiff wall is never pinned since the presence of the external field gives rise to a term proportional to L in (2.87). Therefore, if we cut off the sublinear growth at the so-called *collective pinning length* L_c (Larkin, 1970; Brazovskii and Nattermann,

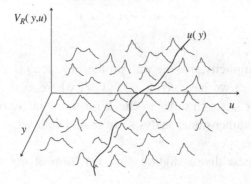

Figure 2.16 Random potential

2004) and allow for small distortions of the wall, its segments of that specific length will be independently pinned. In order to estimate L_c we write (2.87) in an approximate form for a wall of length L and $H = 0$ as

$$E_w(u, L) = b\frac{k_w}{2}\frac{u^2}{L} + bv_R\sqrt{\frac{L}{\zeta}}. \qquad (2.89)$$

Now, minimizing this expression with respect to L for $u \approx \zeta$ we get

$$L_c = \left(\frac{k_w\zeta}{v_R}\right)^{2/3}\zeta, \qquad (2.90)$$

which implies that for pinning energies much weaker than the typical elastic energy scale we have $L_c >> \zeta$. This means that in the absence of an external field the wall finds its optimum profile accommodating long segments of length L_c independently through the valleys of the landscape resulting from the presence of the pinning centers as in Fig. 2.16. It is not hard to imagine that there must be numerous configurations with very similar energies that the wall could accommodate. In other words, there is no reason to think the optimum configuration should be unique, and this leads us again to the concept of metastability. Therefore, our next step is to estimate the energy barriers between these possible configurations, which can easily be done by evaluating $\sqrt{\langle E_w^2(L)\rangle_R}$ in (2.88) at $L = L_c$. This procedure results in

$$U_c = \left(b^3 v_R^2 k_w\zeta\right)^{1/3}. \qquad (2.91)$$

Now, if we turn on the external field H, the random potential energy landscape becomes tilted and there will be a tendency to depin the wall. We can also evaluate the critical depinning field, equating the last term of (2.87) for a wall of length L_c to the energy barrier U_c, which gives us

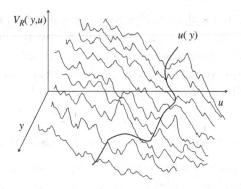

Figure 2.17 Random potential for $H \neq 0$

$$H_{dep} = \left(\frac{v_R}{k_w \zeta} \right)^{1/3} \left(\frac{v_R}{\zeta^2} \right) \frac{1}{M}, \tag{2.92}$$

where M stands for the saturation magnetization of the system. Notice that we could also have estimated the depinning length as a function of the external field, $L_c(H)$, had we performed the minimization of (2.89) for finite H instead. This length would shorten for increasing H and could be used to evaluate an effective potential energy barrier $U_c^{(eff)}(H)$ which must vanish at $H = H_{dep}$. Therefore, as one reaches the depinning field value, the wall moves to accommodate new metastable configurations of lower energy of the tilted random potential energy landscape as shown in Fig. 2.17.

Once again, the same reasoning which led us to argue in favor of depinning for $H < H_{dep}$ can also be used here. As one approaches the depinning field it becomes easier for the pinned segments of length $L_c(H)$ to be, for example, thermally activated over the effective potential barrier $U_c^{(eff)}(H)$. The whole wall ends up in a more favorable energy configuration due to the existence of relaxation effects, which will help it to get rid of the excess energy it should have carried on from the preceding state, and the same process starts over again. The subsequent motion is of the same kind as already analyzed in the periodic strong pinning situation, namely, thermally activated creep. The difference between the present case and that previously studied lies in the mechanism by which the wall hops from one metastable configuration to another. In the strong pinned situation, once a segment of a critical length jumps to the lower-energy minimum, it drags the whole line after itself to the new configuration whereas in the present case, there is a sort of sluggish diffusive motion of the wall to lower-energy configurations.

At any rate, the important conclusion is that in both cases we can talk about critical lengths, fields and energy barriers which are the desiderata to implement the general theory of thermal activation for either case. Actually, this is not only applicable to thermal fluctuations. If one considers now the full (non-dissipative) dynamics of the wall at very low temperatures, one is led to ask about quantum mechanical fluctuations (tunneling) once the appropriate parameters turn out to form an effective action not overwhelmingly larger than $\hbar$.

2.4 Macroscopic quantum phenomena in magnets

Up to now we have been discussing many examples of magnetization dynamics in different models and geometries of ferromagnetic systems. Basically what we have done is employ appropriate approximations to each of our examples in order to reduce a very intricate classical dynamical problem to something that could be represented by a more treatable and intuitive mechanical problem. At least we hope to have succeeded in the three examples we have given above, where we have been able to simplify complex situations and reduce them to particle mechanics or classical field dynamics. Moreover, we have always neglected dissipative terms in our dynamical analysis merely because their effect would only be important in the case where we consider the presence of fluctuations in the phenomenon in which we are interested. Our first step was solely to understand the dynamics of the magnetization under specific circumstances, and damping was only mentioned in order to explain the approach of certain field configurations to either stable or metastable equilibrium states. Actually, relaxation mechanisms also play a very important role when one considers the possibility of thermal activation of our dynamical variable (here collectively representing the magnetization vector) over the potential energy barriers present in the models we have introduced above. We shall return to these issues later in this book.

What we want to discuss now is the possibility we have been loosely mentioning of observing quantum mechanical effects in any of the examples with which we have been dealing. Indeed, since the equations of motion we have can, in the absence of dissipative terms, be deduced within a Hamiltonian formulation, we can employ the canonical quantization procedure directly (see, for example, Merzbacher (1998)) and write, for the particle mechanics problem, a Schrödinger equation as in (2.71). For general field theoretical models, we have to use more appropriate tools to replace the latter. Nevertheless, it is again the straightforward application of the canonical quantization procedure to the "coordinate" of one fictitious particle (or field) representing a very complex quantum mechanical problem.

In principle there is nothing to prevent us from doing this. There are many examples of quantization of approximate equations of motion, or even truncated systems, which we often see in quantum mechanical textbooks. The resulting quantum mechanical effects originating from this procedure must represent a reduced quantum dynamics restricted to a particular subspace of the full Hilbert space of our system. However, even if we boldly assume that the results so obtained do indeed present a signature of quantum effects of the underlying complex system, one should be able to explain what is really going on in terms of the genuine quantum variables (spins or magnetic moments in those cases).

Let us start with the example of the magnetic particle analyzed above in terms of the polar angle θ. In terms of this angular variable our conclusions could be drawn directly from the motion of a fictitious particle in a potential (2.70). However, as we have already mentioned, we are treating here a small particle of a hard magnetic material. Quantum mechanically what can be very useful for analyzing this system is the problem of addition of N spins $1/2$.

This problem is well known and can be found in any textbook of quantum theory (see, for example, Merzbacher (1998)). What we look for is a simultaneous eigenstate of $S^2 = (\sum_k \mathbf{S}_k)^2$ and $S^{(z)} = \sum_k S_k^{(z)}$, where k refers to any given spin of the system. The eigenstates of this problem are labeled by $|S, M_S\rangle$, where $S = N/2, (N-2)/2, (N-4)/2, \ldots, 1/2$ or 0 depending on whether N is odd or even, respectively, and $M_S = S, (S-1), (S-2), \ldots, -(S-1), -S$.

For very high external magnetic fields,[2] the lowest-energy eigenstates of our magnetic particle are approximately $|N/2, M_S\rangle$ and its ground state is then $|N/2, N/2\rangle$, which we denote by $| \uparrow, \uparrow, \ldots, \uparrow \rangle$ where N entries are present. All the other states are excited states of the system from which we can build a spin-coherent state (Takagi, 2002) representing the quantized precession of the whole magnetic moment as a spinning top.

As we reduce the external field the demagnetization term in either (2.43) or (2.63) can no longer be neglected and therefore $| \uparrow, \uparrow, \ldots, \uparrow \rangle$ ceases to be the exact ground state of the system. However, as long as $0 < H < H_c$, one can still use it as a good approximate ground state of the magnetic particle until $H = 0$ is reached and the situation changes drastically. When this happens, we have $| \uparrow, \uparrow, \ldots, \uparrow \rangle$ and $| \downarrow, \downarrow, \ldots, \downarrow \rangle$ completely degenerate and, as the demagnetization term has non-vanishing matrix elements between these two states, we can form a symmetric and an anti-symmetric linear combination of them which gives rise to the two lowest-lying energy eigenstates of the system, with the ground state being

$$|G\rangle = \frac{1}{\sqrt{2}} (| \uparrow, \uparrow, \ldots, \uparrow \rangle + | \downarrow, \downarrow, \ldots, \downarrow \rangle). \tag{2.93}$$

This is an example of a *maximally entangled* state, also known in the literature as the GHZ state (Greenberger *et al.*, 1989). Notice that it is also a quantum

---

[2] Notice that by high fields we mean fields stronger than the demagnetization or anistropy fields but always much weaker than $J\hbar^2/\mu$. Owing to this latter condition, multiplets with $S < N/2$ have very high energy.

mechanical superposition of two states involving a macroscopically large number of spins all pointing in opposite directions, which makes them macroscopically distinguishable. In other words this is a *Schrödinger cat*-like state.

Therefore, as our system was initially pointing along the external field direction, namely $+\hat{z}$, it now evolves in time as

$$|\psi(t)\rangle = a(t)|\uparrow, \uparrow, \ldots, \uparrow\rangle + b(t)|\downarrow, \downarrow, \ldots, \downarrow\rangle, \qquad (2.94)$$

where $a(t)$ and $b(t)$ are complex numbers such that $a(0) = 1$ and $|a(t)|^2 + |b(t)|^2 = 1$. As we are neglecting dissipative effects, the system oscillates forever between $|\uparrow, \uparrow, \ldots, \uparrow\rangle$ and $|\downarrow, \downarrow, \ldots, \downarrow\rangle$. However, as we shall show later on, relaxation effects will tend to destroy this coherent oscillation transforming this pure state into a mixture.

Now, if we turn the field direction to $-\hat{z}$, but still with $0 < |H| < H_c$, the state $|\uparrow, \uparrow, \ldots, \uparrow\rangle$ becomes an excited (metastable) state of the system whereas $|\downarrow, \downarrow, \ldots, \downarrow\rangle$ is its new approximate ground state. Assuming the presence of dissipation only to account for the relaxation of any excited state to the new ground state, we expect that

$$|\psi(t)\rangle = e^{-\Gamma t/2}|\uparrow, \uparrow, \ldots, \uparrow\rangle + \sqrt{1 - e^{-\Gamma t}}|\downarrow, \downarrow, \ldots, \downarrow\rangle, \qquad (2.95)$$

at least for very small damping, mimics the transition from $|\uparrow, \uparrow, \ldots, \uparrow\rangle$ to $|\downarrow, \downarrow, \ldots, \downarrow\rangle$. In the above equation Γ is the transition (tunneling) rate to another excited state of the system with the same energy as the initial metastable configuration. We should stress here that (2.95) is only an approximate form of the decaying state, which in reality cannot even be written as a pure state.

Notice that we have assumed that the system takes Γ^{-1} to leave the metastable configuration $|\uparrow, \uparrow, \ldots, \uparrow\rangle$, make a transition to an intermediate (degenerate) state and only then relaxes (almost instantaneously) to the new approximate ground state $|\downarrow, \downarrow, \ldots, \downarrow\rangle$. We have also assumed that this relaxation occurs within $\tau = \gamma^{-1} << \Gamma^{-1}$. In the latter expression, τ is the relaxation time of the system which, although assumed to be long (compared, for example, with the magnetization precession period), is still much shorter than the tunneling decay time Γ^{-1}. Possible effects of dissipation on the transition rate must be encapsulated in Γ and will be treated explicitly later on.

If we now further increase the external field along $-\hat{z}$ such that $|H| >> H_c$, $|N/2, -N/2\rangle$ becomes the new approximate ground state of the system and the remaining states of $|N/2, M_S\rangle$ are again its excited states. The difference between this case and the one for positive H is that the energy of the excited states now increases with M_S.

Having done the foregoing analysis, we can easily understand what happens in the case of homogeneous nucleation. All we have to do is apply the same reasoning to the case of an infinite (very large) magnet. In so doing we have to think about the critical droplet as a magnetic particle subject to an external field. When the field happens to be pointing along the direction opposite to the magnetization of the sample, as we have seen above, there is a probability that all the spins in a given region collectively make a quantum transition (see (2.95)) to a configuration with reversed spin direction. Although this transition might be quantum mechanical, the subsequent motion of the spins configuration is purely classical and depends on whether the energy reduction achieved by flipping the spins of that region wins over the energy increase due to the presence of the wall. Therefore, depending on the size of the droplet, it either shrinks or expands converting the whole sample to the more stable configuration. Here it should be realized that this reasoning only makes sense if the external field is finite. When $H = 0$ there is no way this transition could take place because, since $| \uparrow, \uparrow, \ldots, \uparrow \rangle$ and $| \downarrow, \downarrow, \ldots, \downarrow \rangle$ are degenerate, there would be no energy reduction in flipping any number of spins from one configuration to another. Only the distortion energy would be gained in this process and any droplet created would shrink, except if it has the size of the whole sample, which makes the transition probability vanishingly small. In other words, there is no cat state in an infinite system!

Finally, let us turn our analysis to the quantum mechanical motion of the wall. Once again the mechanism by which it takes place is very similar to the one just addressed in the case of quantum nucleation. The difference is that the idea of a droplet must be replaced by the enlargement of the more stable spin configuration through the displacement of finite line segments of the wall. This distorted line clearly enlarges the area occupied by, say, spins pointing down in the presence of an external field along $-\hat{z}$ by the same mechanism of (2.95). Whether this region expands or shrinks depends again, in the particular case of strong pinning, on the balance between the energy reduction in this expansion versus the additional distortion to create the kink–anti-kink pair due to the displacement of a finite line segment. From this point onwards the analysis follows that of the nucleation case. In the case of weak pinning there is no fast expansion of the critical segment and the enlargement of the more stable region takes place diffusively as the wall is accommodated along valleys with lower energy.

Summarizing, all we have done in this section is employ a more quantum mechanical point of view of the previously introduced examples to which semi-classical dynamics in terms of some effective – particle or field – variable could be applicable. Therefore, we have developed a deeper understanding of what this semi-classical approach means exactly in terms of the genuine quantum mechanical entities of those systems, namely the elementary magnetic moment.

3

Elements of superconductivity

Superconductivity was discovered by Onnes (1911), who observed that the electrical resistance of various metals dropped to zero when the temperature of the sample was lowered below a certain critical value T_c, the *transition temperature*, which depends on the specific material being dealt with.

Another equally important feature of this new phase was its *perfect diamagnetism* which was discovered by Meissner and Ochsenfeld (1933), the so-called Meissner effect. A metal in its superconducting phase *completely* expels the magnetic field from its interior (see Fig. 3.1). The very fact that many molecules and atoms are repelled by the presence of an external magnetic field is quite well known, as we have already seen in the preceding chapter. The difference here lies in the perfect diamagnetism, which means that it is the whole superconducting sample that behaves as a giant atom!

This effect persists (for certain kinds of metal) until we reach a critical value of the external magnetic field, $H = H_c(T)$, above which the superconducting sample returns to its normal metallic state. Moreover, at fixed temperature, this effect is completely reversible, suggesting that the superconducting phase is an equilibrium state of the electronic system. The temperature dependence of the critical field is such that $H_c(T_c) = 0$ and $H_c(0)$ attains its maximum value as shown in Fig. 3.2.

Despite the great effort of physicists in those days to explain these results in terms of the conventional perfect conductivity and (ordinary) perfect diamagnetism, by which we respectively mean electric conductivity $\sigma = \infty$ and magnetic permeability $\mu = 0$, those attempts failed to account for the experimental hallmark characterizing materials in their superconducting phase, as carefully exposed in London (1961).

3.1 London theory of superconductivity

In order to account for both the resistanceless flow and the more subtle perfect diamagnetism of the new phase the London brothers (London, 1961) proposed a

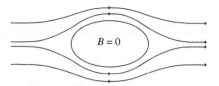

Figure 3.1 Meissner effect

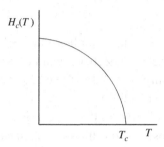

Figure 3.2 Critical field

modification of some of the equations of electrodynamics in material media. These new equations read

$$\mathbf{E} = \frac{\partial}{\partial t}(\Lambda \mathbf{J}_s), \tag{3.1}$$

$$\mathbf{h} = -c\nabla \times (\Lambda \mathbf{J}_s), \tag{3.2}$$

where $\Lambda = 4\pi \lambda_L^2/c^2$ and $\mathbf{J}_s$ is the so-called *supercurrent density*. Notice that $\mathbf{J}_s$ is the source of the microscopic field $\mathbf{h}$ which is related to the magnetic induction through $\mathbf{B}(\mathbf{r}) = \langle \mathbf{h}(\mathbf{r}) \rangle_{\mathbf{r}}$, as mentioned below (2.2).

The first of these equations means that even for $\mathbf{E} = 0$ there can be a constant value of $\mathbf{J}_s$ or, in other words, the resistanceless flow of electrons. The second London equation is such that, when combined with the curl of the Maxwell equation

$$\nabla \times \mathbf{h} = \frac{4\pi}{c} \mathbf{J}_s, \tag{3.3}$$

it gives us the equation that describes the Meissner effect, namely

$$\nabla^2 \mathbf{h} = \frac{\mathbf{h}}{\lambda_L^2}, \tag{3.4}$$

which shows that, given a normal–superconductor interface, the external field will decay exponentially into the superconducting region within a length λ_L, the *London penetration depth*, as shown in Fig. 3.3.

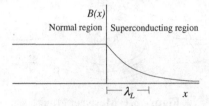

Figure 3.3 The penetration depth

This spatial variation of the field implies that (3.1) *cannot* describe perfect conductivity in the sense that the electrons would be accelerated by an external electric field. This would only be valid if the electronic mean free path $\ell = \infty$, which implies an electric conductivity $\sigma = \infty$ only for electric fields filling the whole space, in clear contradiction to the effects due to the presence of an interface between the two phases. A more thorough analysis of this argument can be followed in Tinkham (2004).

In the absence of an external scalar potential we can rewrite (3.1) as

$$\mathbf{E} = \frac{\partial}{\partial t}(\Lambda \mathbf{J}_s) = -\frac{1}{c}\frac{\partial \mathbf{A}}{\partial t},$$

which implies that

$$\mathbf{J}_s = -\frac{1}{c\Lambda}(\mathbf{A} - \mathbf{A}_0), \tag{3.5}$$

where $\mathbf{A}_0(\mathbf{r})$ is constant in time. Taking the curl of this equation and comparing it with (3.2), one realizes that the only way for the latter to hold is when $\nabla \times \mathbf{A}_0(\mathbf{r}) = 0$. Therefore, $\mathbf{A}_0(\mathbf{r}) = \nabla \chi(\mathbf{r})$ where $\chi(\mathbf{r})$ is a function defined within the superconducting region and, consequently, is only single-valued in simply connected samples.

In contrast, since there is no external scalar potential or, more generally, we are interested in the quasi-stationary regime, we can impose the supplementary condition $\nabla \cdot \mathbf{A} = 0$ on the divergence of (3.5), which together with the continuity equation

$$\frac{\partial \rho}{\partial t} + \nabla \cdot \mathbf{J}_s = 0,$$

implies $\nabla \cdot \mathbf{A}_0 = 0$. Therefore, the scalar function we have just defined obeys

$$\nabla^2 \chi(\mathbf{r}) = 0. \tag{3.6}$$

Now, we can always decompose the supercurrent density at the surface of the superconducting sample into components parallel (divergenceless) and perpendicular (irrotational) to the surface itself as $\mathbf{J}_s = \mathbf{J}_{s\parallel} + \mathbf{J}_{s\perp}$ and choose, at this boundary,

$$\mathbf{J}_{s\perp} = \frac{1}{c\Lambda}\mathbf{A}_{0\perp}.$$

For a *simply connected* superconducting sample, these equations for $\mathbf{A}_0$ have a unique solution if $\mathbf{J}_{s\perp}$ is given on the entire surface of the sample. In particular, for an *isolated superconductor*, where $\mathbf{J}_{s\perp} = 0$, we have $\mathbf{A}_0 = (\nabla\chi)_\perp = 0$ which together with (3.6) implies, by the Neumann boundary condition of the potential theory, that $\chi = 0$. This choice is known as the *London gauge* in which we can write

$$\mathbf{J}_s = -\frac{1}{c\Lambda}\mathbf{A}, \tag{3.7}$$

which represents both (3.1) and (3.2) and is also called the *London equation* in the London gauge. It should be noticed that since this equation is not gauge invariant, it is valid only in this chosen gauge which can be summarized as: $\nabla \cdot \mathbf{A}(\mathbf{r}) = 0$, $\mathbf{J}_{s\perp}(\mathbf{r})$ and $\mathbf{A}_\perp(\mathbf{r})$ are related by (3.7) at the surface of the sample and $\mathbf{A}(\mathbf{r}) \to 0$ deep inside the superconducting region. Besides, it must be kept in mind that (3.7) holds only for a simply connected superconducting specimen and must be replaced by the more general form (3.5) otherwise.

A more fundamental reasoning to obtain the London equation in the above-presented form is the application of quantum mechanics to the expression of the canonical momentum of a charged particle subject to an external field, namely

$$\mathbf{p} = m_s\dot{\mathbf{r}} + e_s\frac{\mathbf{A}}{c}, \tag{3.8}$$

where m_s and e_s are respectively the mass and charge of the superconducting particle whatever its composition in terms of the original electrons may be. Despite there being many physical arguments to assume that the state of the superconducting phase of a given material is of quantum mechanical origin (see, for example, London (1961)), we shall take that presented below as suitable for our needs.

Let us start with a theorem proved by Bloch (London, 1961), which states that in the absence of an external field the most stable state of any electronic system carries zero current. In the particular case of no external field, it coincides with $\langle\mathbf{p}\rangle = 0$. If we now switch on an external field $\mathbf{A}(\mathbf{r})$ it is a simple matter to show (London, 1961) that classical statistical mechanics still gives us $\mathbf{J}_s(\mathbf{r}) = 0$, at variance with the London equation (3.7). In order to reproduce the latter we would need to preserve the relation $\langle\mathbf{p}\rangle = 0$ even in the presence of finite fields, which implies

$$\langle\mathbf{v}_s\rangle = -\frac{e_s\mathbf{A}}{m_s c}, \tag{3.9}$$

where $\langle \mathbf{v}_s \rangle$ is the average value of the superconducting velocity field at a given position of the sample. We show next that quantum mechanics can do this job for us. Assuming that the number density of superconducting electrons (or carriers, to be more precise) is $n_s = n_s(T)$, we get

$$\mathbf{J}_s = n_s e_s \langle \mathbf{v}_s \rangle = -\frac{n_s e_s^2 \mathbf{A}}{m_s c} = -\frac{1}{c\Lambda} \mathbf{A}, \tag{3.10}$$

which is again expression (3.7).

From our latest relations (3.10) and the definition of Λ below (3.1) and (3.2), we can express the London penetration depth as

$$\lambda_L = \left(\frac{m_s c^2}{4\pi n_s e_s^2} \right)^{1/2} \tag{3.11}$$

where $n_s(T_c) = 0$ and $n_s(0) \le n$, the total number density of electrons in the system. In this way, at the critical temperature, the system undergoes a transition to a normal metal when the field fully penetrates the sample. The London penetration depth is typically of the order of 10^{-7} m for conventional superconductors.

If we now assume that a superconductor in its ground state ($T = 0$) can be described, in the absence of an external field, by the wave function $\psi_0(\mathbf{r})$, it means by (3.8) and the Bloch theorem that it carries a current density

$$\mathbf{J}_s = \frac{e_s \hbar}{2m_s i} \left[\psi_0^* \nabla \psi_0 - \psi_0 \nabla \psi_0^* \right] = \frac{e_s}{m_s} \operatorname{Re} [\psi_0^* \mathbf{p} \psi_0] = 0. \tag{3.12}$$

When the field is turned on, we must replace $\mathbf{p} \to \mathbf{p} - e_s \mathbf{A}/c$, $\psi_0 \to \psi$ and then

$$\mathbf{J}_s = \frac{e_s \hbar}{2m_s i} \left[\psi^* \nabla \psi - \psi \nabla \psi^* \right] - \frac{e_s^2 \mathbf{A}}{m_s c} \psi^* \psi. \tag{3.13}$$

If we further assume the *rigidity* of the wave function, which means that its form will not be modified by the presence of the external field, we can take $\psi \approx \psi_0$ and the first term on the r.h.s. of (3.13) vanishes due to (3.12). Then, if the wave function is normalized to the total number of carriers rather than unity, we can rewrite (3.13) as

$$\mathbf{J}_s = -\frac{e_s^2 \mathbf{A}}{m_s c} \psi^* \psi = -\frac{n_s e_s^2 \mathbf{A}}{m_s c}, \tag{3.14}$$

which is known as the *diamagnetic current density*. It should be noticed that not only in (3.14), but also in (3.9), (3.10), and (3.11), the phenomenological parameters n_s, m_s, and e_s appear in such a combination that regardless of the number of electrons composing the superconducting unity (for example, two in the Cooper pair case) we can always replace them by the electronic parameters n, m, and e, respectively. Nevertheless, we shall still use n_s with a somewhat different meaning in what follows.

It is important to stress here that it is n which must be used for the number density of carriers in (3.14) instead of a temperature-dependent function $n(T)$. The reason behind this is simply that we are now dealing with the ground state itself, whereas before we made no hypothesis about the temperature of the system. Our previous arguments (actually, Bloch's arguments) leading to (3.10) and (3.11) were completely macroscopic and based on the dissipationless flow and Meissner effect. As a matter of fact we have imposed n as an upper bound on the value of n_s at $T = 0$.

The hypothesis of rigidity is corroborated by many experimental results which indicate that the superconducting ground state is *gapped* in the vast majority of cases (De Gennes, 1999; Tinkham, 2004). Therefore, it is energetically costly to create excitations in the system. This is a suggestive explanation for the rigidity of the ground state, although there is also the possibility of the existence of *gapless* superconductors (De Gennes, 1999). Nevertheless, at finite temperatures it is not true that this hypothesis still holds fully. Appealing to the BCS (Bardeen, Cooper, and Schrieffer) theory – the microscopic theory of superconductivity (see, for example, Tinkham (2004)) – we really see that the superconducting state in thermal equilibrium at finite temperatures is not completely rigid and therefore the first term in (3.13) (*the paramagnetic current density*) also contributes to the total current.

It is not only with regard to the number density of carriers and finite temperature effects that (3.14) might be modified. In reality, the BCS theory also shows that there may be non-local effects in (3.14) depending on the system under investigation, and this result fits nicely into the non-local phenomenological theory of Pippard (De Gennes, 1999; Tinkham, 2004). Despite the importance and interesting physics of these effects (see below), they are not our main goal in this book and we shall return to the discussion on the wave function of the superconducting ground state.

3.2 Condensate wave function (order parameter)

At this point we might ask ourself whether the idea to describe the ground-state wave function of the superconductor by a single-position variable makes any sense at all, once we know that a wave function representing a system with N particles must be represented by an N-variable function. In order to justify this, let us assume that the system can be described by a many-body wave function (Feynman, 1998) $\Psi_0(\mathbf{r}_1, \ldots, \mathbf{r}_k, \ldots, \mathbf{r}_N)$. If we want to think in terms of the microscopic theory, this would be the coordinate representation of the BCS state which consists of a collection of Cooper pairs. But, lo and behold, in so doing we must be careful with the fact that the BCS ground state has only a fixed *average* number of particles and, therefore, we should employ an alternative approach to the BCS theory in a particle-conserving representation (Leggett, 2006). In the absence of an external field, regardless of its microscopic details, this wave function provides us with a state carrying zero current, as we have seen in (3.9). Moreover, since it is possible for a superconductor to carry a dissipationless current, it must be inferred that a current-carrying wave function and the one obtained from $\Psi_0(\mathbf{r}_1, \ldots, \mathbf{r}_k, \ldots, \mathbf{r}_N)$ by a Galilean transformation must be totally indistinguishable.

Let us call this current-carrying wave function $\Psi_\mathbf{K}(\mathbf{r}_1, \ldots, \mathbf{r}_k, \ldots, \mathbf{r}_N)$, where $\hbar\mathbf{K}$ is the momentum associated with the carrier. In the case of a BCS wave function, for example, this is the momentum of the center of mass of the pair. Consequently, we can write this new wave function, at least for low enough velocities (Merzbacher, 1998), as

$$\Psi_\mathbf{K}(\mathbf{r}_1, \ldots, \mathbf{r}_k, \ldots, \mathbf{r}_N) = \exp\{i \sum_k \mathbf{K} \cdot \mathbf{r}_k\} \Psi_0(\mathbf{r}_1, \ldots, \mathbf{r}_k, \ldots, \mathbf{r}_N). \quad (3.15)$$

In the example just treated we are assuming that all the carriers have the same velocity $\mathbf{v} = \hbar\mathbf{K}/m$, a constant current for the electronic system.

Now, we can try to generalize this by introducing a position-dependent velocity (or carrier's momentum) which would be approximately constant in a neighborhood of the position $\mathbf{r}_k$. In this case the current is position dependent as it generally is in ordinary metals and particularly in superconductors. However, we can do much better by defining

$$\Psi(\mathbf{r}_1, \ldots, \mathbf{r}_k, \ldots, \mathbf{r}_N) = \exp\{i \sum_k \theta(\mathbf{r}_k)\} \Psi_0(\mathbf{r}_1, \ldots, \mathbf{r}_k, \ldots, \mathbf{r}_N), \quad (3.16)$$

from which we can get the average number density of electrons

$$n(\mathbf{r}) = \sum_k \int d\mathbf{r}_1 \ldots d\mathbf{r}_k \ldots d\mathbf{r}_N \delta(\mathbf{r} - \mathbf{r}_k)$$
$$\times \Psi^*(\mathbf{r}_1, \ldots, \mathbf{r}_k, \ldots, \mathbf{r}_N) \Psi(\mathbf{r}_1, \ldots, \mathbf{r}_k, \ldots, \mathbf{r}_N), \quad (3.17)$$

as well as the average current density

$$\mathbf{J}(\mathbf{r}) = \sum_k \int d\mathbf{r}_1 \ldots d\mathbf{r}_k \ldots d\mathbf{r}_N \frac{e\hbar}{2mi} \left[\Psi^* \nabla_k \Psi - \Psi \nabla_k \Psi^* \right] \delta(\mathbf{r} - \mathbf{r}_k). \quad (3.18)$$

If we now use the fact that the current associated with Ψ_0 is zero, we obtain

$$\mathbf{J}(\mathbf{r}) = \frac{e\hbar}{m} n(\mathbf{r}) \nabla\theta, \quad (3.19)$$

where the number density (3.17) can be written further as

$$n(\mathbf{r}) = N \int d\mathbf{r}_2 \ldots d\mathbf{r}_k \ldots d\mathbf{r}_N \Psi_0^*(\mathbf{r}, \ldots, \mathbf{r}_k, \ldots, \mathbf{r}_N) \Psi_0(\mathbf{r}, \ldots, \mathbf{r}_k, \ldots, \mathbf{r}_N),$$
(3.20)

regardless of the statistics of the component particles. This can be recognized, for a system known to be in its ground state Ψ_0, as the diagonal element of the coordinate representation of the *1-particle reduced density operator* (see, for example, Thouless (1972)),

$$n_1(\mathbf{r}; \mathbf{r}') = N \int d\mathbf{r}_2 \ldots d\mathbf{r}_k \ldots d\mathbf{r}_N \Psi_0(\mathbf{r}, \ldots, \mathbf{r}_k, \ldots, \mathbf{r}_N) \Psi_0^*(\mathbf{r}', \ldots, \mathbf{r}_k, \ldots, \mathbf{r}_N),$$
(3.21)

which means $n(\mathbf{r}) = n_1(\mathbf{r}; \mathbf{r})$. This operator can be used to compute the average value of any diagonal 1-particle many-body operator, $\hat{\mathcal{O}}_1 \equiv \sum_i \mathcal{O}_1(\mathbf{r}_i)$, as

$$\langle \hat{\mathcal{O}}_1 \rangle = \text{tr}[\hat{n}_1 \hat{\mathcal{O}}_1] = \int d\mathbf{r}\, n_1(\mathbf{r}; \mathbf{r}) \mathcal{O}_1(\mathbf{r}),$$
(3.22)

where the trace is taken in the coordinate representation of the single-particle operator.

Once we have achieved this point there are some conclusions we can draw about (3.20). Firstly, suppose that all the carriers occupy the same single-particle state (non-interacting particles) and the many-body wave function is a product of these single-particle wave functions for different variables. In this case we can integrate all those $N - 1$ variables and (3.20) becomes

$$n(\mathbf{r}) = N \psi^*(\mathbf{r}) \psi(\mathbf{r}),$$
(3.23)

where we have used $\psi(\mathbf{r})$ as the normalized single-particle wave function.

This maneuver is allowed if our carriers are bosons, since only in this case can they all occupy the same single-particle state. That would be the case of systems which undergo Bose–Einstein condensation (BEC), such as alkali atoms in magneto-optical traps (see, for example, Leggett (2006)). For fermions, this is forbidden by the exclusion principle, and we need to modify our arguments.

Despite trying to avoid any incursion into the microscopic theory of superconductivity, it is wise to at least afford a glimpse of its main results at this point. Since the BCS theory tells us that the charge carriers in materials in their superconducting phase are pairs of electrons (Cooper pairs) which are bound to each other by the exchange of phonons (at least in conventional superconductors), we had better apply this standard result to our present discussion.

As we hope to have convinced the reader that average values of 1-particle many-body operators can be evaluated with the help of $\hat{n}_1$, we are now in a position to extend this result and create another operator which is useful if we need to study the effect of 2-particle many-body operators in the system. As we have just mentioned, there is an effective attractive interaction between electrons in superconductors and, therefore, many of their properties can be computed if we evaluate the average value of the existing effective electronic potential. One example would be the ground-state energy of the superconducting system itself.

Let us then compute the average potential energy of the electronic system,

$$\hat{V} = \frac{1}{2} \sum_{i \neq j} V(\mathbf{r}_i - \mathbf{r}_j), \tag{3.24}$$

which reads

$$\langle \hat{V} \rangle = \frac{1}{2} \sum_{i \neq j} \int d\mathbf{r}_1 \ldots d\mathbf{r}_i \ldots d\mathbf{r}_j \ldots d\mathbf{r}_N \, V(\mathbf{r}_i - \mathbf{r}_j)$$
$$\times |\Psi_0(\mathbf{r}_1, \ldots, \mathbf{r}_i, \ldots, \mathbf{r}_j, \ldots, \mathbf{r}_N)|^2$$
$$= \frac{N(N-1)}{2} \int d\mathbf{x} \, d\mathbf{y} \, d\mathbf{r}_3 \ldots d\mathbf{r}_N \, V(\mathbf{x} - \mathbf{y}) \, |\Psi_0(\mathbf{x}, \mathbf{y}, \mathbf{r}_3, \ldots, \mathbf{r}_N)|^2, \tag{3.25}$$

where we have moved $\mathbf{r}_i$ and $\mathbf{r}_j$ to the first and second entries of $|\Psi_0|^2$ and renamed them $\mathbf{x}$ and $\mathbf{y}$, respectively. Notice that the change of sign due to the asymmetry of the wave function is immaterial here once it takes place both in Ψ_0 and Ψ_0^*.

Now, defining the coordinate representation of the matrix element of the *2-particle reduced density operator* (Thouless, 1972) as

$$n_2(\mathbf{x}, \mathbf{y}; \mathbf{x}', \mathbf{y}') \equiv N(N-1) \int d\mathbf{r}_3 \ldots d\mathbf{r}_N$$
$$\times \Psi_0(\mathbf{x}, \mathbf{y}, \mathbf{r}_3, \ldots, \mathbf{r}_N) \Psi_0^*(\mathbf{x}', \mathbf{y}', \mathbf{r}_3, \ldots, \mathbf{r}_N), \tag{3.26}$$

we can rewrite (3.25) as

$$\langle \hat{V} \rangle = \frac{1}{2} \int d\mathbf{x} \, d\mathbf{y} \, n_2(\mathbf{x}, \mathbf{y}; \mathbf{x}, \mathbf{y}) \, V(\mathbf{x} - \mathbf{y}) \tag{3.27}$$

or, more generally, the average value of any diagonal 2-particle many-body operator, $\hat{O}_2 \equiv \frac{1}{2} \sum_{i,j} O_2(\mathbf{r}_i, \mathbf{r}_j)$, can be written as

$$\langle \hat{O}_2 \rangle = \frac{1}{2} \text{tr}[\hat{n}_2 \hat{O}_2] = \frac{1}{2} \int d\mathbf{x} \, d\mathbf{y} \, n_2(\mathbf{x}, \mathbf{y}; \mathbf{x}, \mathbf{y}) O_2(\mathbf{x}, \mathbf{y}). \tag{3.28}$$

We now proceed with $n_2(\mathbf{x}, \mathbf{y}; \mathbf{x}, \mathbf{y})$ in an analogous way to that done with $n_1(\mathbf{r}; \mathbf{r})$ in order to obtain (3.23). If we assume that $\Psi_0(\mathbf{r}_1, \ldots, \mathbf{r}_N)$ is a properly anti-symmetrized product of normalized pairwise wave functions, $\phi(\mathbf{r}_i, \mathbf{r}_j)$, those remaining non-integrated variables in (3.26) will leave us with a product of functions which are anti-symmetric in the variables $\mathbf{x}$ and $\mathbf{y}$:

$$n_2(\mathbf{x}, \mathbf{y}; \mathbf{x}, \mathbf{y}) = N(N-1)\,\phi^*(\mathbf{x}, \mathbf{y})\phi(\mathbf{x}, \mathbf{y}). \tag{3.29}$$

Furthermore, if we replace the variables $\mathbf{x}$ and $\mathbf{y}$ in (3.26) by the center-of-mass and relative coordinates of the pair, defined as $\mathbf{r} \equiv \frac{1}{2}(\mathbf{x}+\mathbf{y})$ and $\mathbf{u} \equiv \mathbf{x}-\mathbf{y}$, respectively, and integrate over $\mathbf{u}$ we can show that

$$n_1(\mathbf{r}; \mathbf{r}') = \frac{1}{N-1}\int d\mathbf{u}\, n_2(\mathbf{r}, \mathbf{u}; \mathbf{r}', \mathbf{u}). \tag{3.30}$$

Now, inserting (3.29) into (3.30) and further assuming that the pair wave function $\phi(\mathbf{r}, \mathbf{u})$ in (3.29) can be written as $\phi(\mathbf{r}, \mathbf{u}) = \psi(\mathbf{r})\chi(\mathbf{u})$, where both wave functions are conveniently normalized, we have once again

$$n(\mathbf{r}) \equiv n_1(\mathbf{r}; \mathbf{r}) = N\,\psi^*(\mathbf{r})\psi(\mathbf{r}), \tag{3.31}$$

as in (3.23). The difference now lies in the interpretation of what we consider the condensate wave function. Whereas in the case of bosons it should be interpreted as the single-particle wave function of a state which is macroscopically occupied, here, in the case of fermions, it is the center-of-mass wave function of an electron pair. Notice that the above-mentioned hypothesis that the center-of-mass and relative coordinates are separable is in agreement with the fact that, for pairs, the superconducting state is translation invariant which means that they can move freely throughout the sample.

Although our arguments are not rigorous, they at least induce us to accept London's ideas and interpret the "superconducting wave function" as one referring to the system as a whole. It can be viewed as the wave function of a *condensate* of carriers whose dynamics obeys quantum mechanics. Therefore, both the particle number density and the current density of a superconductor can be obtained in many circumstances from their one-particle expressions as applied to the single-particle wave function

$$\psi(\mathbf{r}) = \sqrt{n_s(\mathbf{r})}e^{i\theta(\mathbf{r})}. \tag{3.32}$$

Notice that as we are dealing with systems at zero temperature, we have $n_s(\mathbf{r}) = n(\mathbf{r})$.

It is true that these naïve arguments will be even weaker if we take finite-temperature effects into account. Intuitively, one could appeal to those arguments towards the rigidity of the ground-state wave function and assume that even at finite temperatures the quantum mechanical aspects of ψ will still be preserved, despite the unavoidable fact that this ground state must steadily be depleted as the temperature is raised. Those particles that leave the ground state will no longer present the same quantum effects as their partners still in the condensate, and therefore will behave as ordinary metallic carriers. This is the origin of the so-called *two-fluid model* by which one states that the total current density of the system can be written as

$$\mathbf{J}(\mathbf{r}) = en_s(\mathbf{r})\mathbf{v}_s(\mathbf{r}) + en_N(\mathbf{r})\mathbf{v}_N(\mathbf{r}), \tag{3.33}$$

where $n_s(\mathbf{r})$ and $n_N(\mathbf{r})$ are, respectively, the superconducting and normal number densities of carriers whereas the total number density $n(\mathbf{r}) = n_s(\mathbf{r}) + n_N(\mathbf{r})$. In the same way, $\mathbf{v}_s$ and $\mathbf{v}_N$ are, respectively, the superconducting and normal carriers' velocity fields. Both the superconductor number and current densities can be obtained from expressions such as (3.19) and (3.32) if one replaces $n(\mathbf{r})$ by $n_s(\mathbf{r})$, which is now temperature dependent. This superconducting number density must vanish at the transition temperature.

It is indeed strange to consider (3.32) a genuine wave function once it has a temperature-dependent amplitude and, therefore, must always result from a statistical average using the density operator of the system. Actually, a more thorough treatment of this question is presented in Leggett (2006), where the previous approach of employing the one- and two-particle reduced density operators is generalized to statistical mixtures and consequently incorporates temperature-dependent effects. Using general properties of these operators and the hypothesis of *off-diagonal long-range order* (ODLRO) (Yang, 1962), we characterize in a unified way the effects of BEC, for bosonic particles and pseudo-BEC (pairing), for fermionic particles. Once this is done, the resulting object that plays the same role as (3.32) acquires a new status; it is now regarded as the order parameter of the condensed state, either for bosons or fermions.

This order parameter is the central object in one of the most successful phenomenological theories of phase transitions, namely the *Ginzburg–Landau* (GL) theory of superconductivity (see De Gennes (1999), Tinkham (2004) or any other standard textbook on superconductivity). This object carries the collective as well as the quantum mechanical effects of the condensate within it and is extremely important in the study of the thermodynamic and electromagnetic properties of materials in these exotic phases.

As we have already seen in the preceding chapter, the order parameter is a quantity which develops in the thermodynamical state attained by a system which has undergone a phase transition from a less to a more ordered state. It actually quantifies the amount of ordering in this newly developed state and is, in the vast majority of cases, related to the phenomenon of spontaneous symmetry breaking. In the particular case of superconductivity it is argued (Anderson, 1963) that it is the gauge symmetry which is spontaneously broken in the condensed phase. Nevertheless, this point of view, although broadly accepted, is not unanimous in the scientific community (Leggett and Sols, 1991; Leggett, 2006).

All we have been saying can be made much more rigorous if we use the microscopic theory of superconductivity (see, for example, De Gennes (1999) or Tinkham (2004)), but since this is not our main subject here we shall proceed with the phenomenological approach which will prove very useful for our purposes if some care is exercised at convenient points.

3.3 Two important effects

In this section we study two specific effects that will be fundamentally important for us in future chapters.

3.3.1 Flux quantization

Let us consider now a superconducting ring in the presence of an external magnetic field. If we apply (3.13) to (3.32) we have $\hbar\nabla\theta = e\Lambda\mathbf{J} + e\mathbf{A}/c$, which integrated along a path Γ from position 1 to 2 in the superconducting region gives us

$$\int_1^2 \left(e\Lambda\mathbf{J} + \frac{e}{c}\mathbf{A}\right) \cdot d\mathbf{r} = \hbar(\theta_2 - \theta_1), \tag{3.34}$$

because it is an integral of a gradient field. Additionally, as θ is the phase of a wave function which must be single valued, the integral along a closed loop ($1 = 2$ above) must give

$$\frac{e}{c} \oint \left(c\Lambda\mathbf{J} + \mathbf{A}\right) \cdot d\mathbf{r} = 2\pi n\hbar, \tag{3.35}$$

where n is an integer.

Two points must be emphasized here. First of all, we notice that this result is in apparent contradiction to (3.7), which would result in a vanishing integral if it were applied to (3.35). However, as we have mentioned earlier in this section, it is not (3.7) that must be used for multiply connected superconducting samples, but (3.5) instead. Applying the latter to (3.35) we end up with the result

$$\frac{e}{c} \oint \left(c\Lambda\mathbf{J} + \mathbf{A}\right) \cdot d\mathbf{r} = \Delta\chi, \tag{3.36}$$

which is not zero since $\chi(\mathbf{r})$ need not be single valued in this case. Actually, we have just seen that once it is related to the phase of a wave function, $\Delta\chi = 2\pi n$.

The second point refers to the fact that in (3.35) the charge e appears alone rather than in some specific combination as before. Therefore, we should return to (3.8) and remember that what appears in (3.35) is actually e_s and then,

$$\oint \left(c\Lambda \mathbf{J} + \mathbf{A} \right) \cdot d\mathbf{r} = \frac{2\pi n \hbar c}{e_s} \equiv n\phi_0 \tag{3.37}$$

where the *flux quantum* $\phi_0 \equiv hc/e_s \approx 2.09 \times 10^{-7}$ gauss cm^2, or, in the SI (mks) system of units, $\phi_0 \equiv h/e_s \approx 2.09 \times 10^{-15}$ Wb. Notice that this experimental result can be used to determine the actual charge of the superconducting carriers as $e_s = 2e$. For a simply connected superconductor the integration contour of (3.37) can be deformed continuously to a point and then only $n = 0$ results from the integration and (3.7) holds again.

If the ring is thick enough (all its linear dimensions $\gg \lambda_L$) and we integrate (3.37) along a closed path Γ deep into the ring where there is no current, it yields

$$\oint \mathbf{A} \cdot d\mathbf{r} = \int \mathbf{h} \cdot d\mathbf{s} = n\phi_0, \tag{3.38}$$

where $d\mathbf{s}$ is an element of area on any surface bounded by Γ and we have used $\mathbf{B} = \nabla \times \mathbf{A}$ and Stoke's theorem. From the above equation we see that the magnetic flux trapped in a superconducting ring is quantized in units of ϕ_0.

3.3.2 The Josephson effect

Suppose that we have two pieces of superconductor separated by a non-superconducting material of thickness d. We call them 1 and 2. If d is very large the two superconductors do not feel each other's presence and the dynamics of their condensates should obey two decoupled Schrödinger equations

$$i\hbar\dot{\psi}_1 = E_1\psi_1,$$
$$i\hbar\dot{\psi}_2 = E_2\psi_2. \tag{3.39}$$

Now, if d is such that there is a substantial overlap between the two condensate wave functions then it is reasonable to assume that we now have

$$i\hbar\dot{\psi}_1 = E_1\psi_1 + \Delta\psi_2,$$
$$i\hbar\dot{\psi}_2 = E_2\psi_2 + \Delta^*\psi_1. \tag{3.40}$$

This arrangement is known in the literature as a *Josephson junction*.

Let us for simplicity choose $\Delta \in \mathbb{R}$ and $\mathbf{A} = 0$. Then, writing $\psi = \sqrt{n}e^{i\theta}$ for each wave function in (3.40), we can easily get

$$-\hbar\dot{\theta}_1 = \Delta\sqrt{\frac{n_2}{n_1}} \cos(\theta_2 - \theta_1) + E_1, \tag{3.41}$$

a similar equation for θ_2 (where one should exchange $1 \leftrightarrow 2$) and also

$$\dot{n}_1 = \frac{2\Delta}{\hbar}\sqrt{n_1 n_2}\sin(\theta_2 - \theta_1) = -\dot{n}_2. \tag{3.42}$$

As $\dot{n}_1$ gives us the current through the junction, we can write

$$i = i_0 \sin\Delta\theta, \tag{3.43}$$

$$\Delta\dot{\theta} = \frac{2eV}{\hbar}, \tag{3.44}$$

where we have assumed $n_1 \approx n_2$, $\sqrt{n_1 n_2} = n_s$ and

$$\Delta\theta \equiv \theta_1 - \theta_2, \quad i_0 \equiv \frac{2n_s\Delta}{\hbar} \quad \text{and} \quad \frac{E_2 - E_1}{\hbar} = \frac{2eV}{\hbar}. \tag{3.45}$$

What (3.43), (3.44) tell us is that the difference between the phases of the wave functions on each superconductor can adjust itself to allow for the transport of a constant current, without any measurable voltage, up to a critical value i_0. Beyond this value, a voltage develops between the two ends of the junction. This is the celebrated *Josephson effect*, which was originally obtained by a fully microscopic approach (Josephson, 1962). The simple phenomenological method presented here is attributed to Feynman (1998).

These two particular effects will be very important for us in discussing the devices treated later in this book.

3.4 Superconducting devices

In this section, and from now onwards, we shall always adopt the mks system when dealing with circuit or device applications of superconductivity. The reason for this option is not to encumber our expressions with extra factors involving the speed of light c.

3.4.1 Superconducting quantum interference devices (SQUIDs)

Our first device consists of a superconducting ring closed by a Josephson junction (see Fig. 3.4). We shall be particularly interested in the so-called *weak links* (metallic junctions or point contacts).

Let us suppose that our SQUID ring is subject to an external field perpendicular to its plane and we wish to study the dynamics of the total magnetic flux comprised by the ring. If we had an ordinary ring, we have just seen that the total flux would be quantized in units of ϕ_0. However, in this new example this quantization rule will be modified slightly.

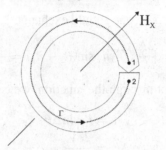

Figure 3.4 SQUID ring

This modification comes about because if the two points 1 and 2 in (3.34) are located right at the two terminals of the junction (see Fig. 3.4), it must be more carefully analyzed. Let us rewrite

$$\int_1^2 \mathbf{J} \cdot d\mathbf{r} = \frac{n_s e \hbar}{m} \int_1^2 \nabla\theta \cdot d\mathbf{r} - \frac{n_s e \hbar}{m} \frac{2\pi}{\phi_0} \int_1^2 \mathbf{A} \cdot d\mathbf{r}. \tag{3.46}$$

The term on the l.h.s. of (3.46) still vanishes if Γ is a path deep into the super-conducting ring. The last integral on the r.h.s., due to the continuity of $\mathbf{A}$, can be approximated by the integral along a closed loop which results in the flux ϕ through the ring. The only term that deserves more attention is the remaining one, which we split as

$$\int_\Gamma \nabla\theta \cdot d\mathbf{r} = \oint \nabla\theta \cdot d\mathbf{r} - \int_2^1 \nabla\theta \cdot d\mathbf{r}, \tag{3.47}$$

and, since θ is a phase, conclude that

$$\int_\Gamma \nabla\theta \cdot d\mathbf{r} = 2\pi n - \Delta\theta \tag{3.48}$$

where $\Delta\theta \equiv \theta_1 - \theta_2$. Consequently, we can write the new quantization relation for the SQUID as

$$\phi + \frac{\phi_0}{2\pi}\Delta\theta = n\phi_0. \tag{3.49}$$

It should be noticed that when $\Delta\theta = 0$, we recover the usual flux quantization in a uniform ring.

Now let us analyze the behavior of the total magnetic flux inside the SQUID ring. As we know from elementary electrodynamics, the total flux can be written as

$$\phi = \phi_x + Li, \tag{3.50}$$

where ϕ_x is the flux due to the external field H_x perpendicular to the plane of the ring, L is the self-inductance of the ring and i its total current. The latter can be decomposed, in the so-called *resistively shunted junction* (RSJ) model (Likharev, 1986), as follows.

Josephson current. This is the current component due to the tunneling of Cooper pairs through the link and is given by (3.43),

$$i_s = i_0 \sin \Delta \theta. \tag{3.51}$$

Normal current. This originates from the two-fluid model (3.33) and obeys Ohm's law,

$$i_N = \frac{V}{R}, \tag{3.52}$$

where V is the voltage across the junction and R the resistance of the material in its normal phase.

Polarization current. This last bit comes from the fact that there is a finite junction capacitance, C, and reads

$$i_c = C\dot{V}. \tag{3.53}$$

Assuming that the total current is given by the sum of these three components, we get

$$i = i_0 \sin \Delta \theta + \frac{V}{R} + C\dot{V} \tag{3.54}$$

which, inserted in (3.50), reads

$$\frac{\phi_x - \phi}{L} = i_0 \sin \frac{2\pi \phi}{\phi_0} + \frac{\dot{\phi}}{R} + C\ddot{\phi} \tag{3.55}$$

where we used the fact that $V = -\dot{\phi}$. This is the equation of motion of a particle of "coordinate" ϕ in a potential (see Fig. 3.5)

$$U(\phi) = \frac{(\phi - \phi_x)^2}{2L} - \frac{\phi_0 i_0}{2\pi} \cos \frac{2\pi \phi}{\phi_0} \tag{3.56}$$

and (3.55) becomes

$$C\ddot{\phi} + \frac{\dot{\phi}}{R} + U'(\phi) = 0 \tag{3.57}$$

where, as usual, $U'(\phi) \equiv dU/d\phi$.

In reality we should add a white-noise fluctuating current $I_f(t)$ to the r.h.s. of (3.57) in order to properly account for the thermodynamical equilibrium properties

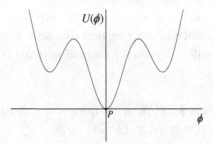

Figure 3.5 Potential energy for $\phi_x = 0$

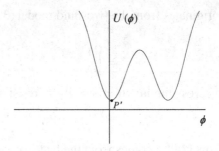

Figure 3.6 Potential energy for $\phi_x = \phi_0/2$

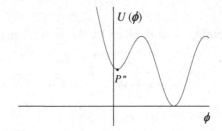

Figure 3.7 Potential energy for $\phi_x = \phi_0$

of the system. With this additional term (3.57) is nothing but the well-known classical *Langevin equation* for the *Brownian motion* (Wax, 2003).

In order to appreciate the diversity of physical phenomena in (3.56), it is worth analyzing that potential in some detail.

The minima of $U(\phi)$ are the solutions, ϕ_m, of $U'(\phi) = 0$ subject to $U''(\phi_m) > 0$. This leads us to two distinct cases:

(i) $2\pi Li_0/\phi_0 > 1 \Rightarrow$ several minima;
(ii) $2\pi Li_0/\phi_0 \le 1 \Rightarrow$ only one minimum.

In Figs 3.5, 3.6, and 3.7 we sketch the form of $U(\phi)$ for three distinct values of the external flux ϕ_x for case (i) above.

Suppose that at $t = 0$ there is no external field and the superconducting current is zero. In this case the equilibrium value of the flux inside the SQUID is also zero (point P in Fig. 3.5). As we slowly turn on the external field H_x, the potential $U(\phi)$ changes accordingly and the equilibrium value of ϕ follows its potential well adiabatically until it becomes an inflection point of $U(\phi)$. In this example, we clearly see the initial equilibrium position moving from $P \rightarrow P' \rightarrow P''$. It is worth noticing that the adiabatic approximation is valid only when dH_x/dt is much smaller than any typical frequency resulting from (3.57), namely, $[U''(\phi_m)/C]^{1/2}$ or $1/2RC$. In this way it is possible to see a broad region of values of H_x throughout which the initial flux changes from a stable to a bistable and, finally, a metastable configuration.

The realization that the dynamics of ϕ is really Brownian comes through the study of the decay of its metastable configuration by thermal fluctuations (Kurki-järvi, 1972). Here, as H_x increases, the value of ϕ that evolves adiabatically from $\phi = 0$ might jump to its neighboring minimum before H_x reaches H_x^*, the field at which the initial local minimum would become an inflection point of $U(\phi)$. The mechanism responsible for this transition is thermal activation over the potential barrier that keeps this configuration metastable. The probability of decaying from the metastable state in the SQUID ring nicely fits the results obtained by regarding ϕ as the coordinate of a Brownian particle.

The observation of the above-mentioned phenomena was made at temperatures of about 4 K (for a Nb SQUID). However, if the temperature is lowered below 1 K thermal fluctuations are not intense enough to trigger this process. Interestingly, for SQUID parameters such as $C \simeq 10^{-12}$ F, $L \simeq 5 \times 10^{-10}$ H and $i_0 \simeq 10^{-5}$ A, one has $T_0 \equiv \hbar\omega/k_B \simeq 1$ K, where ω is one of the natural frequencies of the SQUID. This means that for $T \leq 1$ K quantum effects are important and therefore one should ask about the possibility of observing the same transition driven by another mechanism: for example, quantum tunneling, like in the magnetic examples we have seen before. In this case one should start by neglecting the dissipative term and proceeding with the canonical quantization applied to the Hamiltonian in terms of the flux variable, which allows us to write a Schrödinger equation for the fictitious particle represented by the coordinate ϕ,

$$i\hbar\frac{\partial\psi(\phi)}{\partial t} = -\frac{\hbar^2}{2C}\frac{\partial^2\psi(\phi)}{\partial\phi^2} + U(\phi)\psi(\phi). \qquad (3.58)$$

From then onwards we can apply, for example, the standard WKB method to describe the tunneling of the lowest-energy state initially prepared about the metastable minimum of the potential $U(\phi)$ (see Fig. 3.7).

This procedure would not be useful for studying just the phenomenon of decay of a metastable configuration, but we could also study the coherent tunneling between two bistable configurations or the level structure within a stable potential well. In the former case we restrict ourselves to the two-dimensional Hilbert space spanned by the two lowest-energy eigenstates of the potential of Fig. 3.6. Here, we can also apply approximation methods to describe the coherent tunneling between the two bistable minima of that potential once the initial state is on a given side of the potential barrier, as we will see later on.

We shall address these issues shortly, and their interpretation in future sections.

3.4.2 Current-biased Josephson junctions (CBJJs)

Suppose we now have a weak link coupled to a current source which provides a constant current I_x to the system. In this case the variable of interest is the phase difference, $\Delta\theta$, across the weak link and here the RSJ model can once again be applied. Nevertheless, we shall get the desired equation of motion for the phase difference as a natural limiting process from the equations already deduced for the SQUID ring.

We can easily get the dynamics of $\Delta\theta$ using (3.57) and the fact that the current-biased junction is an extreme case of a huge SQUID where $L \to \infty$, $\phi_x \to \infty$ but $\phi_x/L = I_x$, the current bias. Moreover, since the flux inside the ring ceases to make any sense for the junction with a current bias, we must rewrite it in terms of the phase difference $\Delta\theta$ as (see (3.49))

$$\phi = -\frac{\phi_0}{2\pi}\,\Delta\theta \equiv \frac{\phi_0}{2\pi}\,\varphi \tag{3.59}$$

and (3.49) becomes

$$\frac{\phi_0}{2\pi}\,C\,\ddot{\varphi} + \frac{\phi_0}{2\pi R}\,\dot{\varphi} + \tilde{U}'(\varphi) = 0, \tag{3.60}$$

where

$$\tilde{U}(\varphi) = -I_x\,\varphi - i_0\,\cos\varphi. \tag{3.61}$$

Now it is φ which is representing the "coordinate" of a Brownian particle in the so-called washboard potential (Fig. 3.8). In order to recover a function $U(\varphi)$ with dimension of energy we must multiply the equation of motion (3.60) by $\phi_0/2\pi$ and then the potential energy becomes

$$U(\varphi) = -I_x\,\frac{\phi_0}{2\pi}\,\varphi - E_J\,\cos\varphi, \tag{3.62}$$

where we have just defined the *Josephson coupling energy* as

$$E_J \equiv \frac{\phi_0 i_0}{2\pi}. \tag{3.63}$$

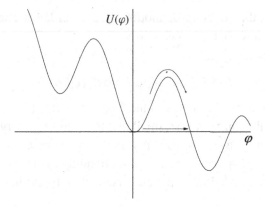

Figure 3.8 Washboard potential

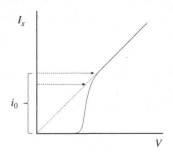

Figure 3.9 Voltage–current characteristic of a CBJJ

When the external current is zero, there is no phase difference across the junction. As the external current is increased, the phase across the junction adjusts itself so the current can cross the junction without the generation of any finite voltage. But, when $I_x \geq i_0$ (see (3.43) and (3.44)), a finite voltage develops across the junction. In our mechanical analogy this means that the maxima and minima of the washboard potential coalesce at $I_x = i_0$ and beyond this value our particle runs downhill, which means that φ varies with time. The current–voltage characteristic for a CBJJ is shown in Fig. 3.9.

However, a finite voltage can be generated even before i_0 is reached because, once again, thermal fluctuations could drive our fictitious particle out of its equilibrium position by overcoming the potential barrier. Now, if the temperature is not high enough to create intense thermal fluctuations, we could also in this case enquire about the possibility of tunneling-induced voltage across the junction. This means that our fictitious particle of coordinate φ tunnels through the potential barrier provided by the washboard potential. Once again we must neglect the dissipative term in the equation of motion of φ in order to apply the canonical quantization procedure to this problem as well.

3.4.3 Cooper pair boxes (CPBs)

The CPB is exactly the same device as the CBJJ with the only difference that now the capacitance of the junction is so small that the charging energy becomes much larger than the Josephson coupling energy. Let us see what results from this fact.

In the preceding example we have referred implicitly to the Hamiltonian operator of the fictitious particle when dissipation is neglected. Its explicit form reads

$$\mathcal{H}_0 = \frac{Q^2}{2C} + U(\varphi) \qquad \text{where} \qquad Q = -i\hbar\frac{\partial}{\partial(\phi_0\varphi/2\pi)}. \qquad (3.64)$$

The charge Q and the variable $\phi_0\varphi/2\pi$ are canonically conjugate variables, or else, $P_\varphi = \phi_0 Q/2\pi$ is the momentum canonically conjugated to φ. This is the operator we use to describe the above-mentioned quantum mechanical effects and which generates the equation of motion (3.60) in the classical limit (for $R \to \infty$). It also allows us to define another important energy scale for the problem, namely, the *charging energy* which is given by

$$E_C = \frac{e^2}{2C}. \qquad (3.65)$$

In the case of an unbiased junction we have a perfectly periodic potential $U(\varphi)$ and therefore the solution of the quantum mechanical problem requires knowledge of the appropriate boundary conditions. When $E_J \gg E_C$, this is not so important because the quantum states are well localized about the minima of the potential (fixed phase), reminding us of localized orbitals in solid-state physics. In this case, a current bias would be the analogue of an electric field tilting a periodic potential. However, for capacitances such that $C \lesssim 10^{-15}$ F one has $E_C \gg E_J$ and the description in terms of localized states is no longer appropriate. Therefore, we should borrow the well-known expression of Bloch's theorem for a particle moving in a one-dimensional potential and write the eigenstates of the system as

$$\psi_{nq}(\varphi) = \exp\left\{i\left(\frac{q}{2e}\right)\varphi\right\} u_n(\varphi), \qquad (3.66)$$

where φ represents the coordinate of the fictitious particle, $u_n(\varphi)$ is a 2π-periodic wave function and $q/2e$ plays the same role as the quasi-momentum of that particle. We refer to q as the *quasi-charge* which is supplied to the junction when it is subject to an external field or current such that $q(t) = q_0 + Q_x(t)$, where

$$Q_x(t) = \int_{t_0}^{t} dt' I_x(t') \qquad (3.67)$$

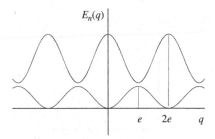

Figure 3.10 Energy bands

and "the first Brillouin zone" of this system extends over the range $-e \leq q \leq e$. This is clearly analogous to the "nearly free electron" approximation in solids. In practice, the set-up of a CPB is more subtle than that we are presenting here and we shall return to this issue later in the book.

The way to tackle the problem described by the Hamiltonian (3.64) is to apply it to (3.66) which, within the adiabatic approximation, results in a new eigenvalue problem given by

$$\mathcal{H}_q u_n(\varphi) = \frac{(Q+q)^2}{2C} u_n(\varphi) + U(\varphi)u_n(\varphi) = E_n(q)u_n(\varphi)$$

$$\text{where } Q = -2ie\frac{\partial}{\partial\varphi}, \tag{3.68}$$

subject to

$$\psi_n(\varphi + 2\pi) = \psi_n(\varphi). \tag{3.69}$$

As we know from solid-state physics, this system will present an energy band structure as in Fig. 3.10 where the lowest-energy gaps open up in the neighborhood of $q = (2p+1)e$, where p is an integer. So, Cooper pair tunneling allows the system to decrease its energy as more charge is fed into the junction (Schön and Zaikin, 1990). We return to the discussion of this point in the last section of this chapter. However, before interpreting physically what is really taking place in the junction, we can foresee very interesting effects in this new situation (such as Bloch oscillations, if we still feed charge very slowly to our circuit or, for slightly faster rates, the occurrence of Zener tunneling between the two lowest-energy bands) (Schön and Zaikin, 1990).

3.5 Vortices in superconductors

Let us return now to the bulk properties of superconductors.

So far we have been dealing with superconductors such that, for external magnetic fields below the critical value H_c, the total field inside the superconducting

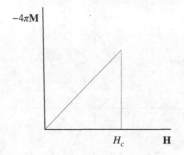

Figure 3.11 Magnetization in a type I superconductor

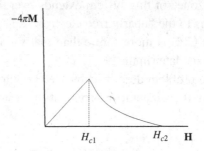

Figure 3.12 Magnetization in a type II superconductor

sample is zero, which implies $\mathbf{H} = -4\pi\mathbf{M}$ where $\mathbf{M}$ is the magnetization of the specimen. Above H_c, the field fully penetrates the material as shown in Fig. 3.11. These are examples of materials we call *type I superconductors*. They are usually very clean pure metallic compounds.

However, the great majority of substances that present superconductivity do not behave in this way. They indeed expel the magnetic field from the interior of their samples, but only until the external field reaches H_{c1}. As the external field is further increased, it starts to penetrate the sample until it reaches H_{c2} and a full penetration is accomplished as can be seen in Fig. 3.12. These are the so-called *type II superconductors*, which are usually metallic alloys and do not need to be very clean.

A very interesting characteristic of this new kind of superconductor is the way they allow the magnetic field to penetrate their interior before they become a normal metal. In order to make things simple, let us imagine a very long superconducting cylinder subject to a magnetic field parallel to its axis, say, along the $\hat{z}$ direction. When the field is increased past H_{c1}, tubes of magnetic field enter the sample as in Fig. 3.13 and therefore increase the total field inside it. As we go on increasing the field, more and more lines (tubes) penetrate the specimen

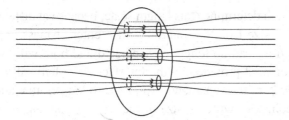

Figure 3.13 Flux tubes in a general geometry

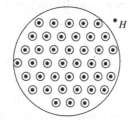

Figure 3.14 Abrikosov lattice

giving rise to an ordered hexagonal structure of vortices, the *Abrikosov vortex lattice* or *Schubnikov phase* (see, for example, De Gennes (1999) and Fig. 3.14). This ordered structure appears because, due to vortex–vortex interaction, it is the most stable spatial configuration of vortex lines possible. On reaching a second critical value, H_{c2}, the vortices coalesce and the external field finally penetrates the cylinder fully.

Despite the importance of the study of vortex lattices, let us return to the case of a single line and try to understand the energetic balance which is responsible for its very appearance. In order to do that we shall start by analyzing the issue of the normal–superconducting interface again.

As we have seen in Section 3.1, the phenomenological London equation (3.2) together with the Maxwell equation (3.3) suffice to explain the space dependence of the magnetic field on crossing an interface between a normal and a superconducting material. We have also seen in Section 3.2 that inside the superconducting material there exists a condensate wave function (order parameter) which can be interpreted (at least at $T = 0$) as the wave function of the center-of-mass coordinate of a Cooper pair. However, this interpretation was based on the separation of center-of-mass and relative coordinate wave functions of the pair, a procedure which is clearly not valid close to the interface separating the two materials. Since the order parameter must vanish in the normal region, there must be a length scale within which it changes from its maximal value (deep inside the superconductor) to zero. It is intuitively plausible to assume that it is the relative coordinate wave function which determines this length scale, because at distances of the same order as the pairs' size from the interface, the center-of-mass and relative coordinate wave functions can no longer be separated and both will vanish according to the external potential provided by the interface. We shall call this distance the *coherence length*, ξ, of the superconductor. At zero temperature and for pure metals, this length coincides with the size of the Cooper pair, ξ_0, as we have argued above. It is then the typical distance over which the electronic velocities are correlated. When $T \neq 0$ it becomes temperature dependent and serves as a measure of the characteristic length of the velocity–velocity correlation function in the superconducting phase.

This whole argument about the existence of an additional length scale in the superconducting phase naturally leads us to estimate the size of the Cooper pair. Although we might think that we would need a full microscopic formulation to do this, we show next that it can be done with very little microscopic insight.

Let us start with a gas of non-interacting electrons at $T = 0$. It is well known that the ground state of the system is described by the Fermi sphere completely filled with two electrons, with opposite spins, for each allowed value of momentum up to the so-called Fermi momentum, p_F, from which we define the Fermi velocity, $v_F \equiv p_F/m$. Therefore, the energetic cost to create an excitation in the system is zero, no matter what it is; an extra electron, a hole, or an electron–hole pair.

However, for superconductors, we have already discussed that the rigidity of the superconducting wave function implies the existence of an energy gap, Δ, in the spectrum of excitations of this phase, which is also corroborated by a plethora of experimental results (see, for example, De Gennes (1999)). In this case, electrons at the Fermi surface experience attractive interaction and deplete the formerly plane-wave energy eigenstates within the energy shell

$$E_F - \Delta < \frac{p^2}{2m} < E_F + \Delta, \tag{3.70}$$

in order to form bound pairs. In (3.70), E_F is the Fermi energy. Assuming that $\Delta \ll E_F$, this equation implies that only electrons within the momentum shell $\delta p = 2\Delta/v_F$ participate in the formation of wave packets representing the Cooper pairs which, by the uncertainty principle, has width

$$\delta x \propto \frac{\hbar}{\delta p} = \frac{\hbar v_F}{2\Delta}. \tag{3.71}$$

The proportionality constant implied above is, for convenience of comparison with the microscopic theory, given by $1/\pi$ and then

$$\xi_0 = \frac{\hbar v_F}{\pi \Delta}. \tag{3.72}$$

At finite temperatures, when we really have to reinterpret the condensate wave function as an order parameter, the coherence length must be replaced by a temperature-dependent expression, $\xi(T)$, which must diverge at the transition temperature. In other words, the pairs unbind at T_c as the superconducting fraction of electrons vanishes. Actually, the temperature dependence of both $\lambda(T)$ (the temperature-dependent penetration depth) and $\xi(T)$ is the same, which means that the characteristics of different superconductors depend only on the ratio λ_L/ξ_0. Analysis of the physical parameters of several materials reveals that pure metals have $\lambda_L/\xi_0 < 1/\sqrt{2}$, whereas metallic alloys have $\lambda_L/\xi_0 > 1/\sqrt{2}$ (the factor $1/\sqrt{2}$ comes from the Ginzburg–Landau theory (De Gennes, 1999; Tinkham, 2004)). In the former case, we see that the hypothesis of locality implicitly assumed in (3.7) does not make sense any more since the spatial variation of the vector potential is fast within the superconducting coherence length. Therefore, the supercurrent density $\mathbf{J_s(r)}$ must be obtained from an average of the vector potential $\mathbf{A(r')}$ over a region such that $|\mathbf{r} - \mathbf{r'}| < \xi_0$. This is the essence of the Pippard phenomenological theory for type I superconductors (De Gennes, 1999; Tinkham, 2004) that we have already mentioned earlier in this chapter but will not explore in this book.

In the latter case above, the variation of the vector potential is, on the contrary, very slow over ξ_0 and, consequently, equation (3.7) holds fully. It is interesting to notice that it is for alloys or dirty metals that (3.7) provides us with an appropriate description of the superconducting state. However, the expression of the penetration depth given by (3.11) must be properly modified in these cases (De Gennes, 1999; Tinkham, 2004). For our present purposes, we do not need to go that far into the phenomenology of superconductivity and will always rely on equation (3.7) with a given phenomenological penetration depth larger than ξ_0.

Once we have these two lengths, the energetic balance due to the existence of an interface between a normal metal and a superconductor can be elaborated using simple thermodynamic arguments. This is readily done in De Gennes (1999), and the general reasoning goes as follows.

If we assume that both the order parameter and the magnetic field change abruptly at the interface, it can be shown that the free energy difference between the normal and superconducting phases, which is called the *condensation energy*, is given by

$$F_N - F_S = \frac{H_c^2}{8\pi} \tag{3.73}$$

where F_N and F_S are, respectively, the Helmholtz free energy densities of the normal and superconducting phases. Nevertheless, when we consider the finite values of both lengths, this free energy balance changes. Owing to the decrease in superconducting order parameter as one approaches the interface, there is a reduction of the condensation energy in a shell of thickness ξ about the interface. This gives rise to a surface tension of the order of $H_c^2 \xi / 8\pi$. In contrast, the penetration of the magnetic field in the superconducting region gives rise to another surface tension term proportional to $-H_c^2 \lambda / 8\pi$. Therefore, we see that the net result is that the presence of the interface produces a change in the condensation energy density given by

$$F_N - F_S = \frac{H_c^2}{8\pi} + \frac{H_c^2}{8\pi} \frac{(\lambda - \xi)S}{V} \tag{3.74}$$

where S and V are, respectively, the area of the interface and the volume of the superconducting sample.

If we allow for the presence of vortices inside the sample and make use of these results, it is clear that for type II superconductors it will be energetically more favorable to have as many vortices as possible since the free energy density difference in (3.74) is further increased in this case. The reason why this process of increasing the number of vortices stops is simply because the vortex–vortex interaction is repulsive, as we shall see below. However, since we will mostly be interested in the dynamics of vortices, let us elaborate a bit more on the structure of a single vortex before we discuss the interaction between them.

Since the field penetrates a cylindrical region fully in order to form the vortex, superconductivity must be destroyed within this region and consequently normal electrons fill up the vortex core. Therefore, the superconductor with a vortex is a clear example of a multiply connected superconducting sample, which leads us back to the discussion prior to the expression (3.38) of flux quantization in Section 3.3 above, and makes us conclude that the vortex must carry an integer number of flux quanta ϕ_0. Actually, using the expression

$$\epsilon_l = \frac{1}{8\pi} \int\limits_{r > \xi} dS(\lambda^2 |\nabla \times \mathbf{h}(\mathbf{r})|^2 + h^2(\mathbf{r})) \tag{3.75}$$

for the energy per unit length, ϵ_l, of the vortex line, we can conclude that many vortices carrying one flux quantum are energetically more favorable than a single vortex carrying many flux quanta. In expression (3.75) above, the first term is the total energy due to the presence of persistent currents and the second one refers to the magnetic field energy (De Gennes, 1999). The integration is performed over a flat surface perpendicular to the vortex axis and is valid only if we neglect core effects, which is a good approximation for extreme type II superconductors ($\kappa \equiv \lambda_L / \xi_0 >> 1$) as discussed in De Gennes (1999) and Tinkham (2004). This integral was evaluated in Tinkham (2004) and the final result is

$$\epsilon_l = \epsilon_0 \ln \kappa, \tag{3.76}$$

where we have defined

$$\epsilon_0 = \left(\frac{\phi_0}{4\pi\lambda}\right)^2 \tag{3.77}$$

as the characteristic scale of the vortex linear energy density.

Once we know this expression, it is possible to evaluate the elastic energy functional of a distorted vortex tube as

$$\mathcal{F}_{el} = \int dz\, \epsilon_l \left\{\left[1 + \left(\frac{\partial \mathbf{u}}{\partial z}\right)^2\right]^{1/2} - 1\right\}$$

$$\approx \int dz\, \frac{\epsilon_l}{2}\left(\frac{\partial \mathbf{u}}{\partial z}\right)^2, \tag{3.78}$$

where $\mathbf{u}(z) = [u_x(z), u_y(z)]$ is the transverse displacement of the line from its equilibrium position, when it is placed along the $\hat{z}$ direction.

Another interesting energy term which is usually neglected as being very small compared with the other parameters appearing in the description of the dynamics of the line (Blatter *et al.*, 1994) refers to its inertial mass or, in other words, is the linear density of kinetic energy of the vortex. Its existence is due to a very simple argument, although its precise determination requires a better understanding of the microscopic theory of superconductivity (see Blatter *et al.* (1994), De Gennes (1999)). Nevertheless, since our main goal here is to appeal to phenomenological arguments, we will content ourselves with the more heuristic reasoning given in Blatter *et al.* (1994), which we repeat below.

It is a very reasonable hypothesis (as we have mentioned above) to consider the vortex as a tube of radius ξ (the region where the order parameter vanishes) completely filled with normal electrons. In order for this to happen, a fraction of $2\pi\xi^2 N(E_F)\delta\epsilon$ electrons per unit length must be localized within the vortex tube, which means that $\delta\epsilon \simeq \hbar v_F/\pi\xi$. Here, $N(E_F) = m_e k_F/2\pi^2\hbar^2$ is the usual density of states (per spin) at the Fermi level and m_e, k_F, and v_F are, respectively, the electronic effective mass, Fermi momentum, and Fermi velocity. Since the effective mass of the electrons confined to the core changes by an amount of the order of $m_e\delta\epsilon/E_F$, we have the linear density of mass of the vortex given by

$$m_l = \frac{2}{\pi^3}m_e k_F. \tag{3.79}$$

There are also other contributions to the mass linear density term (Blatter *et al.*, 1994), but these turn out to be even smaller than the one we have just presented. Therefore, we can regard the vortex in a type II superconductor as an elastic string with a very small linear kinetic energy density. Nevertheless, the elastic and the kinetic energy terms are not the only ones acting on the string. There are other very important contributions we shall address from now on.

In order to do that let us return to the problem of two vortex lines. We shall see that in so doing we will not only find the new contributions to the energy of a single vortex but also answer the previous question of the interaction between them.

Following Tinkham (2004), we start with the assumption that since we are dealing with a type II material, the vortex cores are very small and, consequently, we can write the total field at a point $\mathbf{r}$ as

$$\mathbf{h}(\mathbf{r}) = \mathbf{h}_1(\mathbf{r}) + \mathbf{h}_2(\mathbf{r}), \qquad (3.80)$$

where $\mathbf{h}_{1(2)}(\mathbf{r})$ is the field produced by the vortex 1(2) at $\mathbf{r}$. Then, using this expression in (3.75), one has a vortex pair energy per unit length, E_l, given by

$$E_l = \frac{\phi_0}{8\pi} [h_1(\mathbf{r}_1) + h_1(\mathbf{r}_2) + h_2(\mathbf{r}_1) + h_2(\mathbf{r}_2)], \qquad (3.81)$$

where ϕ_0 is the usual flux quantum which is carried by each vortex and we assumed that the field is along the $\hat{\mathbf{z}}$ direction. This expression clearly shows that the total energy of the vortex pair is written as a sum of the individual self-energies of each vortex plus the interaction energy between them, ΔE_l, which is given by

$$\Delta E_l = \frac{\phi_0}{4\pi} h_1(\mathbf{r}_2) \qquad (3.82)$$

since, by symmetry, $h_1(\mathbf{r}_2) = h_2(\mathbf{r}_1)$. Just for the sake of completeness, the expression for $h(\mathbf{r})$ has been evaluated in Tinkham (2004) and reads

$$h(\mathbf{r}) = \frac{\phi_0}{2\pi\lambda^2} K_0\left(\frac{r}{\lambda}\right). \qquad (3.83)$$

$K_0(r)$ is the zeroth-order Hankel function of imaginary argument, which decays exponentially for $r \to \infty$ and has a logarithmic behavior for $\xi \ll r \ll \lambda$. Therefore we have

$$\Delta E_l(r_{12}) = \frac{\phi_0^2}{8\pi^2\lambda^2} K_0\left(\frac{r_{12}}{\lambda}\right), \qquad (3.84)$$

where $r_{12} \equiv |\mathbf{r}_1 - \mathbf{r}_2|$ and the force between the two vortices can now be evaluated easily if we use, for example, that

$$\mathbf{f}(\mathbf{r}_2) = -\nabla_2 \Delta E_l(r_{12}) = -\frac{\phi_0}{4\pi} \nabla_2 h_1(\mathbf{r}_2). \qquad (3.85)$$

In the above, ∇_2 means that the gradient must be applied on $\mathbf{r}_2$. Now, remembering that $h_i(\mathbf{r})$ is actually the component of the magnetic field on each vortex along $\hat{\mathbf{z}}$, we can use (3.3) to express $\nabla_2 h_1(\mathbf{r}_2)$ as $4\pi \mathbf{J}_1(\mathbf{r}_2)/c$, perpendicular to $\hat{\mathbf{z}}$, and write the force per unit length acting on the vortex at $\mathbf{r}_2$ due to that at $\mathbf{r}_1$ as

$$\mathbf{f}(\mathbf{r}_2) = \mathbf{J}_1(\mathbf{r}_2) \times \frac{\Phi_0}{c}, \tag{3.86}$$

where $\Phi_0 \equiv \phi_0 \hat{\mathbf{z}}$, which clearly shows that this force is repulsive, as we anticipated before. Therefore, the tendency for the creation of more vortices is balanced by the repulsive force between them until one reaches H_{c2}, when they coalesce and the field penetrates the sample fully. Before this has been accomplished, but when $H_{c1} < H < H_{c2}$, the vortices reach stable equilibrium in a hexagonal lattice.

Expression (3.86) is actually quite general and can be extended to the case of a single vortex located at $\mathbf{r}$ subject to a current density $\mathbf{J}(\mathbf{r})$. It then reads

$$\mathbf{f}_L(\mathbf{r}) = \mathbf{J}_s(\mathbf{r}) \times \frac{\Phi_0}{c}, \tag{3.87}$$

where $\mathbf{J}_s(\mathbf{r})$ is now the total supercurrent density at $\mathbf{r}$. This is the well-known Lorentz force.

This expression allows us to extract all we need to know about the motion of vortices in a superconductor. For simplicity, let us assume that there is only one single vortex in our specimen, free to move, and subject to a transport current $\mathbf{J}_s(\mathbf{r})$. By (3.87), the vortex starts to move perpendicular to the current density and in so doing, it becomes subject to many possible effects (Blatter *et al.*, 1994).

The first of these is due to its motion relative to the total superfluid velocity imposed by $\mathbf{J}_s(\mathbf{r})$, which results in the famous Magnus force (see Blatter *et al.* (1994) and references cited therein). Once it has acquired a velocity $\mathbf{v}_l$, it feels a force

$$\mathbf{f}_M(\mathbf{r}) = \rho_s [\mathbf{v}_s(\mathbf{r}) - \mathbf{v}_l] \times \frac{\Phi_0}{c}. \tag{3.88}$$

Furthermore, a vortex in motion relative to the underlying lattice of the superconducting material (laboratory frame) implies that

(i) the normal charges inside the vortex core will start to move relative to the ionic lattice and, consequently, tend to relax toward a stationary regime due to the electron–ion interaction, and
(ii) since the magnetic field is now time-varying it generates an electric field in the direction of the motion.

Both facts (i) and (ii) above obviously lead to dissipative and Hall effects. The detailed study of vortex motion is by no means simple (for one of the simplest approaches, see, for example, Bardeen and Stephen (1965)), but, once again, since we are only interested in the more phenomenological approach to the problem, we can write a general equation of motion for an element of a stiff vortex as (Blatter et al., 1994)

$$m_l \dot{\mathbf{v}}_l + \eta_l \mathbf{v}_l + \alpha_l \mathbf{v}_l \times \hat{\mathbf{z}} = \mathbf{f}_L, \tag{3.89}$$

where the dissipative and Hall coefficients are given by

$$\eta_l = \frac{\phi_0}{c} \rho_s \frac{\omega_0 \tau_r}{1 + \omega_0^2 \tau_r^2},$$

$$\alpha_l = \frac{\phi_0}{c} \rho_s \frac{\omega_0^2 \tau_r^2}{1 + \omega_0^2 \tau_r^2}. \tag{3.90}$$

Here, ω_0 and τ_r are, respectively, the frequency of the level separation of electrons (quasi-particles) bound to the vortex core and the scattering relaxation time. Notice that although both parameters tend to zero as $\omega_0 \tau_r \ll 1$, the motion will be dominated by the dissipative term in this case. On the contrary, in the so-called ultra-clean limit, $\omega_0 \tau_r \gg 1$, $\eta_l \to 0$ whereas $\alpha_l \to \phi_0 \rho_s / c$. So, in the latter case, if the total current density is $\rho_s \mathbf{v}_s$ the vortex is dragged by the superfluid velocity, which means $\mathbf{f}_M = 0$. This is what we usually observe in ordinary superfluid flow when there is no lattice effect to which the carriers must relax.

Finally, there is another effect extremely important for the vortex dynamics which is (similarly to what we have seen for magnetic walls) pinning. Owing to lattice defects, impurities, or anisotropies, the vortex line can be trapped within a given region of the material. It will only be released if one has strong enough currents to provide us with a Lorentz force capable of taking the line away from this pinned configuration. So, if the transport current is high but not enough to create a strong Lorentz force, there will be no dissipative effect generated by the vortex motion, which is what we require for practical applications, such as the development of high-field superconducting magnets.

Pinning theory, as in our study of magnetic systems, can be separated into strong and weak (collective) cases. Although the microscopic mechanisms responsible for the creation of the linear potential energy density for vortices in superconducting systems may be very different from those for creating linear or surface potential energy densities for walls in magnets, the resulting mathematical formalism for the dynamics of these objects is basically the same in either case. Suppose, for instance, we relax the condition of stiffness of the vortex line in (3.89) and assume that it is pinned by a linear potential energy density $V_{pin}(\mathbf{u}(z, t))$. Then, we can write

$$m_l \frac{\partial^2 \mathbf{u}(z, t)}{\partial t^2} + \eta_l \frac{\partial \mathbf{u}(z, t)}{\partial t} + \alpha_l \frac{\partial \mathbf{u}(z, t)}{\partial t} \times \hat{\mathbf{z}} - \epsilon_l \frac{\partial^2 \mathbf{u}(z, t)}{\partial z^2} + \frac{\partial V_{pin}(\mathbf{u}(z, t))}{\partial \mathbf{u}(z, t)} = \mathbf{f}_L,$$

$$\tag{3.91}$$

where $\mathbf{u}(z, t)$ is now the time-dependent transverse displacement of the line from its equilibrium position. If we consider the ultra-clean limit mentioned above, we can neglect both α_l and η_l and (3.91) acquires exactly the same form as the equation of motion we have already treated in (2.81). Therefore, the whole analysis made for either strong or weak pinning of magnetic walls can be repeated here, step by step, once we replace: $y \rightarrow z$, $u(y, t) \rightarrow \mathbf{u}(z, t)$, $c_s \rightarrow \epsilon_l/m_l$, $\tilde{V}_{pin} \rightarrow V_{pin}/\epsilon_l$, and $-(H/H_c)1/\zeta \rightarrow \mathbf{f}_L$. Remember, $u(y, t)$ was a displacement along the $\hat{\mathbf{x}}$ direction whereas $\mathbf{u}(z, t)$ is now in the xy plane. Likewise, the equivalent form of the potential energy functional (2.87) becomes

$$\mathcal{H}[\mathbf{u}(z, t)] = \int\limits_{-\infty}^{+\infty} dz \left[\frac{\epsilon_l}{2} \left(\frac{\partial \mathbf{u}(z, t)}{\partial z} \right)^2 + V_{pin}(z, \mathbf{u}(z, t)) - \mathbf{f}_L \cdot \mathbf{u}(z, t) \right],$$

(3.92)

where $V_{pin}(z, \mathbf{u}(z, t))$ can be replaced by $V_R(z, \mathbf{u}(z, t))$, which satisfies (2.83), (2.84), and (2.85) whenever appropriate.

Although it would be pointless to go through all the details of that analysis again (the reader is urged to follow the discussion from (2.81) to the end of that section), we think it would be instructive to review the main conclusions thereof as applied to the present case.

Let us start with the strong pinning situation. Here we also have a quantity we can change in order to release the vortex from its pinning configuration. In the present case, it is the external transport current density which, at a critical value j_d (analogous to H_d in magnetic systems), distorts the effective pinning potential in Fig. 3.15, changing the stable configuration into an unstable one and allowing for what is known as *flux flow*. On the contrary, if we keep the current just below j_d and wait long enough, thermal fluctuations can excite line segments of the vortex over the potential energy barrier and, depending on the length of this segment, the vortex either returns to its original configuration, for $L < L_p$ (the critical length), or is otherwise dragged downhill. This is the case of a single potential minimum. If the pinning potential allows for more than one minimum energy configuration, the excited vortex segment is accommodated in the next potential energy minimum and, for $L > L_p$, subsequently expands, bringing the rest of the vortex to the new configuration. This expansion can be viewed as two distorted parts of the line (kink–anti-kink) moving away from one another, as we also had for magnetic walls. When this phenomenon occurs repeatedly, we have what is called *flux creep* (Tinkham, 2004), which is the vortex motion due to thermal hopping between metastable energy configurations.

In the weak pinning case things are a little more subtle, as we have already seen for magnetic walls. However, the phenomenological approach to

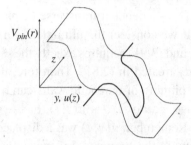

Figure 3.15 Flux flow

the present situation follows exactly the same reasoning we have employed for that case and, once again, we can define a collective pinning length, L_c, a collective pinning barrier, U_c, and a depinning current, j_{dep}, analogously to what we have done in (2.90), (2.91), and (2.92), respectively. The difference here is that the line goes on hopping from one metastable configuration to the next by a diffusive motion through the depinning of segments of length L_c. Therefore, we can also study in this case the effects of flux creep for $j < j_{dep}$ or flux flow for $j > j_{dep}$. For a thorough treatment of this problem to the particular case of vortices in superconductors, we refer the reader to Blatter *et al.* (1994).

If we now compare the above arguments to those previously used for SQUIDs or CBJJs, the next step would naturally be to enquire about the possibility of observing flux creep driven by quantum fluctuations. Here, we can also try to measure finite damping effects or magnetization relaxation in superconducting samples due to the motion of vortex lines at extremely low temperatures when we keep the current density below j_d or j_{dep}. If such a motion exists (once again neglecting dissipative effects) it is bound to take place due to quantum mechanical tunneling of the vortex line, or, in other words, by *quantum creep*.

Before leaving this section we would like to introduce another concept of great importance regarding vortex[1] motion, in particular in relation to the Josephson effect. This is the phenomenon of *phase slip* (Anderson, 1966), which we explain below.

Suppose we have a superconducting sample where a current flows from point 1 to point 2 in Fig. 3.16 and let us, only for simplicity, connect them by a straight line. Now, if a vortex crosses the line joining 1 and 2 we can write

$$\oint \nabla \theta \cdot d\mathbf{l} = \int_1^2 (\nabla \theta)_l \cdot d\mathbf{l} + \int_2^1 (\nabla \theta)_u \cdot d\mathbf{l} = 2\pi, \qquad (3.93)$$

---

[1] We are now referring to the so-called *Josephson vortices* in contrast to the previously introduced *Abrikosov vortices*. For a discussion on the difference between them, see Likharev (1986).

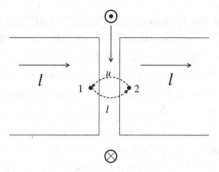

Figure 3.16 Phase slip

where the closed path involves the vortex and contains the endpoints 1 and 2. The integral of $(\nabla\theta)_{l(u)}$ means that we perform it along the lower (upper) part of the closed curve. Therefore,

$$(\theta_1 - \theta_2)_u - (\theta_1 - \theta_2)_l \equiv \Delta\theta_u - \Delta\theta_l = 2\pi. \tag{3.94}$$

So, we can say that every time a vortex crosses the straight line, there must be a phase change of 2π between 1 and 2. If there are N vortices crossing the line per unit time one has

$$\frac{d\Delta\theta}{dt} = 2\pi\frac{dN}{dt}, \tag{3.95}$$

which with the help of the Josephson relation (3.44) becomes

$$V = \phi_0\frac{dN}{dt}. \tag{3.96}$$

Now, if we have n_v vortices per unit area moving with constant speed v_L, perpendicular to the straight-line segment of length d joining 1 and 2, (3.96) becomes

$$V = \phi_0 n_v v_L d. \tag{3.97}$$

Since in the case considered here we expect that the constant vortex speed will be proportional to the imposed current i, this expression clearly shows us that the vortex flow in the system induces a resistive behavior for i.

3.6 Macroscopic quantum phenomena in superconductors

In the preceding section we described three different kinds of superconducting device and also the motion of vortices in bulk superconductors. In all of them we raised some expectation toward the possibility of observing quantum mechanical effects with regard to some specific variables which always played the role of the coordinate of a fictitious particle, or even of an extended object, in a given potential as we had done for magnetic systems. For the latter, we have approached the problem from a more microscopic point of view and have been able to interpret the physics of these effects in terms of approximate quantum mechanical states of those systems. Therefore, it is a natural question to ask now: What do we mean by macroscopic quantum phenomena in the present superconducting systems?

Once again, we should anticipate that there will certainly be problems related to dissipation in all those situations, but we shall postpone the discussion on this issue to future chapters.

As we have seen from the beginning of this chapter until now, the very idea of the existence of a wave function describing the superconducting condensate leads us to the concept of a quantum mechanical macroscopic object. Therefore, something like flux quantization means that we can observe the whole condensate generating screening supercurrents to oppose the external field, and this happens for discrete values of the circulation of the charge carriers (Cooper pairs). Actually, we should mention here that although these states are all metastable, their lifetimes are extremely long. What we conclude from this reasoning is that a superconducting ring behaves as a single giant atom, something like a huge Bohr's atom. It is quantum mechanical but depends on the macroscopic occupation of a single quantum state or the coherent motion of a macroscopic number of carriers.

Another effect that people refer to as a macroscopic quantum effect is the Josephson effect itself. In the way we have introduced it we have made explicit use of the overlap of the macroscopic wave function on different sides of the junction. Nevertheless, this approach to the problem can be reinterpreted as the overlap of the Cooper pair wave functions on each side of the junction. Once again we are back to the concept of a single particle (or pair) state which is macroscopically occupied. The only difference is that in the present case this state refers to a linear combination of single pair states on both sides of the non-superconducting barrier.

The quantum effects we have been referring to are somewhat more subtle. Let us analyze them carefully from the SQUID ring point of view.

The minima of $U(\phi)$ (Fig. 3.5) are reminiscent of the flux quantization in a uniform ring. The weak link actually allows for an easier penetration of the magnetic field in the superconducting material and, consequently, of the flux in the ring. If there were no link, the potential barriers in Fig. 3.5 would be very high, of the order of the critical field for destroying superconductivity. Thus, it is the presence of the link that lowers the energy barriers, allowing for a given amount of flux $\propto \phi_0$ to cross the junction.

Owing to thermal fluctuations one does not need to reach the external field at which the contact becomes normal. Since the junction has dimensions comparable to the penetration depth, it fluctuates between normal and superconducting phases, allowing for the passage of flux every now and then. This is the interpretation of the thermal fluctuations driving the fictitious particle over the barrier.

However, for low enough temperatures, the contact can be assumed in its super-conducting phase until H_x^* is reached and the link becomes normal. So, if it happens that the flux penetrates the ring it must go under a potential barrier or, in other words, it crosses the small superconducting forbidden region. That is the meaning of the flux tunneling.

Since the flux is nothing but $\oint \mathbf{A} \cdot d\mathbf{r}$, we have been discussing the quantum mechanics of the electromagnetic field subject to the boundary conditions imposed by the superconducting material and geometry of the SQUID ring.

If we now remember that the electromagnetic field and charges in the supercon-ductor are coupled fields, we know that the Hilbert space of the composite system is the tensor product of the two Hilbert spaces $\mathcal{E}_A$ of the electromagnetic field and $\mathcal{E}_M$ of matter. Bearing this in mind, we can analyze the total state of the composite system in different flux configurations.

Suppose the flux is in the configuration P'' of Fig. 3.7. In this case there is a finite supercurrent in the ring that creates a field opposing the penetration of addi-tional flux inside it. Therefore, we can write the initial state $|\Phi_i\rangle$ of the composite system as

$$|\Phi_i\rangle = |A_i\rangle \otimes |\psi_i\rangle, \tag{3.98}$$

where $|A_i\rangle$ and $|\psi_i\rangle$ are the initial states of the electromagnetic field and matter, respectively, and $\otimes$ stands, as usual, for the direct product. Remember that $|\psi_i\rangle$ carries a finite current.

Once the flux tunnels to its neighboring value, we must associate another product state with this final configuration. Let us call it $|\Phi_f\rangle$, which is given by

$$|\Phi_f\rangle = |A_f\rangle \otimes |\psi_f\rangle, \tag{3.99}$$

where $|A_f\rangle$ represents the state of the electromagnetic field with some flux ($\sim \phi_0$) inside the ring, whereas the state $|\psi_f\rangle$ now carries a current compatible with the accommodation of that amount of flux inside the ring. We can therefore write the quantum state of the composite system approximately as

$$|\Phi_D(t)\rangle \approx e^{-\frac{\Gamma t}{2}} |A_i\rangle \otimes |\psi_i\rangle + \sqrt{\left(1 - e^{-\Gamma t}\right)} |A_f\rangle \otimes |\psi_f\rangle, \tag{3.100}$$

where Γ is the relaxation frequency (tunneling frequency) for the decay of the initial state to an intermediate excited state of the final configuration, $|\Phi_f^{(*)}\rangle$, with the same energy as the former. Above we assume, as we have done for magnetic systems, that the relaxation time, γ^{-1}, for this excited state to decay to the new ground state of the system is much shorter than Γ^{-1}. Although we have not said anything about dissipation, we should keep in mind that its presence is essential for a transition of the same kind as the one described in (3.100). Notice that this form for $|\Phi_D(t)\rangle$ is not rigorous and great care must be taken when we deal with the more realistic decay of a quantum state, although we use it loosely for our present purposes.

Another interesting situation takes place when $\phi_x \approx \phi_0/2$. The potential $U(\phi)$ is now bistable and the flux inside the ring is bound to oscillate from one minimum to the other. Our product state is then

$$|\Phi_B(t)\rangle \approx a(t)\,|A_i\rangle \otimes |\psi_i\rangle + b(t)\,|A_f\rangle \otimes |\psi_f\rangle, \qquad (3.101)$$

where $|A_{i(f)}\rangle$ and $|\psi_{i(f)}\rangle$ are, respectively, the initial (final) states of the electromagnetic field and matter for this new configuration. The amplitudes $a(t)$ and $b(t)$ are such that $|a(t)|^2 + |b(t)|^2 = 1$. It is this kind of quantum state we are going to study below.

Neglecting dissipative effects we see that this state cannot be written as a product of a state which belongs to $\mathcal{E}_A$ and another belonging to $\mathcal{E}_M$ (actually, neither could $|\Phi_D(t)\rangle$ in (3.100) for any finite time). In other words, it is an entangled state of electromagnetic and matter degrees of freedom. Moreover, it also represents a genuine quantum mechanical superposition of two macroscopically distinct quantum states of matter as in (2.93), in the sense that they can be identified by distinct values of the macroscopic currents they carry. This is again tantamount to the Schrödinger cat state, whose possible existence for superconducting devices (Leggett, 1980) has been discussed extensively in the literature for many years.

It is worth noticing that if we take the coordinate representation of $|\Phi_B(t)\rangle$ it is of the form

$$\Phi_B(\mathbf{r}, \mathbf{r}_1, \ldots, \mathbf{r}_N, t) = a(t)\,A_i(\mathbf{r})\,\psi_i(\mathbf{r}_1, \ldots, \mathbf{r}_N) + b(t)\,A_f(\mathbf{r})\,\psi_f(\mathbf{r}_1, \ldots, \mathbf{r}_N),$$
$$(3.102)$$

which is remarkably different from those we had dealt with beforehand, namely, the condensate wave function and the Josephson effect wave function. The difference from the former is quite obvious, but it is a bit more subtle from the latter. Let us briefly try to stress it in what follows.

The Josephson effect can be understood as a superposition like

$$\varphi(\mathbf{x}_i, \mathbf{y}_i) = a_R^{(i)}\,\varphi_R(\mathbf{x}_i, \mathbf{y}_i) + a_L^{(i)}\,\varphi_L(\mathbf{x}_i, \mathbf{y}_i), \qquad (3.103)$$

where $\varphi_R(\mathbf{x}_i, \mathbf{y}_i)$ and $\varphi_L(\mathbf{x}_i, \mathbf{y}_i)$ are the ith pair wave functions in the superconductor R and L, respectively. Therefore, the total wave function is then

$$\varphi(\mathbf{x}_1, \mathbf{y}_1, \mathbf{x}_2, \mathbf{y}_2, \dots, \mathbf{x}_{N/2}, \mathbf{y}_{N/2}) = \prod_{i=1}^{N/2} \left[a_R^{(i)} \varphi_R(\mathbf{x}_i, \mathbf{y}_i) + a_L^{(i)} \varphi_L(\mathbf{x}_i, \mathbf{y}_i) \right],$$

$$(3.104)$$

which is clearly different in its structure from (3.102). In short, the cat-like state is of the form

$$\phi_C = a\,\phi_1^N + b\,\phi_2^N,$$

$$(3.105)$$

whereas the Josephson state is like

$$\phi_J = (a\,\phi_1 + b\,\phi_2)^{N/2},$$

$$(3.106)$$

where ϕ_1 and ϕ_2 refer to single particle states at macroscopically distinct configurations 1 and 2, respectively. A possible measure of quantification of the degree of how macroscopic these states can be was originally given in Leggett (1980). Therefore, we conclude that the quantum effects we are addressing here are cat-like and, consequently, represent situations in which a fairly large number of particles behave as a single quantum mechanical object in a genuine superposition of states related to macroscopically distinct configurations.

We can also apply the same ideas to the CBJJs. Although the description of the phenomenon in terms of trapped flux does not make sense any more, one can gain some insight into the problem by using the idea of phase slip as explained at the end of the previous section.

Suppose we impose a direct current through a normal junction which we assume to be a tiny region made of the same metal as the superconducting material of the electrodes, in other words, a weak link. Although this restriction is not really necessary, it makes things easier. In order to simplify matters even further, let us assume the link extends along the $\hat{\mathbf{z}}$ direction and connects two superconducting slabs placed on the xz plane as shown in Fig. 3.17. Now, if we keep the current slightly below its critical value i_0 in (3.51), at very low temperatures, there would be no voltage between the terminals of the junction. However, if it happens that there is a transition to the running (finite voltage) state, it could be due to quantum tunneling of the phase of the macroscopic wave function, a phenomenon which is hard to interpret directly. Let us try to analyze this problem in terms of the magnetic field produced by the imposed current.

If the current flows along the $\hat{\mathbf{x}}$ direction, one has vertical field lines pointing along $\hat{\mathbf{z}}$ if $y > 0$ and $-\hat{\mathbf{z}}$ if $y < 0$. This configuration is clearly metastable since its energy can be reduced once it makes a transition to the running state. However, when this transition takes place, the flux of the magnetic field generated by the current through a given area on the xy plane increases on both sides of the junction (we are assuming positive areas for $y > 0$ and negative ones for $y < 0$).

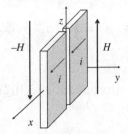

Figure 3.17 Very long junction

This can easily be accomplished if a given amount of flux ϕ is allowed to cross the junction from $y < 0$ to $y > 0$, which means that the number of field lines (negative for $y < 0$ and positive for $y > 0$) increases on both sides of the junction. As we know that flux can only be carried through a superconducting region by the motion of vortices, this transition occurs when a flux tube carrying a quantum of flux, ϕ_0, moves across the junction which, as we have seen, gives rise to a change in the phase of the superconducting wave function (phase slip). This is an unstable situation since as more vortices are allowed to cross the junction the energy of the system is reduced indefinitely, if not for the presence of the dissipative effects we have been neglecting so far. Therefore, remembering that we are dealing with a system at very low temperatures, it is plausible to attribute the triggering transition of the first vortex to quantum mechanical tunneling through a region that, albeit classically forbidden, can be represented by a reasonably small energy barrier. Notice that, for higher temperatures, fluctuations of the amplitude of the superconducting wave function drive the junction normal, which allows for the classical motion of vortices from one side of the junction to the other. This would equally well explain the same kind of transition by thermal fluctuations, a mechanism to which we have previously referred. Summarizing, we can fully reinterpret the dynamics of the phase of the superconducting wave function through a Josephson junction by thermal or quantal phase slippage. Moreover, since this transition involves current states which are macroscopically distinguishable, we can also interpret this "phase tunneling" as a macroscopic quantum phenomenon exactly as in (3.100).

Since we have been addressing the issue of vortex dynamics, let us quickly revisit the phenomenon of vortex depinning (creep or flow) from the point of view of the electronic motion. As we have mentioned above, there is the possibility of a vortex leaving a metastable configuration once an external current density exceeds a critical value j_c. Moreover, even before this value is reached, the vortex could, at very low temperatures, hop to the next stable pinning configuration by what we have called quantum creep. This, in turn, was due to quantum mechanical

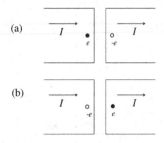

Figure 3.18 Charge configurations (a) and (b) are fully degenerate in energy

tunneling of the vortex line. Here, again, it is the very large number of carriers which keep the flux in the vortex constant and equal to ϕ_0 and, therefore, if a line segment of the tube tunnels so do a reasonably large number of carriers, changing collectively their state of motion. This means that we can regard quantum depinning as a macroscopic quantum effect of the kind proposed in (3.100), with the difference that now the states are macroscopically distinguishable locally. In other words, it is the supercurrent density which is changing its macroscopic configuration by quantum tunneling of the vortex. Once we have learnt that we can return to the CBJJ problem.

Finally, let us analyze the last remaining case, namely, the CPBs. As explained before, this is nothing but a CBJ with extremely low capacitance, which makes it more appropriate to represent by charge instead of phase variables. The analogy is perfect with a particle in a periodic potential which can be represented in position or momentum, depending on whether its kinetic energy is lower or higher than the potential energy. It is the latter which is of interest for the CPB, if we reinterpret it in terms of the electromagnetic variables.

As charge is fed into the system by (3.67), one sees in Fig. 3.18 that when it reaches the electronic charge e, the state of the junction becomes degenerate with the state of charge e minus one Cooper pair. In order to visualize what happens, let the charge on the left plate of the capacitor formed by the junction be e and on the right plate, $-e$. Notice that we are using the standard convention of the current transporting positive charges. This state is degenerate with that having charge $-e$ on the left plate and e on the right plate, which can be obtained from the former by allowing a Cooper pair of charge $2e$ to tunnel from the left plate to the right plate in Fig. 3.19. As this tunneling process costs the Josephson coupling energy (3.63), the two above-mentioned states, which we denote $|0\rangle = |e, 0\rangle_L \otimes |-e, 0\rangle_R$ and $|1\rangle = |e, -2e\rangle_L \otimes |-e, 2e\rangle_R$, can be used in the linear combination forming the eigenstates of the CPB as

$$|\pm\rangle = |0\rangle \pm |1\rangle. \tag{3.107}$$

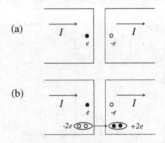

Figure 3.19 Charge configuration (b) of Fig. 3.18 is obtained by a Cooper pair tunneling from the left to the right plate in configuration (a)

These states have energies $\mp E_J$ relative to the degeneracy point. Notice the resemblance of this situation to that of a Bragg-reflected electron in a periodic potential within the reduced band scheme.

Therefore, when the external charge fed into the system reaches e, the eigenstates of the CPB represent simply the even and odd superpositions of a Cooper pair on the left and right terminals of the junction, which obviously does not characterize what we are calling here a macroscopic quantum phenomenon. It has indeed the same status as the ordinary Josephson effect, as it should have. In the end, the two phenomena are merely distinct because of their different representation. So, a constant phase difference in the CBJ is equivalent to the Bloch oscillation in the CPB and what plays the role of the transition to the running state in the former is the Zener tunneling phenomenon in the latter.

4

Brownian motion

In the two preceding chapters of this book we have analyzed many interesting physical phenomena in magnetic and superconducting systems which could adequately be described by phenomenological dynamical equations in terms of collective classical variables. One unavoidable consequence of this approach is that, as we are always dealing with variables that describe only part of the whole system, the interaction with the remaining degrees of freedom shows up through the presence of non-conservative terms which describe the relaxation of those variables to equilibrium. Those phenomenological equations are able to describe a very rich diversity of physical phenomena, in particular, those which can be studied in the context of quantum mechanics. Since these are genuine dynamical equations, there is no reason why they should be restricted to classical physics. However, as we do not yet know how to treat dissipative effects in quantum mechanics, we have deliberately neglected those terms when trying to describe quantum mechanical effects of our collective variables.

In this chapter we will describe the general approach to dealing with dissipation in quantum mechanics. However, before we embark on this enterprise we should spend some time learning a little bit about the classical behavior of dissipative systems. In this way we can develop some intuition on how systems evolve during a dissipative process and, hopefully, this will be useful later on when we deal with quantum mechanical systems.

The immediate problem we have to face concerns the choice of dissipative system to be studied. As we have already seen, there are many ways to introduce dissipative terms in the phenomenological equations of motion. Nevertheless, one of them is of particular interest; namely, dissipative terms which depend linearly on the rate of change (speed) of the collective variable under investigation. Although this kind of dissipation is not the only possibility, as we have already stressed, the reason why we have chosen to study it is twofold: firstly, because it is quite ubiquitous among all dissipative systems and secondly, for its simplicity. Equations of motion containing phenomenological terms of this kind turn out to be the basis for the study of *Brownian motion* – a paradigm of dissipative systems – and are known in the literature as *Langevin equations*. Next, we start our study by reviewing the theory of classical Brownian motion.

4.1 Classical Brownian motion

In the nineteenth century, the English botanist R. Brown observed that small particles immersed in a viscous fluid exhibited an extremely irregular motion. If no external force is applied on the particle, its average velocity $\langle \vec{v} \rangle = 0$ and its variance $\langle v^2 \rangle$ is finite. The averages are taken over an ensemble of identically prepared systems. This phenomenon has since been known as Brownian motion.

The theoretical approach to treat this problem is through the so-called Langevin equation, which reads

$$M\ddot{q} + \eta\dot{q} + V'(q) = f(t), \tag{4.1}$$

where $f(t)$ is a fluctuating force such that $\langle f(t) \rangle = 0$ and $\langle f(t)f(t') \rangle = 2\eta k_B T \delta(t - t')$, where M is the mass of the particle, η is the damping constant, $V(q)$ is an external potential, and k_B is the Boltzmann constant. This equation is a good description of the phenomenon only if

- the mass M of the Brownian particle is much larger than the mass m of the molecules composing the viscous fluid, and
- we are interested in the behavior of the particle for time intervals much longer than the average time τ between molecular collisions.

Although we have used a classical example to introduce the idea of Brownian motion, we should now recall that we have already been through this kind of equation in previous chapters when we wrote, for example, the equation describing the dynamics of the flux inside a SQUID ring (see (3.57)). In order to reproduce the correct Brownian dynamics all one has to do is insert a fluctuating current $I_f(t)$ on the r.h.s. of (3.57), which now becomes

$$C\ddot{\phi} + \frac{\dot{\phi}}{R} + U'(\phi) = I_f(t), \tag{4.2}$$

where $\langle I_f(t) \rangle = 0$ and $\langle I_f(t)I_f(t') \rangle = 2k_B T R^{-1} \delta(t - t')$. Actually, the Langevin equation can generally be used to describe the dynamics of the electromagnetic variables in most electric circuits.

The dynamical variables of conventional circuits usually exhibit a purely classical motion all the way down to extremely low temperatures. However, as we have seen before, in SQUID rings things can be very different on reaching the appropriate domain of the circuit parameters when quantum mechanics comes into play.

In this circumstance we have to deal with dissipation, fluctuations, and quantum effects on the same footing, and it is our main goal in this book to tie all these effects together. We shall return to this point in the next section.

Although we could jump directly to the development of our strategy to tackle the above-mentioned problem, we will, for the sake of completeness, introduce some key concepts in the classical theory of Brownian motion.

4.1.1 Stochastic processes

The Langevin equation involves the concept of *fluctuating forces*, which introduces a probabilistic character into the dynamics of the variable $q(t)$. One way to understand this process is by creating a statistical ensemble of equally prepared Brownian particles. After a time t each particle will occupy a different position due to the fact that it is being subject to a random force. Therefore, we can define the probability $P(q, t)dq$ to find the Brownian particle within the interval $[q, q + dq]$ at time t, which allows us to compute the average value of any function $g(q)$. Let us then start by introducing some useful concepts of stochastic processes (Reichl, 2009).

Let $y(t)$ be a variable which can assume any value in the interval $-\infty < y(t) < \infty$. The probability density that the variable $y(t)$ has the value y_1 at time t_1 is $P_1(y_1, t_1)$. This concept can be generalized to the case of n events through the probability that $y(t)$ has the value y_1 at time t_1, y_2 at t_2, ..., y_n at t_n, which is described by the *joint probability density*

$$P_n(y_n, t_n; y_{n-1}, t_{n-1}; \ldots; y_1, t_1). \tag{4.3}$$

The functions P_n are normalized as

$$\int_{R^n} P_n(y_n, t_n; y_{n-1}, t_{n-1}; \ldots; y_1, t_1) \, dy_1 \, dy_2 \ldots dy_n = 1, \tag{4.4}$$

and if we integrate them with respect to one of its variables, it reduces to P_{n-1},

$$\int P_n(y_n, t_n; y_{n-1}, t_{n-1}; \ldots; y_1, t_1) \, dy_n = P_{n-1}(y_{n-1}, t_{n-1}; \ldots; y_1, t_1). \tag{4.5}$$

The moment of the random variable $y(t)$ at times $t_1, t_2, \ldots, t_n$ is

$$\mu_n(t_1, t_2, \ldots, t_n) \equiv \langle y(t_1) \ldots y(t_n) \rangle$$
$$= \int_{R^n} y_1 \, y_2 \ldots y_n \, P_n(y_n, t_n; y_{n-1}, t_{n-1}; \ldots; y_1, t_1) \, dy_1 \, dy_2 \ldots dy_n. \tag{4.6}$$

The first moment is called the *average* or *mean*, $\mu_1(t) = \langle y(t) \rangle$. The second moment at time $t_1 = t_2 = t$ is given by $\mu_2(t) = \langle y^2(t) \rangle$, and the *variance* or *dispersion* of the probability distribution is defined as $\sigma^2(t) = \langle (y(t) - \langle y(t) \rangle)^2 \rangle = \mu_2 - \mu_1^2$.

A stochastic process is said to be *stationary* if

$$P_n(y_n, t_n; y_{n-1}, t_{n-1}; \ldots; y_1, t_1) = P_n(y_n, t_n + \tau; y_{n-1}, t_{n-1} + \tau; \ldots; y_1, t_1 + \tau),$$
(4.7)

which implies that

$$P_1(y_1, t_1) = P_1(y_1) \qquad \text{and} \qquad P_2(y_2, t_2; y_1, t_1) = P_2(y_2, t_2 - t_1; y_1, 0).$$

Another important concept is that of *conditional probability*. It is the probability that the variable $y(t)$ assumes the value y_2 at t_2 given that it was y_1 at t_1, and its form reads

$$P_{11}(y_2, t_2 | y_1, t_1) = \frac{P_2(y_2, t_2; y_1, t_1)}{P_1(y_1, t_1)}.$$
(4.8)

Integrating this expression with respect to y_1 we obtain

$$P_1(y_2, t_2) = \int dy_1 \, P_{11}(y_2, t_2 | y_1, t_1) P_1(y_1, t_1),$$
(4.9)

which can be integrated again with respect to y_2 to give the normalization of the conditional probability

$$\int dy_2 \, P_{11}(y_2, t_2 | y_1, t_1) = 1.$$
(4.10)

We can generalize this concept by introducing

$$P_{lk}(y_{k+l}, t_{k+l}; \ldots; y_{k+1}, t_{k+1} | y_k, t_k; \ldots; y_1, t_1) \equiv \frac{P_{k+l}(y_{k+l}, t_{k+l}; \ldots; y_1, t_1)}{P_k(y_k, t_k; \ldots; y_1, t_1)},$$
(4.11)

which represents the probability that $y(t)$ assumes the values $y_{k+l}, \ldots, y_{k+1}$ at $t_{k+l}, \ldots, t_{k+1}$ given that they were $y_k, \ldots, y_1$ at $t_k, \ldots, t_1$.

Two examples of stochastic processes are particularly important in physics. The first is the *independent process*, where

$$P_n(y_n, t_n; y_{n-1}, t_{n-1}; \ldots; y_1, t_1) = \prod_{i=1}^{n} P_n(y_i, t_i)$$
(4.12)

and hence there is no correlation between the values of $y(t)$ at t_i and t_{i+1}. The second is the *Markovian process*, where

$$P_{1,n-1}(y_n, t_n | y_{n-1}, t_{n-1}; \ldots; y_1, t_1) = P_{11}(y_n, t_n | y_{n-1}, t_{n-1})$$
(4.13)

which means that y_n is only correlated with y_{n-1} and independent of the previous values assumed by the stochastic variable at earlier times. Since a Markovian process is fully determined by $P_{11}(y_2, t_2|y_1, t_1)$ and $P_1(y_1, t_1)$, we can write for $t_1 < t_2 < t_3$

$$P_3(y_3, t_3; y_2, t_2; y_1, t_1) = P_{12}(y_3, t_3|y_2, t_2; y_1, t_1) P_2(y_2, t_2; y_1, t_1)$$
$$= P_{11}(y_3, t_3|y_2, t_2) P_{11}(y_2, t_2|y_1, t_1) P_1(y_1, t_1).$$

Now, integrating over y_2 we have

$$P_2(y_3, t_3; y_1, t_1) = \int dy_2\, P_3(y_3, t_3; y_2, t_2; y_1, t_1)$$
$$= P_1(y_1, t_1) \int dy_2\, P_{11}(y_3, t_3|y_2, t_2) P_{11}(y_2, t_2|y_1, t_1).$$

Dividing the above equation by $P_1(y_1, t_1)$ and using (4.8), we end up with

$$P_{11}(y_3, t_3|y_1, t_1) = \int dy_2\, P_{11}(y_3, t_3|y_2, t_2) P_{11}(y_2, t_2|y_1, t_1) \qquad (4.14)$$

which is the *Chapman–Kolmogorov equation*. Therefore, a Markovian process is fully determined by (4.9) and (4.14). Here are two examples of this kind of process.

(i) The Wiener–Lévy process.
By setting

$$P_{11}(y_2, t_2|y_1, t_1) = \frac{1}{\sqrt{2\pi(t_2 - t_1)}} \exp -\frac{(y_2 - y_1)^2}{2(t_2 - t_1)}, \qquad (4.15)$$

we can show that (4.14) holds for $t_2 > t_1$. If we choose $P_1(y_1, 0) = \delta(y_1)$ in (4.9), a non-stationary process results:

$$P_1(y, t) = \frac{1}{\sqrt{2\pi t}} \exp -\frac{y^2}{2t}. \qquad (4.16)$$

This was originally used to describe the stochastic behavior of the position of a Brownian particle in the theory of Brownian movement proposed by Einstein and Smoluchowsky.

(ii) The Ornstein–Uhlenbeck process.
We now set

$$P_{11}(y_2, t_2|y_1, t_1) = \frac{1}{\sqrt{2\pi(1 - \exp(-2\tau))}} \exp -\frac{(y_2 - y_1 \exp(-\tau))^2}{2(1 - \exp(-2\tau))}, \qquad (4.17)$$

where $\tau \equiv t_2 - t_1$. This also obeys (4.14) and together with

$$P_1(y_1, t_1) = \frac{1}{\sqrt{2\pi}} \exp - \frac{y_1^2}{2},$$ (4.18)

maintains this same form of $P(y, t)$ for any time t. This represents, for example, the velocity distribution of Brownian particles in the theory proposed by Ornstein and Uhlenbeck. Besides being Markovian, this process is also stationary. Actually, there is a theorem by J. L. Doob (see Wax (2003)) which asserts that this is the only process that is stationary, Gaussian, and Markovian.

4.1.2 The master and Fokker–Planck equations

We are now interested in studying the time evolution of the probability density $P(y, t)$ and will, therefore, deduce a couple of dynamical equations that will prove to be very useful. Our starting point is equation (4.9), which can be rewritten as $(P_1 \equiv P)$

$$P(y, t + \tau) = \int dy_1 P_{11}(y, t + \tau | y_1, t) P(y_1, t).$$ (4.19)

Using the fact that

$$\frac{\partial P(y, t)}{\partial t} \equiv \lim_{\tau \to 0} \frac{P(y, t + \tau) - P(y, t)}{\tau}$$ (4.20)

in (4.19), we have

$$\frac{\partial P(y, t)}{\partial t} = \lim_{\tau \to 0} \frac{1}{\tau} \int dy' [P_{11}(y, t + \tau | y', t) P(y', t) - P_{11}(y, t | y', t) P(y', t)].$$ (4.21)

But, expanding $P_{11}(y, t + \tau | y', t)$ up to first order in τ and keeping the normalization (4.10) up to this same order, we must have

$$P_{11}(y, t + \tau; y', t) \approx \frac{\delta(y - y') + \tau W_t(y, y')}{1 + \tau \int W_t(y, y') \, dy}$$

$$\approx \delta(y - y') \left[1 - \tau \int W_t(y'', y') \, dy'' \right] + \tau W_t(y, y'),$$ (4.22)

where we have used the fact that $P_{11}(y, t | y', t) = \delta(y - y')$ and defined $W_t(y, y') \equiv \partial P_{11}(y, t | y', t') / \partial t |_{t=t'}$. Inserting (4.22) in (4.21) we finally get

$$\frac{\partial P(y, t)}{\partial t} = \int dy' W_t(y, y') P(y', t) - \int dy' W_t(y', y) P(y, t),$$ (4.23)

which is the famous *master equation* of the stochastic processes. We can further write this equation in a more appropriate form if we make the following change of variables. Define $\xi \equiv y - y'$ and

$$W_t(y, y') = W_t(y' + \xi, y') \equiv W(\xi, y') = W(\xi, y - \xi),$$
$$W_t(y', y) = W_t(y - \xi, y) \equiv W(-\xi, y). \tag{4.24}$$

In terms of these new variables and functions, (4.23) becomes

$$\frac{\partial P(y, t)}{\partial t} = \int d\xi\, W(\xi, y - \xi) P(y - \xi, t) - P(y, t) \int d\xi\, W(-\xi, y), \tag{4.25}$$

where we have used the fact that these integrals extend from $-\infty$ to $+\infty$ when we write them in terms of the newly defined quantities. Now, assuming that only small jumps occur, that is, $W(\xi, y)$ is a sharply peaked function of ξ about $\xi = 0$ but varies slowly with y, and that $P(y, t)$ is also a slowly varying function of y, one can expand the product $W(\xi, y - \xi) P(y - \xi, t)$ in a Taylor series in ξ and rewrite (4.25) as

$$\frac{\partial P(y, t)}{\partial t} = \sum_{n=1}^{\infty} \frac{(-1)^n}{n!} \frac{\partial^n}{\partial y^n} [\alpha_n(y) P(y, t)], \tag{4.26}$$

where

$$\alpha_n(y) \equiv \int d\xi\, \xi^n\, W(\xi, y) = \lim_{\tau \to 0} \frac{1}{\tau} \int dy'(y - y')^n\, P_{11}(y, t + \tau | y', t). \tag{4.27}$$

Notice that the dependence of $W(\xi, y)$ on its first argument ξ has been fully kept since an expansion in this variable would not be allowed as W varies rapidly with it.

This is the *Kramers–Moyal expansion* and can easily be shown to be valid for the conditional probability distributions $P_{11}(y, t | y_0, t_0)$ as well. The only difference between its application to either case resides in the establishment of the appropriate initial conditions to the specific case we choose to deal with.

In the case of a multidimensional variable $\mathbf{y} \in \mathbb{R}^N$, this equation can be generalized to

$$\frac{\partial P(\mathbf{y}, t)}{\partial t} = -\sum_{i=1}^{N} \frac{\partial}{\partial y_i} [\alpha_i(\mathbf{y}) P(\mathbf{y}, t)] + \frac{1}{2} \sum_{i,j=1}^{N} \frac{\partial^2}{\partial y_i \partial y_j} [\alpha_{ij}(\mathbf{y}) P(\mathbf{y}, t)] + \ldots, \tag{4.28}$$

where

$$\alpha_i(\mathbf{y}) \equiv \int d\xi_i\, \xi_i\, W(\xi, \mathbf{y}) \quad \text{and} \quad \alpha_{ij}(\mathbf{y}) \equiv \int d\xi_i\, d\xi_j\, \xi_i\, \xi_j\, W(\xi, \mathbf{y}), \quad \text{etc.} \tag{4.29}$$

In the following, we will apply the equations above to some specific examples of Brownian motion.

Examples

(i) Free Brownian particle.

In this case the equation of motion for the particle is given by

$$M\frac{dv}{dt} + \eta v = f(t), \tag{4.30}$$

where $\langle f(t)\rangle = 0$ and $\langle f(t)f(t')\rangle = 2D_{pp}\,\delta(t-t')$ with $D_{pp} = \eta k_B T$.

This equation, when integrated over a time interval Δt, gives us

$$\Delta v = -\frac{\eta v}{M}\Delta t + \frac{1}{M}\int_{t}^{t+\Delta t} d\xi\, f(\xi), \tag{4.31}$$

which together with $\langle f(t)\rangle = 0$ and $\langle f(t)f(t')\rangle = 2D_{pp}\,\delta(t-t')$ allows us to compute

$$\alpha_1 \equiv \lim_{\Delta t \to 0}\frac{\langle \Delta v\rangle}{\Delta t} = -\frac{\eta v}{M},$$

$$\alpha_2 \equiv \lim_{\Delta t \to 0}\frac{\langle (\Delta v)^2\rangle}{\Delta t} = \lim_{\Delta t \to 0}\left[\frac{1}{M^2}\int_{t}^{t+\Delta t} d\xi \int_{t}^{t+\Delta t} d\lambda\, 2D_{pp}\,\delta(\xi-\lambda)\right]$$

$$= \frac{2D_{pp}}{M^2}, \tag{4.32}$$

and $\alpha_n = 0$ if $n \geq 3$. Then we have

$$\frac{\partial P(v,t)}{\partial t} = \frac{\eta}{M}\frac{\partial}{\partial v}[vP(v,t)] + \frac{D_{pp}}{M^2}\frac{\partial^2}{\partial v^2}P(v,t). \tag{4.33}$$

As we have mentioned before, this equation is also obeyed by the conditional probability distribution $P_{11}(v,t|v_0,t_0)$. In this particular case, it can be shown that

$$P_{11}(v,t|v_0,0) = \frac{1}{\sqrt{2\pi\,\langle(\Delta v(t))^2\rangle}}\exp-\frac{(v - v_0\exp-(\eta t/M))^2}{2\,\langle(\Delta v(t))^2\rangle}, \tag{4.34}$$

where

$$\langle(\Delta v(t))^2\rangle = \frac{k_B T}{M}\left[1 - \exp-\left(\frac{2\eta t}{M}\right)\right], \tag{4.35}$$

which is exactly the Ornstein–Uhlenbeck process we presented in (4.17) with $\tau = 2\eta t/M$ and $y = v$. When $t \to \infty$ this approaches the Maxwell–Boltzmann distribution of velocities of the Brownian particle.

It can also be shown, and we leave it as an exercise for the reader, that in the limit of very long times, when the acceleration term in (4.30) can be neglected, the same procedure that led us to (4.33) will now, with $v = dx/dt$, lead us to

$$\frac{\partial P(x, t)}{\partial t} = D_{xx} \frac{\partial^2}{\partial x^2} P(x, t), \tag{4.36}$$

where $D_{xx} = k_B T/\eta$. This is the well-known diffusion equation and the coefficient D_{xx} is the "spatial" diffusion constant. The solution for $P_{11}(x, t|x_0, 0)$ is

$$P_{11}(x, t|x_0, 0) = \frac{1}{\sqrt{2\pi \sigma^2(t)}} \exp -\frac{(x - x_0)^2}{2\sigma^2(t)}, \tag{4.37}$$

where

$$\sigma^2(t) = \langle (\Delta x(t))^2 \rangle = 2 D_{xx} t, \tag{4.38}$$

a famous formula first derived by Einstein (Wax, 2003). Thus, with the replacements $D_{xx} = 1/4$ and $x = y$, we can also derive, from the same equation (4.33), the Wiener–Lévy process introduced earlier in (4.16).

From these results, we see that the Einstein–Smoluchowsky theory is the long-time limit of the Ornstein–Uhlenbeck theory of Brownian motion, and this can also be checked directly if one writes the full solution for the position of the free Brownian particle described by (4.30) as

$$s(t) = x(t) - x_0 = \frac{M v_0}{\eta} \left[1 - \exp -\left(\frac{\eta t}{M} \right) \right]$$
$$- \int_0^t dt' \int_0^{t'} dt'' \frac{f(t'')}{M} \exp -\frac{\eta(t' - t'')}{M}, \tag{4.39}$$

which gives us

$$\langle s(t) \rangle = \frac{M v_0}{\eta} \left[1 - \exp -\left(\frac{\eta t}{M} \right) \right] \tag{4.40}$$

and

$$\langle s^2(t) \rangle = \frac{2k_B T}{\eta} t + \frac{M^2 v_0}{\eta^2} \left[1 - \exp -\left(\frac{\eta t}{M} \right) \right]^2 - \frac{M k_B T}{\eta^2}$$
$$\times \left[3 - 4 \exp -\left(\frac{\eta t}{M} \right) - \exp -\left(\frac{2\eta t}{M} \right) \right]. \tag{4.41}$$

Studying the long- and short-time limits of the equations above we have, respectively, $\langle s^2(t) \rangle = 2 D_{xx} t$ and $\langle s^2(t) \rangle = v_0^2 t^2$, showing that the motion of the Brownian particle is purely diffusive for long times whereas, for short times, its dynamics is ballistic, in agreement with the fact that in this latter limit the medium where the particle is immersed has not had enough time to act on it.

(ii) Brownian particle in a potential $V(q)$.

Here, it is more appropriate to consider the Hamilton equations of motion

$$\frac{dq}{dt} = \frac{p}{M},$$

$$\frac{dp}{dt} = -\frac{\eta p}{M} - V'(q) + f(t).$$

Integrating these equations in a time interval Δt and proceeding as before, we can consider this problem as one of a two-dimensional stochastic variable $\mathbf{y} = (q, p)$ and compute

$$\alpha_1 = \lim_{\Delta t \to 0} \frac{\langle \Delta q \rangle}{\Delta t} = \frac{p}{M},$$

$$\alpha_2 = \lim_{\Delta t \to 0} \frac{\langle \Delta p \rangle}{\Delta t} = -\frac{\eta p}{M} - V'(q),$$

$$\alpha_{11} = \lim_{\Delta t \to 0} \frac{\langle (\Delta q)^2 \rangle}{\Delta t} = 0,$$

$$\alpha_{12} = \alpha_{21} = \lim_{\Delta t \to 0} \frac{\langle \Delta q \Delta p \rangle}{\Delta t} = 0,$$

$$\alpha_{22} = \lim_{\Delta t \to 0} \frac{\langle (\Delta p)^2 \rangle}{\Delta t} = 2D_{pp},$$

and all the other α vanish. Therefore, using the generalized form (4.28) we have

$$\frac{\partial P}{\partial t} = -\frac{\partial}{\partial q}(pP) + \frac{\partial}{\partial p}\left[\left(\frac{\eta}{M}p + V'(q)\right)P\right] + D_{pp}\frac{\partial^2 P}{\partial p^2}, \tag{4.42}$$

where $P = P(q, p, t)$. Notice that if $V(q) = 0$ and $P = P(p, t)$, we recover (4.33).

Equations (4.33) and (4.42) are examples of the so-called *Fokker–Planck* equation. Its general form is expression (4.28) when all the functions α beyond second order are null. These equations will be very useful for comparison with some quantum mechanical expressions we will develop in future chapters.

To summarize this section we can say that in order to describe the dynamics of stochastic variables we can either work directly with their Langevin equations, which carry all their statistical characteristics through the fluctuating forces, or determine their statistical distribution as a function of time from which all their moments can be determined. We have also given the recipe to link these two approaches by defining the coefficients α of the Fokker–Planck equation (or Kramers–Moyal expansion). So, what we have presented in this section would be a starting point to deal with the dynamics of the systems we have introduced so far, but only once we can make sure they operate in their classical regime.

4.2 Quantum Brownian motion

As we have seen in Chapter 3, for example, the equations of motion of the variables of interest of the superconducting devices we have introduced are all dissipative. Although they have their origin in quantum mechanisms their dynamics is, for a vast range of parameters, purely classical. No wonder this is so, because they are macroscopic variables. Nevertheless, we have already argued that these devices, when properly built, could present quantum effects at very low temperatures. These would be due to superpositions of macroscopically distinct quantum states which make the problem interesting by itself. On top of that, there is another very important issue we have to deal with when studying these devices, namely dissipation. Since we are talking about macro- or mesoscopic variables, it is almost impossible to isolate them completely from their environments. This means that dissipative effects are always present in our problem. Actually, it is not even true that these systems might be considered very weakly damped. As a general rule, there is no restriction on dissipative terms in the equations of motion, and they can even be tuned by the experimentalist in such a way that they result in overdamped dynamics. Our problem then reduces to reconciling damped equations of motion with the procedure of quantization.

The origin of this problem lies in the fact that the standard procedures of quantization are based on the existence of either a Hamiltonian or a Lagrangian function for the system in which we are interested. In contrast, it is well known that we cannot obtain a Langevin equation from the application of the classical Lagrange or Hamilton equations to any Lagrangian or Hamiltonian which has no explicit time dependence. The employment of time-dependent functions would allow us to use the standard procedures of quantization directly, but would lead us to face problems with the uncertainty principle which, even if properly circumvented, would not free the theory from being not very physically appealing.

Over many decades, people have tried to solve this problem. In spite of the variety of methods used, including the above-mentioned time-dependent method, all these attempts fall into two main categories: they either look for new schemes of quantization or they use the system-plus-reservoir approach (see Caldeira and Leggett (1983b) and references cited therein). The former approaches always rely on some questionable hypotheses and lead us to results dependent on the method used, besides not being very realistic. The way out of this dilemma is to consider explicitly the fact that the dissipative system is always coupled to a given thermal environment (the second approach above). We know of no dissipative system in

Nature which is not coupled to another system responsible for its losses. Then, before trying to modify the canonical scheme of quantization, we believe that it is wiser to apply the traditional methods to more realistic situations.

Conceptually, the idea is very simple. However, in practice, its implementation requires a little labor. Once we have decided to consider explicitly the coupling of the system of interest to the environment, we must know what sort of system the latter is and how their mutual coupling takes place. This can be a very hard task.

Nevertheless, fundamentally different composite systems, by which we mean systems of interest plus environment, might have the former obeying Brownian dynamics in the classical limit. Although this appears to be an additional complication to our approach, it actually gives us a chance to argue in favor of some simplifying hypotheses. For instance, we can assume that different reservoirs may share some common characteristics such as the behavior of their spectrum of excitations or the way they respond when acted on by an external input. Before we explore this idea further with the specific model we will develop soon, let us outline the general program we will follow throughout the next chapters.

4.2.1 The general approach

Once we have decided on the system-plus-reservoir approach (we will use the words reservoir, environment, and bath interchangeably throughout the book), we need to make two other choices.

The first one we have already touched upon above and has to do with the choice of a specific model which captures the desired physics in the classical limit. If we make the appropriate choice we believe this would be a very good candidate to account for the quantum dynamics of the system we are interested in describing. We shall discuss some possibilities in the next chapter.

The second choice deals with the method we are going to use to treat our composite system. To start with, we should make some connection with the classical approaches in order to establish which one of them will be more useful for our needs. Here, once again, there are two methods. As we have seen earlier in this chapter we can either employ the equation of motion method, through which one must generate a Langevin equation like (4.1), or study the time evolution of probability densities using the Fokker–Planck equation (4.42). Quantum mechanically there are two equivalent ways corresponding to each of them, and these are tantamount to the two well-known pictures of quantum mechanics, namely Heisenberg and Schrödinger.

In order to analyze each of them, let us assume that our system-plus-reservoir is described by the Hamiltonian

$$\mathcal{H} = \mathcal{H}_0(q, p) + \mathcal{H}_I(q, q_k) + \mathcal{H}_R(q_k, p_k), \qquad (4.43)$$

where $\mathcal{H}_0$, $\mathcal{H}_I$, and $\mathcal{H}_R$ are the Hamiltonians of the system, interaction, and reservoir, respectively. Here, (q, p) is the pair of canonically conjugate operators of the system, which means $[q, p] = i\hbar$, where $[*, *]$ stands for the commutator of the two operators, whereas (q_k, p_k) represents the same for the kth particle composing the environment.

In the Heisenberg picture we attribute the time evolution of the whole composite system to the operators through

$$\hat{O}(t) = e^{i\mathcal{H}t/\hbar}\,\hat{O}(0)\,e^{-i\mathcal{H}t/\hbar}, \tag{4.44}$$

keeping the quantum state fixed at, say, $t = 0$. Here, $\hat{O}(t)$ represents an observable which depends on any combination or function of the canonical variables of the composite system. In this way we can write the equation for the time evolution of any operator using the Heisenberg equation of motion

$$i\hbar\frac{d\hat{O}(t)}{dt} = \frac{\partial\hat{O}(t)}{\partial t} + [\hat{O}(t), \mathcal{H}]. \tag{4.45}$$

When we consider that those operators are (q, p) or (q_k, p_k) themselves, we can describe the coupled dynamics between system and environment in a closed form. The solution of this system of equations depends on how complex the coupling we have chosen is. In general, it will not be a very simple matter to solve it. Nevertheless, there are some choices (see next chapter) which allow us to write an equation of motion for the variable of interest $q(t)$ in terms of the environment operators at $t = 0$ and, under certain conditions, this turns out to become a Langevin equation. Therefore, knowledge of the state of the composite system at $t = 0$ would allow us to describe the average value of any observable $\hat{O}(q, p)$ at time t in an analogous way as done above, for example, in (4.40) and (4.41). Even if this strategy cannot be followed so easily, at least in principle one should be able to do so. At any rate, what one has to do is evaluate the average

$$\langle\hat{O}(t)\rangle = \langle\Psi(0)|\hat{O}(t)|\Psi(0)\rangle, \tag{4.46}$$

assuming that the composite system is initially in a pure state $|\Psi(0)\rangle$. However, in dealing with a system-plus-reservoir this initial condition might not be the most appropriate one. Actually, most of the time we will be treating systems at finite temperatures which cannot be represented by pure states. What we do in these cases is represent the state of the composite system by the density operator representing a mixed state, defined as

$$\hat{\rho}(0) \equiv \sum_{\Psi} p_{\Psi}|\Psi(0)\rangle\langle\Psi(0)| \tag{4.47}$$

where p_Ψ is a classical probability labeled by the set of parameters defining the state $|\Psi(0)\rangle$. Averages are then given by

$$\langle \hat{O}(t) \rangle = \text{tr} \left\{ \hat{\rho}(0) \hat{O}(t) \right\} \tag{4.48}$$

where the trace must be performed over a basis of the composite system.

Now let us turn to the Schrödinger picture. If we want to pursue this approach what we need is knowledge of the state of the composite system, in other words, its wave function $\Psi(q, q_k, t)$. Now, it is the state which carries the time dependence and the operators are time independent, except for explicit time dependence. The time evolution of $\Psi(q, q_k, t)$ can be obtained by solving the Schrödinger equation

$$i\hbar \frac{\partial \Psi(q, q_k)}{\partial t} = \mathcal{H}\left(q, q_k, -i\hbar\frac{\partial}{\partial q}, -i\hbar\frac{\partial}{\partial q_k} \right) \Psi(q, q_k), \tag{4.49}$$

which is obviously an extremely hard task except for some nicely chosen models. Nevertheless, once again one should, at least in principle, solve it and compute the desired averages of any operator of the composite system.

As we have argued before, what we must really aim at is not the wave function of the composite system but rather its density operator. Therefore, writing the time evolution of the physical state independently of representation, we have

$$|\Psi(t)\rangle = e^{-i\mathcal{H}t/\hbar}|\Psi(0)\rangle, \tag{4.50}$$

which with the help of (4.47) gives us

$$\hat{\rho}(t) \equiv \sum_\Psi p_\Psi |\Psi(t)\rangle\langle\Psi(t)| = e^{-i\mathcal{H}t/\hbar}\hat{\rho}(0)e^{i\mathcal{H}t/\hbar}. \tag{4.51}$$

If we now take the time derivative of (4.51), we get

$$\frac{d\hat{\rho}(t)}{dt} = \frac{1}{i\hbar}\left[\mathcal{H}, \hat{\rho}(t) \right] \tag{4.52}$$

whose solution can be used to evaluate averages as in the Heisenberg picture by

$$\langle \hat{O}(t) \rangle = \text{tr}\left\{ \hat{\rho}(0)\hat{O}(t) \right\} = \text{tr}\left\{ \hat{\rho}(t)\hat{O}(0) \right\}. \tag{4.53}$$

Although the averages we compute do not depend on the picture in which we choose to work, there are certainly advantages in adopting one or the other picture. In our case, for example, it will be more advantageous to work in the Schrödinger picture since we will address questions related to quantum tunneling and coherence, and these are more suitably approached by studying the dynamics of the state of the system.

Equation (4.52) can still be written in another form if restricted to the subspace of the system of interest. This resulting equation plays the role of a master equation in the classical theory of Brownian motion, and is used extensively in the literature (see, for example, Breuer and Petruccione (2002)). Nevertheless, since the problems we have just mentioned above cannot be solved by the usual techniques, we will develop our method in the next section.

4.2.2 The propagator method

Suppose we have a general system described by the Hamiltonian (4.43). Notice that (q, p) do not necessarily refer to a point particle. If appropriately generalized, this pair of canonically conjugate variables can also represent either a local field variable or one of its normal modes and their respective momenta. In the case of a continuous variable one should remember that the Hamiltonians in (4.43) become functionals of the dynamical variables of the system. Hence, we can evaluate the density operator of the composite system at time t, as we have done in (4.51).

As we are usually interested in operators only referring to the system S, $\hat{O} = \hat{O}(q, p)$, we have

$$\langle \hat{O}(q, p) \rangle = \text{tr}_{RS}\{\hat{\rho}(t)\hat{O}\} = \text{tr}_S\{[\text{tr}_R \hat{\rho}(t)]\hat{O}\} = \text{tr}_S\{\tilde{\rho}(t)\hat{O}\}, \qquad (4.54)$$

where tr_S and tr_R represent the partial traces with respect to S and R, and

$$\tilde{\rho}(t) \equiv \text{tr}_R \hat{\rho}(t) \qquad (4.55)$$

is the reduced density operator of the system of interest. Notice that this is the operator to which we have referred above as the density operator restricted to the subspace of the system of interest. If we take this partial trace over the environmental degrees of freedom of equation (4.52), we end up with the master equation we have mentioned above. Similarly, we can also define

$$\tilde{\rho}_R(t) \equiv \text{tr}_S \hat{\rho}(t) \qquad (4.56)$$

as the reduced density operator of the environment.

If we introduce the general vector $\mathbf{R} = (R_1, \ldots, R_N)$, where R_k is the value assumed by q_k, we can write the coordinate representation of the total $\hat{\rho}(t)$ in (4.51) as

$$\hat{\rho}(x, \mathbf{R}, y, \mathbf{Q}, t) = \int \int \int \int dx' dy' d\mathbf{R}' d\mathbf{Q}' \, K(x, \mathbf{R}, t; x', \mathbf{R}', 0)$$
$$\times K^*(y, \mathbf{Q}, t; y', \mathbf{Q}', 0) \, \hat{\rho}(x', \mathbf{R}', y', \mathbf{Q}', 0),$$

$$(4.57)$$

where

$$K(x, \mathbf{R}, t; x', \mathbf{R}', 0) = \left\langle x, \mathbf{R} \left| e^{-i\mathcal{H}t/\hbar} \right| x', \mathbf{R}' \right\rangle \tag{4.58}$$

is the quantum mechanical propagator of the whole universe $S + R$ and

$$\hat{\rho}(x', \mathbf{R}', y', \mathbf{Q}', 0) = \left\langle x', \mathbf{R}' | \hat{\rho}(0) | y', \mathbf{Q}' \right\rangle \tag{4.59}$$

is the coordinate representation of its initial state. Taking the trace with respect to the environmental variables means making $\mathbf{R} = \mathbf{Q}$ and integrating over it. Then, assuming that (4.59) is separable (see Smith and Caldeira (1990) for more general choices of initial conditions) which means

$$\hat{\rho}(x', \mathbf{R}', y', \mathbf{Q}', 0) = \hat{\rho}_S(x', y', 0) \, \hat{\rho}_R(\mathbf{R}', \mathbf{Q}', 0), \tag{4.60}$$

we reach

$$\tilde{\rho}(x, y, t) = \int \int dx' dy' \, \mathcal{J}(x, y, t; x', y', 0) \, \tilde{\rho}(x', y', 0), \tag{4.61}$$

where

$$\mathcal{J}(x, y, t; x', y', 0) = \int \int \int d\mathbf{R}' d\mathbf{Q}' d\mathbf{R} \left\{ K(x, \mathbf{R}, t; x', \mathbf{R}', 0) \right.$$
$$\left. \times K^*(y, \mathbf{R}, t; y', \mathbf{Q}', 0) \, \tilde{\rho}_R(\mathbf{R}', \mathbf{Q}', 0) \right\}. \tag{4.62}$$

Notice that, in these expressions, we have already replaced $\hat{\rho}(x', y', 0)$ and $\hat{\rho}_R(\mathbf{R}', \mathbf{Q}', 0)$ by $\tilde{\rho}(x', y', 0)$ and $\tilde{\rho}_R(\mathbf{R}', \mathbf{Q}', 0)$, respectively, which can easily be shown for product states. Here we should also notice that if S and R are decoupled then $K(x, \mathbf{R}, t; x', \mathbf{R}', 0) = K_0(x, t; x', 0) K_R(\mathbf{R}, t; \mathbf{R}', 0)$, which inserted in (4.62) yields

$$\mathcal{J}(x, y, t; x', y', 0) = K_0(x, t; x', 0) K_0^*(y, t; y', 0), \tag{4.63}$$

where we have used the fact that $\mathrm{tr}_R \left\{ \hat{\rho}_R(t) \right\} = 1$. Thus, $\mathcal{J}$ acts as a superpropagator of the reduced density operator of the system.

From (4.61) and (4.62) we see that all we need to do is evaluate the total propagator $K(x, \mathbf{R}, t; x', \mathbf{R}', 0)$, since $\hat{\rho}(0)$ is assumed to be known. The evaluation of the propagator of the whole universe is, in general, a formidable task and the reason for that is twofold. Firstly, the environment itself might be a very complex system for which very little can be done as far as the standard mathematical approximations are concerned. As we have already anticipated, we shall try to model it in such a way that this difficulty can be circumvented. Secondly, even if we have a simple environment, the evaluation of the whole propagator can still be quite laborious since we are dealing with wave functions of a huge number of variables. The latter can be handled properly by employing the *Feynman path integral* (Feynman and Hibbs, 1965) representation of quantum mechanics, which will prove extremely useful to deal with the quantum propagators throughout future sections of this book. Therefore, we shall concentrate on the choice of our model in the next chapter.

5

Models for quantum dissipation

As we have stressed already, the treatment of a realistic model for the environment
to which our system of interest is coupled, is not necessarily the best path to take. It
would only encumber the intermediate steps of the calculation and hide the essence
of the important physics of the problem. In what follows we will not attempt to
justify the use of one specific model. We will choose a *minimal model* which, under
certain conditions, reproduces Brownian motion in the classical regime. Thus, the
justification for the choice of the model will be provided *a posteriori*. However, it
is worth mentioning that the employment of detailed microscopic models for some
environments may show different quantum mechanical behavior, which turns out
be very important in some cases.

Once we have done that, we will consider some specific extensions of the minimal model to treat more general problems which we must learn how to handle in
order to deal with the realistic physical examples given before. In so doing, we will
see that it is possible to describe a broader range of dissipative systems through
general phenomenological models.

5.1 The bath of non-interacting oscillators: Minimal model

Let us suppose we want to study a *classical* composite system described by the
Lagrangian

$$L = L_S + L_I + L_R + L_{CT},\tag{5.1}$$

where

$$L_S = \frac{1}{2} M \dot{q}^2 - V(q),\tag{5.2}$$

$$L_I = \sum_k C_k q_k q,\tag{5.3}$$

$$L_R = \sum_k \frac{1}{2} m_k \dot{q}_k^2 - \sum_k \frac{1}{2} m_k \omega_k^2 q_k^2, \tag{5.4}$$

$$L_{CT} = -\sum_k \frac{1}{2} \frac{C_k^2}{m_k \omega_k^2} q^2 \tag{5.5}$$

are respectively the Lagrangians of the system of interest, interaction, reservoir, and counter-term (see below). The reservoir consists of a set of non-interacting harmonic oscillators of coordinates q_k, masses m_k, and natural frequencies ω_k. Each of them is coupled to the system of interest by a coupling constant C_k. A fairly general justification of this model was presented in Caldeira and Leggett (1983a).

Initially, we shall study the classical equations of motion resulting from (5.1). Writing the Euler–Lagrange equations of the composite system, one has

$$M\ddot{q} = -V'(q) + \sum_k C_k q_k - \sum_k \frac{C_k^2}{m_k \omega_k^2} q, \tag{5.6}$$

$$m_k \ddot{q}_k = -m_k \omega_k^2 q_k + C_k q. \tag{5.7}$$

Taking the Laplace transform of (5.7), one gets

$$\tilde{q}_k(s) = \frac{\dot{q}_k(0)}{s^2 + \omega_k^2} + \frac{s\, q_k(0)}{s^2 + \omega_k^2} + \frac{C_k\, \tilde{q}(s)}{m_k\, (s^2 + \omega_k^2)}, \tag{5.8}$$

which after the inverse transformation can be taken to (5.6), yielding

$$M\ddot{q} + V'(q) + \sum_k \frac{C_k^2}{m_k \omega_k^2} q = \frac{1}{2\pi i} \int_{\varepsilon - i\infty}^{\varepsilon + i\infty} \sum_k C_k \left\{ \frac{\dot{q}_k(0)}{s^2 + \omega_k^2} + \frac{s\, q_k(0)}{s^2 + \omega_k^2} \right\} e^{st}\, ds$$

$$+ \sum_k \frac{C_k^2}{m_k} \frac{1}{2\pi i} \int_{\varepsilon - i\infty}^{\varepsilon + i\infty} \frac{\tilde{q}(s)}{s^2 + \omega_k^2} e^{st}\, ds. \tag{5.9}$$

Thus, using the identity

$$\frac{1}{s^2 + \omega_k^2} = \frac{1}{\omega_k^2} \left\{ 1 - \frac{s^2}{s^2 + \omega_k^2} \right\}, \tag{5.10}$$

we can show that the last term on the r.h.s. of (5.9) generates two other terms, one of which exactly cancels the last one on its l.h.s., and the resulting equation is

$$M\ddot{q} + V'(q) + \sum_k \frac{C_k^2}{m_k \omega_k^2} \frac{1}{2\pi i} \int\limits_{\varepsilon-i\infty}^{\varepsilon+i\infty} \frac{s^2 \tilde{q}(s)}{s^2 + \omega_k^2} e^{st} ds$$

$$= \frac{1}{2\pi i} \int\limits_{\varepsilon-i\infty}^{\varepsilon+i\infty} \sum_k C_k \left\{ \frac{\dot{q}_k(0)}{s^2 + \omega_k^2} + \frac{s q_k(0)}{s^2 + \omega_k^2} \right\} e^{st} ds. \qquad (5.11)$$

Then we see that the inclusion of L_{CT} in (5.1) was solely to cancel one extra harmonic contribution that would come from the coupling to the environmental oscillators.

The last term on the l.h.s. of (5.11) can be rewritten as

$$\frac{d}{dt} \left\{ \sum_k \frac{C_k^2}{m_k \omega_k^2} \frac{1}{2\pi i} \int\limits_{\varepsilon-i\infty}^{\varepsilon+i\infty} \frac{s \tilde{q}(s)}{s^2 + \omega_k^2} e^{st} ds \right\} \qquad (5.12)$$

which, with the help of the convolution theorem and the Laplace transform of $\cos \omega_k t$, reads

$$\frac{d}{dt} \left\{ \sum_k \frac{C_k^2}{m_k \omega_k^2} \int\limits_0^t \cos\left[\omega_k (t - t') \right] q(t') dt' \right\}. \qquad (5.13)$$

In order to replace $\sum_k \longrightarrow \int d\omega$, let us introduce the spectral function $J(\omega)$ as

$$J(\omega) = \frac{\pi}{2} \sum_k \frac{C_k^2}{m_k \omega_k} \delta(\omega - \omega_k), \qquad (5.14)$$

which allows us to write

$$\sum_k \frac{C_k^2}{m_k \omega_k^2} \cos\left[\omega_k (t - t') \right] = \frac{2}{\pi} \int\limits_0^\infty d\omega \frac{J(\omega)}{\omega} \cos\left[\omega (t - t') \right]. \qquad (5.15)$$

The function $J(\omega)$ is, from linear response theory, nothing but the imaginary part of the Fourier transform of the retarded dynamical susceptibility of the bath of oscillators, namely

$$J(\omega) = \operatorname{Im} \mathcal{F} \left[\theta(t - t') \left\langle \left\{ \sum_k C_k q_k(t), \sum_{k'} C_{k'} q_{k'}(t') \right\}_{PB} \right\rangle \right], \qquad (5.16)$$

where $\{ *, * \}_{PB}$ stands for the Poisson brackets of the involved variables and the average $\langle * \rangle$ is taken over the equilibrium state of the non-interacting oscillators.

A simpler way to see this is by evaluating, directly from (5.7), the Fourier transform of the susceptibility (response function) of $q_k(t)$ to a stimulus $q(t)$, which reads

$$\tilde{q}_k(\omega) = -\frac{C_k}{m_k(\omega^2 - \omega_k^2)}\tilde{q}(\omega) \tag{5.17}$$

and allows us to write for the collective coordinate $\sum_k C_k q_k(t)$ of the environment

$$\chi_{env}(\omega) = -\sum_k \frac{C_k^2}{m_k(\omega^2 - \omega_k^2)}, \tag{5.18}$$

which can be rewritten as

$$\chi_{env}(\omega) = \sum_k \left(\frac{C_k^2}{2m_k\omega_k(\omega + \omega_k)} - \frac{C_k^2}{2m_k\omega_k(\omega - \omega_k)} \right). \tag{5.19}$$

The imaginary part of $\chi_{env}(\omega)$ comes from the usual replacement for causal responses, $\omega \pm \omega_k \to \omega \pm \omega_k + i\epsilon$ ($\epsilon \to 0$), and the identity

$$\frac{1}{(\omega \pm \omega_k) + i\epsilon} = \mathcal{P}\left(\frac{1}{\omega \pm \omega_k} \right) - i\pi\delta(\omega \pm \omega_k), \tag{5.20}$$

which gives us

$$\mathrm{Im}\chi_{env}(\omega) \equiv \chi_{env}''(\omega) = \frac{\pi}{2}\sum_k \frac{C_k^2}{m_k\,\omega_k}[\delta(\omega - \omega_k) - \delta(\omega + \omega_k)]. \tag{5.21}$$

Therefore, since ω and $\omega_k > 0$, we have $J(\omega) = \chi_{env}''(\omega)$.

Now, assuming that

$$J(\omega) = \begin{cases} \eta\omega & \text{if} \quad \omega < \Omega \\ 0 & \text{if} \quad \omega > \Omega, \end{cases} \tag{5.22}$$

where Ω is a microscopic high-frequency cutoff and $\gamma \equiv \eta/2M$ the macroscopic relaxation frequency of the problem (the factor $2M$ is introduced for later convenience), we can rewrite (5.15) as

$$\sum_k \frac{C_k^2}{m_k\,\omega_k^2} \cos\left[\omega_k\,(t - t')\right] = \frac{2}{\pi}\int_0^\Omega d\omega\,\eta\,\cos\left[\omega\,(t - t')\right] = 2\,\eta\,\delta\,(t - t'), \tag{5.23}$$

where we took $\Omega \to \infty$, which means that we are considering $\Omega \gg \gamma$. This result allows us to rewrite (5.13) as

$$\frac{d}{dt}\int_0^t 2\,\eta\,\delta\,(t - t')\,q(t')dt' = \eta\,\dot{q} + 2\,\eta\,\delta\,(t)\,q(0). \tag{5.24}$$

Finally, the r.h.s. of (5.11) can be interpreted as a force $\tilde{f}(t)$ depending on the initial conditions imposed on the oscillators of the bath. Performing the inverse Laplace transform involved on the r.h.s. of (5.11), we have

$$\tilde{f}(t) = \sum_k C_k \left\{ \frac{\dot{q}_k(0)}{\omega_k} \sin \omega_k t + q_k(0) \cos \omega_k t \right\}. \tag{5.25}$$

Now, suppose that the equilibrium position of the kth environmental oscillator is about $\bar{q}_k \equiv C_k q(0)/m_k \omega_k^2$. Thus, subtracting the term

$$\sum_k \frac{C_k^2 q(0)}{m_k \omega_k^2} \cos \omega_k t = \sum_k C_k \bar{q}_k \cos \omega_k t$$

from both sides of (5.11) we can show, with the help of (5.14), that in the limit $\Omega \to \infty$ it exactly cancels the term $2 \eta \delta(t) q(0)$ on its l.h.s. (see (5.24)) whereas, on its r.h.s., there appears a new forcing term given by

$$f(t) = \sum_k C_k \left\{ \frac{\dot{q}_k(0)}{\omega_k} \sin \omega_k t + (q_k(0) - \bar{q}_k) \cos \omega_k t \right\}. \tag{5.26}$$

In order to study the statistical properties of (5.26) we can use the equipartition theorem which, in the classical limit, states that

$$\langle q_k(0) \rangle = \bar{q}_k \quad \text{and} \quad \langle \dot{q}_k(0) \rangle = \langle \dot{q}_k(0) \Delta q_k(0) \rangle = 0,$$

$$\langle \dot{q}_k(0) \dot{q}_{k'}(0) \rangle = \frac{k_B T}{m_k} \delta_{kk'},$$

$$\langle \Delta q_k(0) \Delta q_{k'}(0) \rangle = \frac{k_B T}{m_k \omega_k^2} \delta_{kk'}, \tag{5.27}$$

where $\Delta q_k(0) \equiv q_k(0) - \bar{q}_k$ and the averages are taken over the classical distribution of non-interacting harmonic oscillators in equilibrium about the positions $\bar{q}_k$. Using these relations and (5.26), it can be shown that

$$\langle f(t) \rangle = 0 \qquad \text{and}$$

$$\langle f(t) f(t') \rangle = 2 \eta k_B T \delta(t - t'). \tag{5.28}$$

In conclusion, after all these maneuvers equation (5.11) becomes

$$M \ddot{q} + \eta \dot{q} + V'(q) = f(t), \tag{5.29}$$

where $f(t)$ satisfies (5.28), which is the well-known Langevin equation (see (4.1)) for the classical Brownian motion.

So, we see that the hypothesis that the environmental oscillators are in equilibrium about $\bar{q}_k$ was essential for us to get rid of the spurious term $2 \eta \delta(t) q(0)$

in (5.24). Another possibility to reach (5.29) is to assume that the oscillators are in equilibrium independently of the initial position of the external particle, which means $\langle q_k(0) \rangle = 0$, and drop that term by considering that the particle motion (with the appropriate initial conditions) starts at $t = 0^+$. A more thorough analysis of this problem with the justification for the latter procedure can be found in Rosenau da Costa *et al.* (2000).

Another point we should emphasize here is that the choice we have made for the behavior of $J(\omega)$ in (5.22) – the so-called *ohmic spectral function* – was crucial for the attainment of the term proportional to the velocity in (4.1). Actually, since $\gamma = \eta/2M$ is the relaxation frequency of the macroscopic motion of the particle, and we also expect the appearance of another macroscopic frequency scale from the external potential $V(q)$, we should argue that it is the low-frequency behavior of $J(\omega)$ which is important for describing the long-time dynamics of the particle. Any typical frequency of the macroscopic motion should be much smaller than the microscopic cutoff Ω we have introduced before. The latter sets the time scale of the microscopic motion of the components of the environment.

Bearing this in mind, there is no reason not to deviate from the choice made in (4.1). In general we can choose

$$J(\omega) = \begin{cases} A_s \omega^s & \text{if} \quad \omega < \Omega \\ 0 & \text{if} \quad \omega > \Omega, \end{cases} \tag{5.30}$$

where $[A_s] = MT^{s-2}$. The cases in which $0 < s < 1$ or $s > 1$ are respectively called *subohmic* or *superohmic* spectral functions. For example, the classical equation of motion of an electron interacting with its own electromagnetic field can be deduced from our minimal model with a superohmic ($s = 3$) spectral function (Barone and Caldeira, 1991). This is the well-known *Abraham–Lorentz equation* which has a damping term proportional to $\dddot{q}(t)$, a direct consequence of the low-frequency behavior of the spectral distribution of photons in a cavity. So, much regarding the dissipative term in the classical equation of motion of the particle can be inferred from the low-frequency behavior of the spectral function.

There is one important remark we would like to make about the minimal model and it has to do with alternative forms to write it. The first one is accomplished if we replace the interaction Lagrangian (5.3) by

$$\tilde{L}_I = \sum_k \tilde{C}_k q \, \dot{q}_k, \tag{5.31}$$

and omit L_{CT}. In this way we can define the canonical momenta p_k as

$$p_k = \frac{\partial L}{\partial \dot{q}_k} = m_k \dot{q}_k + \tilde{C}_k q, \tag{5.32}$$

and write

$$\tilde{\mathcal{H}} = p\dot{q} + \sum_k p_k \dot{q}_k - L$$

$$= \frac{p^2}{2M} + V(q) + \sum_k \left\{ \frac{1}{2m_k} \left(p_k - \tilde{C}_k q \right)^2 + \frac{1}{2} m_k \omega_k^2 q_k^2 \right\}. \quad (5.33)$$

Now, performing a canonical transformation $p \to p, \, q \to q, \, p_k \to m_k \omega_k q_k$, $q_k \to p_k/m_k \omega_k$ and defining $C_k \equiv \tilde{C}_k \omega_k$ we have

$$\mathcal{H} = \frac{p^2}{2M} + V(q) - \sum_k C_k q_k q$$

$$+ \sum_k \left\{ \frac{p_k^2}{2m_k} + \frac{1}{2} m_k \omega_k^2 q_k^2 \right\} + \sum_k \frac{C_k^2}{2m_k \omega_k^2} q^2, \quad (5.34)$$

which has (5.1) as its corresponding Lagrangian. Therefore, the electromagnetic-like Lagrangian with $\tilde{L}_I$ replacing L_I in (5.1) and no counter-term is completely equivalent to (5.1) itself.

Actually, there is a third way to describe the model (5.1) if we rescale the coordinates and momenta of the bath variables in (5.34), replacing $q_k \to C_k q_k/m_k \omega_k^2$ and $p_k \to C_k p_k/m_k \omega_k^2$, which allows us to write

$$\mathcal{H} = \frac{p^2}{2M} + V(q) + \sum_k \left\{ \frac{p_k^2}{2\mu_k} + \frac{1}{2} \mu_k \omega_k^2 (q_k - q)^2 \right\}, \quad (5.35)$$

where $\mu_k \equiv C_k^2/m_k \omega_k^4$. It is worth noting that for $V(q) = 0$ this form is manifestly translation invariant, which was very useful for the exact diagonalization of the problem of the free Brownian particle (Hakim and Ambegaokar, 1985).

One should keep in mind that applying alternative forms for the model (5.1), the spectral density $J(\omega)$ must be redefined as a function of $\tilde{C}_k$ or μ_k, accordingly.

Now that we know a treatable model that generates the classical Brownian motion for the variable of interest ($q(t)$ in the present case), we can study the quantum mechanics of the composite system and extract from it only the part referring to the system of interest by applying the general strategy described by (4.61) and (4.62) to our model, and this is what we are going to do in a later chapter. However, there is something we can already say about the quantum mechanics of the Brownian particle from what we have developed so far.

The quantum Langevin equation. If we consider that we are dealing directly with the minimal model in terms of quantum mechanical variables, everything that has been done within the Lagrangian formalism for the classical model can easily be repeated here just by writing exactly the same equations of motion for the operators

$\hat{q}(t)$ and $\hat{q}_k(t)$ which result from the application of the Heisenberg equations of motion to (5.34). Only in two places, where we have explicitly used the fact that we are dealing with classical variables, do we have to be more cautious

The first place is in equation (5.16), where we have defined the response function of the bath in terms of Poisson brackets and performed the averages over the equilibrium state of non-interacting classical oscillators. Nevertheless, we have shown that the spectral function $J(\omega)$ could also be attained directly from the equations of motion of the reservoir variables, and it is actually independent of the oscillator equilibrium distribution. This means that the previous analysis follows exactly as before until we reach equations (5.26) and (5.29), written in terms of operators.

The second place is in equation (5.27), where the equipartition theorem was used and, consequently, a classical equilibrium distribution of oscillators was assumed. Therefore, if we want to generalize these relations all we have to do is evaluate the same averages with a quantum mechanical equilibrium distribution of a set of non-interacting oscillators. This can easily be done, and yields

$$\frac{\{\hat{f}(t), \hat{f}(t')\}}{2} = \frac{\hbar}{\pi} \int_0^\infty d\omega\, \eta\, \omega \coth \frac{\hbar\omega}{2k_B T} \cos \omega(t - t'). \qquad (5.36)$$

Owing to the non-commuting character of the bath operators involved in (5.26) (remember to replace $\dot{\hat{q}}_k(0) \rightarrow \hat{p}_k(0)/m_k$), we have also replaced the product $f(t)f(t')$ by its symmetrized form $\{\hat{f}(t), \hat{f}(t')\}/2$ where $\{*, *\}$ stands for the anti-commutator of the two operators. The latter obviously reduces to the former in the classical regime as the r.h.s. of (5.36) reproduces the white noise expression (5.28) when $k_B T >> \hbar\tilde{\omega}$, where $\tilde{\omega}$ is a typical frequency scale of the particle motion.

Therefore, our minimal model has also been useful for deriving an equation of motion for the position operator $\hat{q}(t)$ exactly like (4.1), with the fluctuating force operator $\hat{f}(t)$ obeying a colored noise relation given by (5.36). This is what is called the *quantum Langevin equation* for Brownian motion.

Notice that although this equation is instantaneous, or Markovian, as far as the damping term is concerned, the colored noise introduces a non-Markovian character to the full quantum mechanical (low-temperature) dynamics of the particle. This will be seen more clearly later on, when we deduce the master equation for the reduced density operator of the system of interest. In that case, a Markovian equation will only result for very special cases.

5.2 Particle in general media: Non-linear coupling model

It has been shown extensively in the literature that, within the range of interest, other approaches to dealing with dissipative systems described by a single dynamical variable always furnish us with the same results as those obtained by the bath of non-interacting oscillators with a properly chosen spectral function. This is the case, for example, in the employment of microscopic models to describe more realistic systems such as a charged particle in a metallic environment (Guinea, 1984; Hedegård and Caldeira, 1987; Weiss, 1999). However, despite its success there are certain dissipative systems for which the minimal model can be shown to be inappropriate to account for the physics we expect. Let us consider, for instance, the dynamics of two Brownian particles.

Although this example does not apply directly to any of the realistic situations presented in the first two chapters of this book, there are two things we should mention to defend our choice. The first is that a composition of devices or, generally speaking, dissipative structures does exist and can be very important from the point of view of either fundamental physics or applications. Second, the model we have introduced is not bound to be applicable only to meso- or nanoscopic devices or collective dynamical variables. Actually, we expect that we can also apply it to the motion of particles (macroscopic or not) in general media, in other words, to transport problems. Although the latter will not be our main goal in this book, sometimes examples thereof will be timely. Besides, the motion of particles in material media is the paradigm of Brownian motion. We hope to have convinced the reader that our analysis will not be purely academic.

Suppose we immerse two independent (but identical) particles in the same medium where each of them would separately behave as a Brownian particle. This hypothesis implies that the coupling constants of the individual particles to the common bath are the same. Since for each individual particle we could describe the effect of the medium with a bath of oscillators, it would be very natural to try to generalize the model to cope with the presence of these two particles. This generalization is quite straightforward and the only point where we need to be a bit cautious is when introducing the well-known counter-term (5.5) in the generalized model. In order to do that unambiguously, we have to employ an equivalent model for the system–bath Lagrangian where the interaction is replaced by a velocity–coordinate coupling (5.31) and perform the usual canonical transformations to achieve the desired coordinate–coordinate coupling with the appropriate counter-term.

Once again, we can obtain the equations of motion for each particle under the influence of the environment using the Laplace transform method as in the preceding section. These are now coupled Langevin equations which decouple when written in terms of the center of mass and relative coordinates, q and u, of the two particles. However, the equation for u presents very strange behavior; it obeys a free particle equation of motion, which is quite unexpected.

In this section we propose an extension of the usual model of a bath of oscillators in order to remedy this deficiency.[1] On top of succeeding in so doing, we will also be able to show that this extension provides us with an effective coupling between the two particles mediated by the presence of the bath and the resulting interaction potential depends on the specific form of the spectral function of the environmental oscillators. This effect reminds us of the formation of Cooper pairs or bipolarons in material systems due to the electron–phonon interaction. We will also establish the conditions under which one can approximate the common bath by two independent ones. A trivial solution to this problem would be to assume that the coupling constants of each particle to the environment are different from each other. However, this is not a very reasonable assumption since we have been assuming that the particles are identical and the reservoir is common to both of them. Another possibility is to consider a position-dependent coupling constant, but this has already been done in Caldeira and Leggett (1983a) and gives rise to a position-dependent damping. Nevertheless, we will show below that if we make a specific choice of this coupling function based on very general arguments, it is possible to generate a constant damping again.

In the generalization of the minimal model, the system of interest is represented by the free particle Lagrangian

$$L_S = \frac{1}{2} M \dot{x}^2, \tag{5.37}$$

whereas the heat bath is now described as a symmetrized collection of independent harmonic modes,

$$L_R = \frac{1}{2} \sum_k m_k (\dot{R}_k \dot{R}_{-k} - \omega_k^2 R_k R_{-k}), \tag{5.38}$$

and for the coupling term we assume the interaction Lagrangian

$$L_I = \sum_k \tilde{C}_k(x) \dot{R}_k, \tag{5.39}$$

where $C_k(x)$ is no longer a linear function of x. Notice that this form of writing the model Lagrangian for the environment is quite general. Actually, it is the form appropriate to describing the Fourier transform of the coupling of the particle to a real scalar field $\phi(x, t)$, such as the deformation field of a one-dimensional solid.

We can rather use the Hamiltonian formulation and perform the canonical transformation, $P_k \rightarrow m_k \omega_k R_k$ and $R_k \rightarrow P_k / m_k \omega_k$, to show that the coupling term, after being properly symmetrized, transforms into

[1] In the remainder of this section we shall be following Duarte and Caldeira (2006) closely.

$$\mathcal{H}_I = \frac{1}{2}\sum_k (C_{-k}(x)R_k + C_k(x)R_{-k}) - \sum_k \frac{C_k(x)C_{-k}(x)}{2m_k\omega_k^2}, \qquad (5.40)$$

where $C_k(x) \equiv \tilde{C}_k(x)\omega_k$ is the new coupling constant. In order to represent the effect of a local interaction of the particle with a spatially homogeneous environment, we choose

$$C_k(x) = \kappa_k e^{ikx}, \qquad (5.41)$$

which means that the entire system is translationally invariant and, moreover, the coupling to the environment is homogeneous as we show next.

If the particle is displaced by a distance d, the coupling term transforms into $C_{-k}(x + d)R_k = C_{-k}(x)e^{-ikd}R_k$, which suggests the definition of a new set of canonical variables as $\tilde{R}_k = e^{-ikd}R_k$, leaving the total Lagrangian (or Hamiltonian) invariant. It is now a trivial matter to show that, with a coupling like (5.41), the potential renormalization term in (5.40) is constant and therefore does not contribute to the particle dynamics. This coupling appears in many realistic situations such as, for example, the interaction of a particle with a fermionic bath (Hedegård and Caldeira, 1987).

Following the same steps as in the preceding section, we write the Euler–Lagrange equations for all the variables involved in the problem, take their Laplace transforms and eliminate the bath variables, in order to find the following equation of motion for the system of interest:

$$M\ddot{x} + \int_0^t K(x(t) - x(t'), t - t')\dot{x}(t')dt' = F(t). \qquad (5.42)$$

The resulting non-linear dissipation kernel, given by

$$K(x, t) = \sum_k \frac{k^2\kappa_k\kappa_{-k}}{m_k\omega_k^2}\cos kx \cos \omega_k t, \qquad (5.43)$$

shows that the interaction with the thermal bath induces a systematic influence on the system which is non-local and non-instantaneous.

The function $F(t)$, can once again be interpreted as a fluctuating force

$$F(t) = -\frac{\partial}{\partial x}\sum_k \left\{ \left(C_{-k}(x)\tilde{R}_k(0) + C_k(x)\tilde{R}_{-k}(0)\right)\frac{\cos \omega_k t}{2} \right.$$

$$\left. + \left(C_{-k}(x)\dot{R}_k(0) + C_k(x)\dot{R}_{-k}(0)\right)\frac{\sin \omega_k t}{2\omega_k} \right\}, \qquad (5.44)$$

where $\tilde{R}_k = R_k - C_k(x_0)/m_k\omega_k^2$ and x_0 is the initial position of the particle. $F(t)$ depends explicitly on the initial conditions of the bath variables, and its statistical properties are obtained from the initial state of the total system. Notice that when $k \to 0$ and $C_k \equiv k\kappa_k$ we recover the results of the minimal model, as expected.

Now we need to choose a suitable spectral function for the oscillators which, in the continuum limit, allows us to recover the Brownian motion. The kernel in (5.43) can be rewritten as

$$
K(x, t) = \sum_k \int_0^\infty d\omega 2k^2 \kappa_k \kappa_{-k} \frac{\mathrm{Im}\chi_k^{(0)}(\omega)}{\pi\omega} \cos k(x(t) - x(t')) \cos \omega(t - t'),
$$

(5.45)

where $\mathrm{Im}\chi_k^{(0)}(\omega) = \pi\delta(\omega - \omega_k)/2m_k\omega_k$ is, as we have seen before, the imaginary part of the dynamical response $\chi_k^{(0)}(\omega)$ of a non-interacting oscillator with wave number k. This is the point where we slightly deviate from the procedure adopted to define $J(\omega)$ in (5.17)–(5.21). Before, we had only the dependence of the response functions on the frequency ω_k and k played the role of a harmless label. In contrast, now k acts as a wave number and there is an extra summation over its allowed values. However, it will not be an additional complication, as it seems at first sight.

If we assume that the oscillators (or normal modes) in the environment are in fact only approximately non-interacting and replace *solely* their response functions by those of damped oscillators, $\chi_k(\omega)$, we have

$$
\mathrm{Im}\chi_k^{(0)}(\omega) \to \mathrm{Im}\chi_k(\omega) = \frac{\gamma_k \omega}{m_k \left[(\omega^2 - \omega_k^2)^2 + \omega^2\gamma_k^2 \right]},
$$

(5.46)

where γ_k is the relaxation frequency of the kth oscillator. Here we should mention that all the environmental oscillators are assumed to be very weakly damped.

Our main interest is, as before, the study of the system for times longer than the typical time scale of the reservoir and, therefore, we should concentrate on the low-frequency limit of (5.46) in which $\mathrm{Im}\chi_k(\omega) \propto \omega$. This behavior suggests assuming that

$$
\mathrm{Im}\chi_k(\omega) \approx f(k)\omega\theta(\Omega - \omega),
$$

(5.47)

where we have, as usual, introduced a high-frequency cutoff Ω as the characteristic frequency of the bath. A functional dependence like (5.47) for the dynamical response of the bath has been employed in the references (Guinea, 1984; Hedegård and Caldeira, 1987) for fermionic environments. The particular choice of the dynamical susceptibility of the bath allows us to separate, at least at low frequencies, its time and length scales, which results in a Markov dynamics when we replace (5.47) in (5.45) and integrate with respect to ω when $\Omega \to \infty$. With these considerations the equation of motion (5.42) reads

$$M\ddot{x}(t) + \eta\dot{x}(t) = F(t), \tag{5.48}$$

where $\eta \equiv \sum_k k^2 \kappa_k \kappa_{-k} f(k)$. Notice that within this new model we obtain a relation between the damping constant and some microscopic parameters of the oscillator bath. Assuming that the bath is initially in thermal equilibrium it is easy to show that, for high temperatures, the fluctuating force $F(t)$ satisfies the relations $\langle F(t) \rangle = 0$ and $\langle F(t)F(t') \rangle = 2\eta k_B T \delta(t - t')$, which are characteristic of white noise. Notice that this is valid only if we assume a classical distribution of oscillators as the initial state of the bath.

Therefore, the system–reservoir model with non-linear coupling presented here allows us to reproduce the result we would have obtained by coupling the particle of interest bilinearly in coordinates to a bath of non-interacting harmonic oscillators with spectral function $J(\omega) = \eta\omega$.

We are now in a position to deal safely with the problem of two particles coupled to a dissipative environment. In this case the Lagrangian of the system of interest is

$$L_S = \frac{1}{2}M\dot{x}_1^2 + \frac{1}{2}M\dot{x}_2^2, \tag{5.49}$$

and the coupling term

$$L_I = -\frac{1}{2}\sum_k \left[(C_{-k}(x_1) + C_{-k}(x_2)) R_k + (C_k(x_1) + C_k(x_2)) R_{-k} \right]. \tag{5.50}$$

Notice that there is no counter-term in (5.50) since, as we have argued above, our system is translationally invariant. The equations of motion for this Lagrangian are then

$$M\ddot{x}_i + \int_0^t K(x_i(t) - x_i(t'), t - t')\dot{x}_i(t')dt'$$

$$+ \int_0^t K(x_i(t) - x_j(t'), t - t')\dot{x}_j(t')dt'$$

$$+ \frac{\partial(V(x_i(t) - x_j(t)))}{\partial x_i} = F_i(t), \tag{5.51}$$

where $i \neq j = 1, 2$, the fluctuating force $F_i(t)$ has the form given by (5.44) replacing $x(t)$ by $x_i(t)$, and

$$V(r(t)) = -\sum_k \int_0^\infty d\omega 2\kappa_k \kappa_{-k} \frac{\mathrm{Im}\chi_k^{(0)}(\omega)}{\pi\omega} \cos kr(t). \qquad (5.52)$$

In the long-time limit, the second term on the l.h.s. of (5.51) can be written as $\eta \dot{x}_i$, providing us with the usual dissipative term. The third term represents a cross-dissipative term that depends on the velocity of the second particle and the relative distance between them, reminding us of a current–current interaction between charged particles. The last term is an effective interaction induced by coupling with the common environment.

The explicit forms of these terms in the long-time regime are obtained through evaluation of (5.45) and (5.52) with the replacement introduced in (5.46) and further employment of the approximation (5.47). Performing the frequency integrals still leaves us with summations over k which can also be transformed into integrations with the help of a function $g(k)$ defined as

$$\eta g(k) \equiv \sum_{k'} \kappa_{k'} \kappa_{-k'} f(k') \delta(k - k'), \qquad (5.53)$$

where we are assuming k' to be discrete whereas k is a continuous variable. Integrating this expression over k and using the definition of η just below (5.48), we see that $g(k)$ satisfies the condition $\int_0^\infty g(k)k^2 dk = 1$. Notice it is at this point that the dimensionality of the bath comes into play, and it does so because of the appearence of the integral over k. We have been speaking loosely about k as a simple scalar, but there is nothing to prevent us from considering it as a vector in D dimensions. All we have to do is change the integration measures accordingly.

Let us for simplicity work with a one-dimensional environment and choose $g(k)$ as

$$g(k) = Ae^{-k/k_0}, \qquad (5.54)$$

where $A = 1/(2k_0^3)$ is the normalization constant and k_0 determines the characteristic length scale of the environment. This choice was inspired by the case of fermionic environments where $\mathrm{Im}\Pi^{(0)}(\mathbf{k}, \omega)$ (the imaginary part of the Fourier transform of the retarded polarization function) has, at low frequencies, a very similar behavior to our susceptibility in (5.47) and k_F plays the same role as k_0 (see Fetter and Walecka (2003) for details). We should bear in mind that the choice of $g(k)$ is guided either by knowledge of the microscopic details of the environment or by some phenomenological input about the macroscopic motion of the two particles, in the same way as $J(\omega)$ in the minimal model.

With (5.54) the kernel of the third term in (5.51) for high temperatures is

$$K(x_1(t) - x_2(t'), t - t') = 2\delta(t - t')\eta(u(t)), \qquad (5.55)$$

where $u(t) = x_1(t) - x_2(t)$,

$$\eta(u(t)) = \eta \left(\frac{1}{\left(k_0^2 u^2 + 1\right)^2} - \frac{4u^2 k_0^2}{\left(k_0^2 u^2 + 1\right)^3} \right), \tag{5.56}$$

and the effective potential reads

$$V(u(t)) = -\frac{2\Omega\eta}{\pi k_0^2 \left(k_0^2 u^2(t) + 1\right)}. \tag{5.57}$$

The strength of the effective potential depends on the characteristic length and time scales of the environment. Therefore, the contribution of this coupling to the dynamics of the Brownian particles is only important for times longer than the typical time scale of the reservoir and distances not much longer than its characteristic length k_0^{-1}. A simple hypothesis about the relation between the frequency and wave-number cutoffs can lead us to a much more suggestive expression for (5.57). Suppose Ω and k_0 are related by an expression like the dispersion relation of a non-relativistic particle of mass m, namely $\Omega = \hbar k_0^2 / 2m$. Using this expression in (5.57) we see that the strength of the potential $V(u)$ becomes proportional to $\hbar\gamma M/m$, where m is a typical mass scale of the environmental oscillators. In other words, it is expressible in terms of the phenomenological relaxation constant of the particles but independent of the frequency and length cutoffs we have introduced.

The fluctuating forces still satisfy the white noise properties, but now they present an additional property associated with the distance between the particles. The forces $F_1(t)$ and $F_2(t)$ are also spatially correlated, that is

$$\langle F_1(t) F_2(t') \rangle = 2\eta(u(t)) k_B T \delta(t - t'). \tag{5.58}$$

In Fig. 5.1 we see the noise correlation strength as a function of the distance between the particles. For short distances the noise correlation has the standard Brownian behavior. However, for longer distances the correlation function becomes negative and this *anti-correlation* induces an anomalous diffusive process in the system which will ultimately tend to normal diffusion once the particles are infinitely far apart.

In terms of the relative and center-of-mass coordinates, the equations of motion read

$$M\ddot{u}(t) + \eta\dot{u}(t) - \eta(u(t))\dot{u} + V'(u(t)) = F_u(t) \tag{5.59}$$

and

$$M\ddot{q}(t) + \eta\dot{q}(t) + \eta(u(t))\dot{q} = F_q(t), \tag{5.60}$$

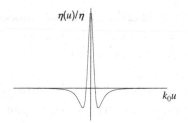

Figure 5.1 Force–force correlation

where $F_u(t) = F_1(t) - F_2(t)$ and $F_q(t) = (F_1(t) + F_2(t))/2$. From the form of $V(u(t))$ and $\eta(u(t))$ and the statistical properties of the fluctuating forces, we can write

$$\langle F_u(t) \rangle = 0 \quad \text{and} \quad \langle F_u(t) F_u(t') \rangle = 4k_B T (\eta - \eta(u)) \delta(t - t'),$$

$$\langle F_q(t) \rangle = 0 \quad \text{and} \quad \langle F_q(t) F_q(t') \rangle = k_B T (\eta + \eta(u)) \delta(t - t'). \tag{5.61}$$

It is evident that at large distances the equations of motion for the relative and center-of-mass coordinates represent the motion of Brownian particles with reduced mass $M/2$ and total mass $2M$, respectively. In general, the dissipation depends on the relative distance between the particles and for distances such that $uk_0 \ll 1$, we have up to first order in $u(t)$, $V'(u) \propto -u(t)$. In this approximation both dissipation and fluctuations are negligible and then we have an undamped oscillatory motion for $u(t)$. This is equivalent to the ballistic limit of a single Brownian particle at very short times. In other words, the internal motion of this small Brownian "molecule" is insensitive to the presence of the environment.

It should be stressed here that all the above results and conclusions are strongly dependent on the hypothesis we have made in (5.47) and the model employed in (5.54). If we modify these choices we are certainly bound to end up with something very different from the results presented above. On the level of treating dissipative effects in quantum systems, the model we adopt must be guided by the phenomenology of the specific problem we face.

So, we have presented in this section a system-plus-reservoir model with a non-linear coupling in the system coordinates which, under appropriate conditions, allows us to reproduce the phenomenological results known for the dynamics of a Brownian particle. As the model is manifestly translation invariant it turns out to become the natural candidate to describe the dissipative motion of a particle immersed in a homogeneous material medium. Actually, it is generally observed that the exponential form employed here is quite ubiquitous in the interaction of a moving particle with a realistic many-body system (see, for example, Hedegård and Caldeira (1987)). Moreover, the dissipation coefficient is still expressible in terms of a few parameters of the thermal bath that can be measured experimentally.

It should also be emphasized that the traditional bilinear model is the linearized version of the one we have just presented here, and, therefore, is more appropriate to describe the motion of a single particle about a localized region of the environment.

Besides improving the microscopic description of Brownian motion, this generalization is still capable of predicting an effective coupling between two Brownian particles mediated by the bath, which depends on the choice of its dynamical susceptibility. Although this kind of coupling is quite common in condensed matter physics (formation of Cooper pairs and bipolarons or RKKY interaction, to name just a few cases) there is no reason we should expect it to take place between two ordinary Brownian particles. However, it was the choice made for the behavior of the bath response function in order to reproduce the Brownian motion which naturally generated this effective coupling. Actually, it is the separable k-dependent part of the susceptibility that gives the specific form of the static potential we have obtained here. Retarded effects will show up as one can no longer separate the k and ω dependences in that function. These non-local effects are also relevant for the single-particle case, but only in a non-Markovian approximation.

5.3 Collision model

As we have seen in the last two sections, the models introduced therein were able to account for the classical motion of a Brownian particle if we make very general assumptions about the response of the environment to its presence. Moreover, we have also been able to extend what we have called the minimal model to cope with the presence of two Brownian particles without any ambiguity. In this sense, our approach has been absolutely phenomenological so far. We have never really bothered about the microsocopic details underlying the dynamics of the systems in which we are interested, although we have given some examples of realistic systems where, for instance, the specific form of a non-linear coupling can be applied. Nevertheless, there are some situations in which a partial microscopic knowledge of the coupling to the environment is so evident that we had best take it into account even if we later realize that, once again, the resulting dynamics can be described by some generalization of the minimal model.

This is the case, for example, with a system composed of an external particle coupled via a very localized potential to identical non-interacting and non-relativistic point particles (Caldeira and Castro Neto, 1995). In other words, the external particle would collide with the particles of the medium and consequently lose energy after multiple scattering processes. This is the most intuitive model for Brownian motion, although not as simple as the minimal model. As we show below, even the classical analysis of the motion of the external particle is not as straightforward as in the latter case.

Another important example which presents a very similar physics is that resulting from the application of the *collective coordinate method* (Rajaraman, 1987; Castro Neto and Caldeira, 1992, 1993) to quantize the motion of solitons in field theoretical or many-body models. In this approach, the center of the soliton acquires the status of a quantum mechanical operator coupled to the linearized excitations of the system. The coupling happens to be through the collision of those excitations with the deformation of the medium (potential) provided by the presence of the localized solitonic excitation. Therefore, the quantum soliton motion can be viewed as the quantum version of the model we have mentioned above, the only difference being the fact that the excitations of the medium need not be massive as before. A straightforward application of this model to our needs would be the study of the motion of localized collective excitations (solitons or kinks), such as domain walls or vortices in magnetic or superconducting media, respectively.

Although we are not going to develop this model along the same lines as we have done for previous ones, and the reason for that will become apparent shortly, let us for the sake of completeness at least sketch what we would need to do in order to address the problem classically.

Let us start from the Hamiltonian

$$\mathcal{H}(q, p, q_i, p_i) = \frac{p^2}{2M} + V(q) + \sum_{i=1}^{N} U(q_i - q) + \sum_{i=1}^{N} \frac{p_i^2}{2m}, \tag{5.62}$$

which describes the coupling of the external particle to environmental ones through the potential $\sum_{i=1}^{N} U(q_i - q)$ and also takes into account the possibility of the existence of a potential $V(q)$ acting only on the external particle.

It is an easy exercise to show that, under the transformation $q_i - q \rightarrow Q_i$ and $p + p_i \rightarrow P$, where P and p_i are canonically conjugate to q and Q_i, respectively, the new Hamiltonian

$$\mathcal{H}'(q, P, Q_i, p_i) = \frac{1}{2M} \left(p - \sum_{i=1}^{N} p_i \right)^2 + \sum_{i=1}^{N} \left(\frac{p_i^2}{2m} + U(Q_i) \right) \tag{5.63}$$

generates the same equations of motion as before, namely

$$M\ddot{q} + V'(q) - \sum_{i=1}^{N} U'(q_i - q) = 0$$

and $\quad m\ddot{q}_i + \sum_{i=1}^{N} U'(q_i - q) = 0. \tag{5.64}$

Since the equation of motion for $q_i(t)$ is not linear in this variable, the method of the Laplace transform we have been employing so far is no longer useful to describe the effective equation of motion of the external particle coordinate $q(t)$ as before. In other words, with the set of canonical variables just described we would not be able to generate an effective Langevin equation for the particle coordinate in the appropriate limit. However, there is a very subtle canonical transformation from the old (p_i, Q_i) to a new set of variables which would do this job. Although this can be done classically, it is not trivial and its motivation and implementation are much simpler if we treat the quantum mechanical problem directly with the Hamiltonian (5.63). Therefore, we will postpone the discussion of this model for a while until we start with the full quantum mechanical treatment of our models.

5.4 Other environmental models

All the models of system–bath interaction we have treated in this chapter, on top of being phenomenological, have been defined through their classical counterparts. Actually, it has been through the classical effective equations of motion for the variables of interest, whenever available, that we have been able to extract the appropriate dependence of the spectral function of the environment at low frequencies which is ultimately responsible for the long-time behavior of the particle dynamics. However, there are situations in which the most suitable description is inevitably quantum mechanical. In these cases, we can also propose phenomenological models in terms of appropriate quantum operators in order to reproduce some expected relaxation processes. Although examples thereof may not necessarily be useful for describing the quantum mechanics of collective variables, we think some of them deserve a few words at the end of this chapter. Let us present two in particular.

Rotating wave approximation (RWA). As its name indicates, it is actually an approximation by which a general Hamiltonian of, say, an electromagnetic mode in a cavity acquires a very simplified form (Louisell, 1990). Generally, it is written as

$$\mathcal{H}_{RWA} = \hbar\omega_a a^\dagger a + \sum_k \{V_k a^\dagger b_k + V_k^* a\, b_k^\dagger\} + \sum_k \hbar\omega_k b_k^\dagger b_k, \qquad (5.65)$$

where a and b_k are respectively the annihilation operators of the electromagnetic field mode and the kth excitation mode of the cavity, ω_a and ω_k their corresponding frequencies, and V_k the coupling between them. The generalization of the model to many electromagnetic modes is straightforward and we only need to consider the presence of more modes of the form a_l and couplings V_{lk}.

As a matter of fact, this model has a very broad range of applicability. It not only appears in quantum optics but also in several examples of condensed matter systems, in particular those dealing with the coupling of elementary excitations (phonons, magnons, etc.) in solids (Kittel, 1987).

If $V_k = V_k^*$, (5.65) can also be viewed as a modified minimal model (5.34) applied to a particle in a harmonic potential written in terms of creation and annihilation operators of the particle of interest and bath oscillators which read

$$a = \sqrt{\frac{M\omega_a}{2\hbar}} \left(q + i\frac{p}{M\omega_a} \right) \quad \text{and}$$

$$b_k = \sqrt{\frac{m_k\omega_k}{2\hbar}} \left(q_k + i\frac{p_k}{m_k\omega_k} \right) \tag{5.66}$$

with their corresponding adjoints $a^\dagger$ and $b_k^\dagger$.

The difference between the two is that the counter-term (last term on the r.h.s. of (5.34)) and the counter-rotating terms, $a\, b_k$ and $a^\dagger b_k^\dagger$, should be dropped from (5.34) in order to reproduce (5.65). Now, since its dependence on both creation and annihilation operators is quadratic, the equations of motion for all of them are linear, which allows us to use once again the method of Laplace transforms as in (5.6)–(5.9) to write down an effective time evolution equation for the variable of interest a (or $a^\dagger$) in terms of the bath operators $b_k(0)$ or $b_k^\dagger(0)$. Moreover, similarly to what we have done in (5.14), we can also define for (5.65) another spectral function given by

$$S(\omega) = \frac{2\pi}{\hbar^2} \sum_k |V_k|^2 \delta(\omega - \omega_k), \tag{5.67}$$

which, if applied to the minimal model $\left(V_k = \hbar C_k / 2\sqrt{M\,\omega_a\, m_k\, \omega_k}\right)$, can be related to (5.14) by $J(\omega) = 2M S(\omega)\omega_a$.

However, contrary to what we have seen for the minimal model, a Markovian equation for a and $a^\dagger$ is always possible to obtain if we consider a very weak coupling between the system and its environment in such a way that we can treat the effect of the latter within first-order perturbation theory. This means operationally that in all the resulting frequency integrals we must consider $S(\omega) \approx S(\omega_a) = \gamma$ and $n(\omega) \approx n(\omega_a)$, where

$$n(\omega) = \frac{1}{\exp\left(\frac{\hbar\omega}{k_B T}\right) - 1} \tag{5.68}$$

is the thermal occupation number for a harmonic mode of frequency ω. Notice that the choice made for $S(\omega)$ just above (5.68) coincides, as it should, with the exponential relaxation obtained from the application of Fermi's golden rule to the problem of the decay of an excited atomic level (see, for example, Merzbacher (1998)) and whose rate reads

$$w = 2\gamma = \frac{2\pi}{\hbar^2} \sum_{k \neq a} |V_k|^2 \delta(\omega_a - \omega_k) = \frac{2\pi}{\hbar^2} |V_a|^2 \rho(\omega_a), \qquad (5.69)$$

where $\rho(\omega_a)$ is the density of modes evaluated at the natural frequency ω_a.

Another point worth emphasizing here concerns the absence of the counter-term in the present model, which implies the existence of a real energy shift ΔE to the unperturbed term $\hbar\omega_a$. This issue is important and usually causes some confusion in the literature.

The Hamiltonian (5.65) represents a system whose dynamical behavior is known and which we couple to an external bath. Consequently, the effects of energy shift and excited state decay are physical and there is nothing we can do about them except neglect one or the other depending on their strength or the physical question we wish to answer. On the contrary, the Hamiltonian (5.34) was proposed to reproduce some phenomenological input provided by the classical (damped) equation of motion for a given variable of interest and the existence of the counter-term is due to the arbitrary choice made for the bath degrees of freedom suitable for our needs. The two models address different questions and there is no ambiguity in that.

If we are still unhappy with the former explanation there is an alternative way to exemplify the different physical situations. Suppose we insert a diatomic and very massive molecule in the harmonic bath with the non-linear coupling introduced in the last section. Our previous results show us that the motion of its center-of-mass is dissipative, but no external potential is provided by the bath. It behaves as a damped doubly massive free particle. On the contrary, the relative coordinate behaves as a particle having reduced atomic mass performing a damped motion in a potential composed of the original interatomic potential and an extra potential mediated by the bath. In other words, we now see a dissipative motion in a modified effective potential. Therefore, if we want to model each of these two situations separately as being the result of different minimal models applied to the center-of-mass and relative coordinates, the former would be of the kind (5.34) whereas the latter, if treated in the harmonic approximation, would be mimicked by (5.65). Notice that we are not saying that the bath-mediated interaction could be simply expressed as a harmonic correction to the original potential. Our message is that the physics introduced by the harmonic correction might be a real phenomenon in this case. We hope to have settled this issue with these two brief explanations.

Since the RWA has been studied extensively in the literature, in particular in the context of quantum optics (Louisell, 1990), we will not pursue it any further. For an exact diagonalization of the model we refer the reader to Rosenau da Costa *et al.* (2000).

Two-state system (TSS) bath. Suppose now that our external particle either inter-acts magnetically with a system composed of magnetic moments ($s = 1/2$) with different gyromagnetic factors or collides with different atoms, each of which has only one allowed electronic transition during the inelastic scattering process. Then, we can describe the reservoir as a set of non-interacting two-state systems which are coupled to the particle through the Hamiltonian

$$\mathcal{H}_{TSS} = \mathcal{H}_0 + \mathcal{H}_I + \mathcal{H}_R \tag{5.70}$$

where

$$\mathcal{H}_I = -\sum_k F_k \, q \, \sigma_k^{(x)}, \tag{5.71}$$

$$\mathcal{H}_R = \sum_k \hbar \omega_k \, \sigma_k^{(z)}, \tag{5.72}$$

and $\mathcal{H}_0(q, p)$ is the Hamiltonian of the particle when isolated. In the above expres-sions, F_k are the new coupling constants and $\sigma_k^{(\alpha)}$ ($\alpha = x, y, z$) are the Pauli matrices. This can be viewed as a bath of harmonic oscillators which are projected onto their two lowest-lying energy eigenstates. In this way we can compare (5.70) and (5.34), defining

$$\sigma_k^{(+)} = [\sigma_k^{(-)}]^\dagger \equiv \frac{\sigma_k^{(x)} + i \, \sigma_k^{(y)}}{2} \tag{5.73}$$

and realizing that for the two lowest-lying states

$$q_k = \sqrt{\frac{\hbar}{2m_k\omega_k}} \left(b_k + b_k^\dagger\right) = \sqrt{\frac{\hbar}{2m_k\omega_k}} \left(\sigma_k^{(-)} + \sigma_k^{(+)}\right) = \sqrt{\frac{\hbar}{2m_k\omega_k}} \sigma_k^{(x)} \tag{5.74}$$

which gives us

$$F_k = \sqrt{\frac{\hbar}{2m_k\omega_k}} C_k. \tag{5.75}$$

Now we can define a new spectral function for this model as

$$J_{TSS}(\omega) \equiv \pi \sum_k F_k^2 \, \delta(\omega - \omega_k), \tag{5.76}$$

which relates to (5.14) through $J_{TSS}(\omega) = \hbar J(\omega)$. The relaxation process origi-nated from this model obviously depends on the choice we make for $J_{TSS}(\omega)$. As we will see below, the previous choice we have made for $J(\omega)$ in (5.22) will no longer give us the same result for the effective motion of the external particle. In order to recover the Brownian-like behavior, another choice needs to be made. We shall return to this point shortly.

A simple variation of (5.71), in fact its RWA, is very useful for treating, for example, a harmonic mode coupled to the TSS bath. It reads

$$\mathcal{H}_I = - \sum_k \left(G_k \sigma_k^{(+)} a + G_k^* \sigma_k^{(-)} a^\dagger \right) \tag{5.77}$$

where a and $a^\dagger$ are, respectively, annihilation and creation operators of the harmonic mode. The full Hamiltonian (5.70) with (5.77) replacing (5.71) is known in the literature as the *Jaynes–Cummings model* (Jaynes and Cummings, 1963), which is another form very useful in quantum optics.

If we once again borrow the parameters from a mechanical oscillator for $\mathcal{H}_0 = \hbar \omega_a a^\dagger a$, F_k and G_k become related through (assuming without loss of generality that $G_k = G_k^*$)

$$G_k = \sqrt{\frac{\hbar}{2M\omega_a}} \, F_k, \tag{5.78}$$

in terms of which another spectral function defined as

$$S_{JC}(\omega) \equiv \pi \sum_k G_k^2 \, \delta(\omega - \omega_k) \tag{5.79}$$

can also be related to the previous ones in (5.14) and (5.76) by

$$S_{JC}(\omega) = \frac{\hbar}{2M\omega_a} J_{TSS}(\omega) = \frac{\hbar^2}{2M\omega_a} J(\omega). \tag{5.80}$$

As a final remark about the models we have just introduced, we must emphasize that we have worried about defining their corresponding spectral functions and always relate them to that of the minimal model, namely $J(\omega)$, only to provide some sort of guidance to the reader in comparing how these different models influence the damping and fluctuation terms. In treating one of these specific models, for example in the Heisenberg picture (Merzbacher, 1998), to write the effective equation of motion for the operator of the system of interest, we must always use the spectral function of the corresponding model to replace the summations over k by integrals over frequencies and establish the desired form of the relaxation rate or the conditions under which the Markovian approximation applies. Although we have done that explicitly for the minimal and non-linear coupling models, we deviate from this approach from now onwards. In the next chapter we will develop the strategy we will adopt in the remainder of the book to attack general quantum mechanical problems in dissipative systems.

6

Implementation of the propagator approach

Now that we have decided on the method to evaluate the reduced density operator of the system and also introduced some reasonably simple phenomenological models which account for some properties of given dissipative systems, let us develop the explicit forms of that operator for some of them.

6.1 The dynamical reduced density operator

6.1.1 The minimal model case

As the reservoir consists of a set of N non-interacting oscillators, we can now evaluate (4.61) and (4.62) analytically. However, it is still quite laborious to go on with their evaluation through conventional approaches such as the explicit use of the environmental wave functions. Our approach is, as we have mentioned before, Feynman's representation for K in terms of functional integrals which will prove more suitable than many other ways to deal with the same problem.

It can be shown (Feynman and Hibbs, 1965) that in the functional integral representation the total propagator reads (see also Appendix A)

$$K(x, \mathbf{R}, t; x', \mathbf{R}', 0) = \int_{x'}^{x} \int_{\mathbf{R}'}^{\mathbf{R}} \mathcal{D}x(t') \, \mathcal{D}\mathbf{R}(t') \, \exp\left\{\frac{i}{\hbar} S[x(t'), \mathbf{R}(t')]\right\}, \quad (6.1)$$

where

$$S[x(t'), \mathbf{R}(t')] = \int_0^t L(x(t'), \mathbf{R}(t'), \dot{x}(t'), \dot{\mathbf{R}}(t')) \, dt' \quad (6.2)$$

is the action of the composite system $R + S$. Those integrals in (6.1) must be evaluated over all the geometric paths $x(t')$ and $\mathbf{R}(t')$ such that $x(0) = x'$, $x(t) = x$, $\mathbf{R}(0) = \mathbf{R}'$, and $\mathbf{R}(t) = \mathbf{R}$.

Inserting (6.1) in (4.62) we reach

$$
\mathcal{J}(x, y, t; x', y', 0) = \int_{x'}^{x} \int_{y'}^{y} \mathcal{D}x(t') \mathcal{D}y(t') \exp\left\{\frac{i}{\hbar}\tilde{S}_0[x(t')]\right\}
$$

$$
\times \exp\left\{-\frac{i}{\hbar}\tilde{S}_0[y(t')]\right\} \mathcal{F}[x(t'), y(t')], \qquad (6.3)
$$

where $\tilde{S}_0$ is the action of the system of interest when isolated, S_0, plus the counter-term action, S_{CT}, and

$$
\mathcal{F}[x(t'), y(t')] = \int \int \int d\mathbf{R}' \, d\mathbf{Q}' \, d\mathbf{R} \, \rho_R(\mathbf{R}', \mathbf{Q}', 0) \int_{\mathbf{R}'}^{\mathbf{R}} \int_{\mathbf{Q}'}^{\mathbf{R}} \mathcal{D}\mathbf{R}(t') \mathcal{D}\mathbf{R}(t')
$$

$$
\times \exp\left\{\frac{i}{\hbar}\Big[S_I[x(t'), \mathbf{R}(t')] - S_I[y(t'), \mathbf{Q}(t')] + S_R[\mathbf{R}(t')] - S_R[\mathbf{Q}(t')]\Big]\right\}
$$

$$
(6.4)
$$

is the so-called influence functional (Feynman and Vernon Jr., 1963; Feynman and Hibbs, 1965). This functional is the average of the product of two time evolutions over the initial state of the environment. One of them is the time evolution of the environment when acted on by the system of interest and the other is its time-reversed partner. One needs these two histories to describe the time evolution of the reduced density operator of the system.

It is clear that this approach can be applied to several systems, but it is unlikely to be exactly soluble in the great majority of cases. That is where our model (5.1) comes into play. Since our bath is harmonic, we can easily compute its propagator when acted on by the coordinate of the particle of interest (forced harmonic oscillators) and perform the average in (6.4) by assuming that the environment is in thermal equilibrium at a given temperature T (Feynman and Hibbs, 1965; Feynman, 1998).

Although the integrals in (6.4) are all Gaussians, we are not going to write the intermediate steps of their evaluation explicitly.

In order to compute the influence functional (6.4) we shall make explicit use of the expression for the propagator of the kth environmental oscillator when it is acted on by an external force $C_k x(t)$ which reads (Feynman and Hibbs, 1965)

$$
K_{RI}^{(k)} = \sqrt{\frac{m_k \omega_k}{2\pi i \hbar \sin\omega_k t}} \exp\frac{i}{\hbar} S_{cl}^{(k)}, \qquad (6.5)
$$

where

$$S_{cl}^{(k)} = \frac{m_k \omega_k}{2 \sin \omega_k t} \left[\left(R_k^2 + R_k'^2 \right) \cos \omega_k t - 2 R_k R_k' \right.$$

$$- \frac{2 C_k R_k}{m_k \omega_k} \int_0^t x(t') \sin \omega_k t' dt' - \frac{2 C_k R_k'}{m_k \omega_k} \int_0^t x(t') \sin \omega_k (t - t') dt'$$

$$\left. - \frac{2 C_k^2}{m_k^2 \omega_k^2} \int_0^t dt' \int_0^{t'} dt'' x(t') x(t'') \sin \omega_k (t - t') \sin \omega_k t'' \right]. \qquad (6.6)$$

This and its time-reversed counter-part must be multiplied by one another for all k and averaged over the initial state of the environment. In our particular problem we will assume that the environment is initially in thermal equilibrium at temperature T independently of the position of the particle of interest, and therefore its density operator can be written as

$$\rho_R(\mathbf{R'}, \mathbf{Q'}, 0) = \prod_k \rho_R^{(k)}(R_k', Q_k', 0) = \prod_k \frac{m_k \omega_k}{2 \pi \hbar \sinh\left(\frac{\hbar \omega_k}{k_B T}\right)}$$

$$\times \exp - \left\{ \frac{m_k \omega_k}{2 \hbar \sinh\left(\frac{\hbar \omega_k}{k_B T}\right)} \left[\left(R_k'^2 + Q_k'^2 \right) \cosh\left(\frac{\hbar \omega_k}{k_B T}\right) - 2 R_k' Q_k' \right] \right\}. \qquad (6.7)$$

Since all the integrals in (6.4) are Gaussians, we can readily evaluate them. Substituting (6.5)–(6.7) in (6.4) and the resulting expression in (6.3), we get

$$\mathcal{J}(x, y, t; x', y', 0) = \int_{x'}^x \int_{y'}^y \mathcal{D}x(t') \mathcal{D}y(t') \exp \frac{i}{\hbar} \left\{ \tilde{S}_0[x(t')] - \tilde{S}_0[y(t')] \right.$$

$$\left. - \int_0^t \int_0^\tau d\tau d\sigma [x(\tau) - y(\tau)] \alpha_I(\tau - \sigma)[x(\sigma) + y(\sigma)] \right\}$$

$$\times \exp - \frac{1}{\hbar} \int_0^t \int_0^\tau d\tau d\sigma [x(\tau) - y(\tau)] \alpha_R(\tau - \sigma)[x(\sigma) - y(\sigma)], \qquad (6.8)$$

where

$$\alpha_R(\tau - \sigma) = \sum_k \frac{C_k^2}{2 m_k \omega_k} \coth \frac{\hbar \omega_k}{2 k_B T} \cos \omega_k (\tau - \sigma) \qquad (6.9)$$

and

$$\alpha_I(\tau - \sigma) = -\sum_k \frac{C_k^2}{2 m_k \omega_k} \sin \omega_k (\tau - \sigma). \qquad (6.10)$$

For later convenience let us define the damping function

$$\eta(\tau - \sigma) = 2M\gamma(\tau - \sigma) \equiv \sum_k \frac{C_k^2}{2m_k\omega_k^2} \cos\omega_k(\tau - \sigma), \qquad (6.11)$$

which is related to $\alpha_I(\tau - \sigma)$ by

$$\alpha_I(\tau - \sigma) = \frac{d\eta(\tau - \sigma)}{d\tau}, \qquad (6.12)$$

and the diffusion function $D(\tau - \sigma)$, which is actually $\alpha_R(\tau - \sigma)$.

Now, using (5.14) and (5.22) we can rewrite these expressions as

$$\alpha_R(\tau - \sigma) = \frac{1}{\pi} \int_0^\Omega d\omega \, \eta\omega \coth\frac{\hbar\omega}{2k_BT} \cos\omega(\tau - \sigma) \qquad (6.13)$$

and

$$\alpha_I(\tau - \sigma) = -\frac{1}{\pi} \int_0^\Omega d\omega \, \eta\omega \sin\omega(\tau - \sigma) = \frac{\eta}{\pi} \frac{d}{d(\tau - \sigma)} \int_0^\Omega d\omega \cos\omega(\tau - \sigma). \qquad (6.14)$$

In order to obtain the final expression for (6.3), we must insert (6.13) and (6.14) into (6.8). These substitutions must be done carefully, in particular the one involving the expression (6.14). The double integral in the imaginary part of (6.8) transforms into

$$\frac{\eta}{\pi} \int_0^t \int_0^\tau \int_0^\Omega [x(\tau) - y(\tau)] \frac{d}{d(\tau - \sigma)} \cos\omega(\tau - \sigma)[x(\sigma) + y(\sigma)]\,d\tau d\sigma d\omega$$

$$= -\frac{\eta\Omega}{\pi} \int_0^t [x^2(\tau) - y^2(\tau)]d\tau + \frac{\eta}{\pi}(x' + y') \int_0^t \frac{\sin\Omega\tau}{\tau}[x(\tau) - y(\tau)]\,d\tau$$

$$+ \frac{\eta}{\pi} \int_0^t \int_0^\tau [x(\tau) - y(\tau)] \frac{\sin\Omega(\tau - \sigma)}{(\tau - \sigma)} [\dot{x}(\sigma) + \dot{y}(\sigma)]\,d\tau d\sigma, \qquad (6.15)$$

whose r.h.s. has been obtained after an integration by parts with respect to σ.

The first term on the r.h.s. of (6.15) comes from the harmonic correction that can be canceled by the counter-term present in $\tilde{S}_0$. The remaining integrals on the r.h.s.

of (6.15) can be simplified further if we use the fact that the time interval t is such that $t \gg \Omega^{-1}$, which allows us to approximate

$$\frac{1}{\pi} \frac{\sin \Omega (\tau - \sigma)}{\tau - \sigma} \approx \delta(\tau - \sigma). \tag{6.16}$$

However, there is a point where we have to use (6.16) with great care. A brief analysis of the integrals on the r.h.s. of (6.15) shows us that when $\Omega t \gg 1$ its second term reads

$$\eta(x' + y') \int_0^t \frac{1}{\pi} \frac{\sin \Omega \tau}{\tau} [x(\tau) - y(\tau)] d\tau \rightarrow \eta(x' + y') \int_0^t \delta(\tau) [x(\tau) - y(\tau)] d\tau, \tag{6.17}$$

which, as a function of its lower limit of integration, is not continuous. Its particular form is

$$\eta(x' + y') \int_{t'}^t \delta(\tau) [x(\tau) - y(\tau)] d\tau = \begin{cases} 0 & \forall \quad t' > 0 \\ \eta(x'^2 - y'^2)/2 & \text{if} \quad t' = 0. \end{cases} \tag{6.18}$$

In contrast, the last integral on the r.h.s. of (6.15) is a continuous function of its lower limit of integration. When $\Omega t \gg 1$ we have

$$\frac{\eta}{2} \int_0^t d\tau \, [x(\tau) - y(\tau)] \int_0^t \delta(\tau - \sigma) [\dot{x}(\sigma) + \dot{y}(\sigma)] d\sigma, \tag{6.19}$$

which could also have been obtained by

$$\lim_{t' \to 0} \frac{\eta}{2} \int_{t'}^t d\tau \, [x(\tau) - y(\tau)] \int_{t'}^t \delta(\tau - \sigma) [\dot{x}(\sigma) + \dot{y}(\sigma)] d\sigma, \tag{6.20}$$

and, therefore, it does not really matter whether we include the lower limit of integration in its evaluation.

We conclude that we must specify what we mean by those integrals from $\tau, \sigma = 0$ to $\tau, \sigma = t$. In order to recover the appropriate boundary conditions to the superpropagator (4.62), namely

$$\mathcal{J}(x, y, 0^+; x', y', 0) = \delta(x - x')\delta(y - y'),$$

we must perform all those integrations within the interval $0 < \tau, \sigma \leq t$ and finally write

$$\mathcal{J}(x, x', t; y, y', 0) = \int_{x'}^{x} \int_{y'}^{y} \mathcal{D}x(t')\mathcal{D}y(t')$$

$$\times \exp\frac{i}{\hbar}\left\{ S_0[x(t')] - S_0[y(t')] - M\gamma \int_0^t (x\dot{x} - y\dot{y} + x\dot{y} - y\dot{x})dt' \right\}$$

$$\times \exp - \frac{2M\gamma}{\pi\hbar} \int_0^\Omega d\omega\,\omega \coth\frac{\hbar\omega}{2k_BT}$$

$$\times \int_0^t \int_0^\tau d\tau d\sigma[x(\tau) - y(\tau)]\cos\omega(\tau - \sigma)[x(\sigma) - y(\sigma)], \qquad (6.21)$$

where $\gamma \equiv \eta/2M$ is the relaxation constant of the system. More generally,

$$\mathcal{J}(x, x', t; y, y', 0) = \int_{x'}^{x} \int_{y'}^{y} \mathcal{D}x(t')\mathcal{D}y(t') \exp\frac{i}{\hbar}\left\{ S_0[x(t')] - S_0[y(t')] \right.$$

$$\left. - \int_0^t \int_0^\tau d\tau d\sigma[x(\tau) - y(\sigma)]M\gamma(\tau - \sigma)[\dot{x}(\tau) + \dot{y}(\sigma)] \right\}$$

$$\times \exp - \frac{1}{\hbar} \int_0^t \int_0^\tau d\tau d\sigma[x(\tau) - y(\tau)]D(\tau - \sigma)[x(\sigma) - y(\sigma)], \qquad (6.22)$$

where $D(\tau - \sigma)$ and $\gamma(\tau - \sigma)$ have been defined in (6.9) and (6.11), respectively. In (6.21) we see that, for ohmic dissipation, $\gamma(\tau - \sigma) \to \gamma\delta(\tau - \sigma)$. Notice that the issue of the limits of integration we have been involved with is strongly dependent on the choice of initial condition made in (4.60). This is related to the problem of the spurious term that appeared in (5.24). As we mentioned in our analysis below (5.29), one possible way to circumvent the problem is to adopt the completely separable initial condition and disregard the spurious term by imposing that the initial value for $q(t)$ is actually established as $q(0^+)$. That is, purely for simplicity, the solution we have adopted here. The other possibility would be to modify (4.60) by imposing that the bath of oscillators is in equilibrium about the initial position of the particle, which requires that we should change expression

(6.7) accordingly. We call this a *conditionally separable* initial condition since it depends on the parameters used to specify the initial condition of the system of interest. In this second approach the spurious $t = 0$ terms cancel out naturally, as we have seen in the method of the equation of motion.

The explanation why the factorizable initial condition still works if it is applied appropriately to our problem is not complicated, albeit a bit subtle. In order to understand what goes on we had best analyze the issue from the form (5.35). Since the whole set of oscillators can be considered as attached to the external particle but assumed to be in equilibrium about $q_k = 0$, they will exert a force on that particle if it is initially displaced by q_0 from the common origin. Moreover, suppose the external particle moves with natural and relaxation frequencies ω_0 and γ such that $\omega_0, \gamma \ll \Omega$. Therefore, fast environmental oscillators are initially very well localized about the origin $\left(\sqrt{\langle q_k^2 \rangle} \ll q_0 \right)$, whereas slow ones are spread over a broad region in space $\left(\sqrt{\langle q_k^2 \rangle} \gg q_0 \right)$. As we allow the system to relax, the fastest oscillators will adjust their motion adiabatically to the very slow motion of the particle and within a time interval $\Delta t \gtrsim \Omega^{-1}$ their positions will be such that $\langle q_k(\Delta t) \rangle \approx q_0$. On the contrary, for the slowest oscillators, it does not matter much where their equilibrium position is since $\left(\sqrt{\langle q_k^2 \rangle} \gg q_0 \right)$. Hence, let us consider them to be in thermal equilibrium about q_0. The preceding reasoning justifies why we have imposed the initial conditions at $t = 0^+$, which actually means $t \approx \Omega^{-1}$ when $\Omega \to \infty$. In other words, at $t \approx \Omega^{-1}$ the reservoir acts as if it were in equilibrium about the position of the particle.

Expression (6.21) is the main result of this section. Given the external potential $V(q)$ to which the particle is subject we can, at least in principle, determine $\mathcal{J}(x, x', t; y, y', 0)$ through the evaluation of that double path integral. As expected, there are few cases where $\mathcal{J}(x, x', t; y, y', 0)$ can be obtained in closed analytical form. In the great majority of cases some methods of approximation must be developed. Potentials of the form $V(q) = aq^n$, where $n = 0, 1, 2$, are very special cases that can be solved exactly. After having evaluated $\mathcal{J}(x, x', t; y, y', 0)$, we can obtain $\tilde{\rho}(x, y, t)$ using (4.61) for some particular initial conditions of the particle of interest. However, let us first investigate a particularly important limit of (6.21), namely the semi-classical limit, where $k_B T \gg \hbar\omega$ for frequencies such that $\omega \ll \Omega$. In other words, we are assuming that most frequencies in this range will be present in the Fourier decomposition of the paths, effectively contributing to the evaluation of $\mathcal{J}(x, x', t; y, y', 0)$.

In this limit, using the fact that $\coth x \approx x^{-1}$ for $x \to 0$ and performing the integration over ω, we can approximate (6.21) by

$$\mathcal{J}(x, x', t; y, y', 0) = \int\limits_{x'}^{x} \int\limits_{y'}^{y} \mathcal{D}x(t')\mathcal{D}y(t')$$

$$\times \exp\frac{i}{\hbar}\left\{ S_0[x(t')] - S_0[y(t')] - M\gamma\int\limits_0^t (x\dot{x} - y\dot{y} + x\dot{y} - y\dot{x})dt' \right\}$$

$$\times \exp - \frac{2M\gamma k_B T}{\hbar^2}\int\limits_0^t d\tau [x(\tau) - y(\tau)]^2. \tag{6.23}$$

Nevertheless, this is not the most appropriate form to study the semi-classical limit of our problem. We had best work directly with the reduced density operator.

Quantum master equations. Actually, it is very convenient to write a differential equation for $\tilde{\rho}(x, y, t)$, not only because of this specific limit we are treating, but also for possible comparison with other approaches to the problem (see, for example, Breuer and Petruccione (2002)). This equation can be obtained if we follow the same steps as Feynman, when he deduced the Schrödinger equation from the path integral representation of $K(x, t; x', t')$ (see Appendix B), and it reads

$$\frac{\partial \tilde{\rho}}{\partial t} = -\frac{\hbar}{2Mi}\frac{\partial^2 \tilde{\rho}}{\partial x^2} + \frac{\hbar}{2Mi}\frac{\partial^2 \tilde{\rho}}{\partial y^2} - \gamma(x - y)\frac{\partial \tilde{\rho}}{\partial x} + \gamma(x - y)\frac{\partial \tilde{\rho}}{\partial y}$$

$$+ \frac{V(x)}{i\hbar}\tilde{\rho} - \frac{V(y)}{i\hbar}\tilde{\rho} - \frac{2M\gamma k_B T}{\hbar^2}(x - y)^2\tilde{\rho}, \tag{6.24}$$

which is the coordinate representation of

$$\frac{\partial \tilde{\rho}}{\partial t} = \frac{1}{i\hbar}[\mathcal{H}_0, \tilde{\rho}] + \frac{\gamma}{i\hbar}[q, \{p, \tilde{\rho}\}] - \frac{D}{\hbar^2}[x, [x, \tilde{\rho}]], \tag{6.25}$$

where $\mathcal{H}_0$ is the Hamiltonian of the system of interest, and $D = D_{pp} = \eta k_B T$. This equation is one possible form of the so-called *non-rotating wave* (NRW) master equations, which are very popular in the literature of quantum open systems (Breuer and Petruccione, 2002). The *rotating wave* (RW) quantum master equation is in the so-called *Lindblad form* (Lindblad, 1975), which reads

$$\frac{\partial \tilde{\rho}}{\partial t} = \frac{1}{i\hbar}[\tilde{\mathcal{H}}, \tilde{\rho}] + \sum_n \left\{ 2A_n^\dagger \tilde{\rho} A_n - \tilde{\rho} A_n A_n^\dagger - A_n A_n^\dagger \tilde{\rho} \right\}, \tag{6.26}$$

where $\tilde{\mathcal{H}}$ is a Hermitian Hamiltonian operator and A_n are operators which depend on the specific problem being considered. It was shown by Lindblad (1975) that this is the only form of a first-order time evolution linear differential equation for $\tilde{\rho}$ that preserves its semi-definite positivity property, which means $\langle \varphi | \tilde{\rho}(t) | \varphi \rangle \geq 0$ for all $|\varphi\rangle$ at any instant t.

As (6.25) is not in the Lindblad form, we should not be surprised if in some circumstances it violates the semi-definite positivity requirement. However, we do not need to worry much about it and the reason is twofold. Firstly, as shown in Ambegaokar (1991) and Munro and Gardiner (1996), this violation is due to very short time transients that only happen once we try to localize the particle initially in a length shorter than its de Broglie wavelength. Simple uncertainty arguments suffice to show that within a very short time scale, positivity is recovered. Another way to remedy the situation is to consider corrections to the approximation $\coth x \sim 1/x$ we used in (6.21) to obtain (6.24). Secondly, as observed in Pechukas (1994), semi-definite positivity is a very particular requirement that only holds necessarily if we assume that the full density operator is separable at any instant along the time evolution of the composite system. This is definitely not our case, although as we have said above, with appropriate measures we can still preserve positivity.

Finally, we should also worry about the meaning of this semi-classical (high-temperature) limit of the quantum evolution equation. In order to study this limit in a coherent manner, we introduce the *Wigner transform* of $\tilde{\rho}$, which is defined as

$$W(q, p, t) \equiv \frac{1}{2\pi\hbar} \int_{-\infty}^{+\infty} \exp\left(\frac{ip\xi}{\hbar}\right) \tilde{\rho}\left(q - \frac{\xi}{2}, q + \frac{\xi}{2}, t\right) d\xi. \tag{6.27}$$

This function is particularly important because in the classical limit ($\hbar \to 0$) it reduces to the phase space probability distribution of the particle. Then, inverting (6.27) we have

$$\tilde{\rho}(x, y, t) = \int_{-\infty}^{+\infty} W\left(\frac{x+y}{2}, p, t\right) \exp\left(\frac{-ip(x-y)}{\hbar}\right) dp, \tag{6.28}$$

which, if taken to (6.24), yields

$$\frac{\partial W}{\partial t} = -\frac{\partial}{\partial q}(pW) + \frac{\partial}{\partial p}\left[(2\gamma p + V'(q)) W\right] + D\frac{\partial^2 W}{\partial p^2}. \tag{6.29}$$

Therefore, we see that, in the classical limit, our model leads us to the Fokker–Planck equation of the Wigner transform of $\tilde{\rho}(x, y, t)$, as expected. It is worth noticing that, although the Wigner function is purely quantal, only its classical limit makes sense in (6.29) once this equation has been deduced in this circumstance. For the general case, we should have a generalization of (6.24), which would lead us to a modified Fokker–Planck equation for $W(q, p, t)$. The advantage of working with the path integral representation of this problem is that we can treat the quantum regime directly, with no need to generalize (6.24) or (6.29). Hence, the dynamical solution of the problem at any temperature boils down to the evaluation of (6.21).

6.1.2 The non-linear coupling case

In the last section we saw that the influence functional (6.4) is the central object to be evaluated in order to describe the time evolution of the reduced density operator of an external particle placed in a general environment. Moreover, as in our model the reservoir has been represented by a set of non-interacting harmonic oscillators bilinearly coupled in coordinates to the system of interest, we have been able to write it in a Gaussian form as shown in (6.21). However, for the non-linear coupling model things work differently.

Although the oscillators are still non-interacting and the coupling to the external particle is a linear function of the bath coordinates, we have assumed an exponential dependence on the particle coordinate itself in (5.41), which no longer allows us to write the superpropagator $\mathcal{J}(x, y, t; x', y', 0)$ in a Gaussian form. In spite of that, and due to the linear coupling to the bath coordinates, we can easily find the analytical structure of this non-Gaussian form by replacing the paths $x(t')$ in the forced harmonic oscillator expression (6.5) with $\kappa_k \exp i k x(t')$ and proceed as before to reach a generalization of (6.8). However, as there might be more complicated forms of the dependence on the bath coordinates in the interaction term, a more general approach to this problem would be desirable. Let us then try to develop our strategy to work with these situations.[1] The influence functional (6.4) can be written compactly in the Hamiltonian formulation as

$$\mathcal{F}\left[x(t'), y(t')\right] = \mathrm{Tr}_R \left(\rho_R U_{RI}^{\dagger}[y(t')] U_{RI}[x(t')] \right), \tag{6.30}$$

where $U_{RI}[x(t')]$, a functional of $x(t')$, is the unitary time evolution operator of the reservoir subject to the influence of the system which evolves through the Hamiltonian $\mathcal{H}_{RI} = \mathcal{H}_R + \mathcal{H}_I(x(t'))$. This means that a given trajectory $x(t')$ of the system for $0 \leq t' \leq t$, or a function thereof, acts as an external source on the environment. The time evolution of the operator $U_{RI}[x]$ obeys

$$i\hbar \frac{dU_{RI}(t)}{dt} = \mathcal{H}_{RI}(t) U_{RI}(t), \tag{6.31}$$

with the initial condition $U_{RI}(0) = 1$, and has the formal solution

$$U_{RI}(t) = \mathcal{T} \exp -\frac{i}{\hbar} \int_0^t dt' \mathcal{H}_{RI}(t'), \tag{6.32}$$

where $\mathcal{T}$ is the well-known time ordering operator which, applied on the exponential, means that it must be applied to all the terms of its series expansion. In the interaction picture (Fetter and Walecka, 2003) this result can be written as

---

[1]   In the remainder of this section we shall be following closely part of Duarte and Caldeira (2009).

$$U_{RI}(t) = e^{-i\mathcal{H}_{Rt}/\hbar} T \exp -\frac{i}{\hbar} \int_0^t dt' \tilde{\mathcal{H}}_I[x(t')], \tag{6.33}$$

where $\tilde{\mathcal{H}}_I[x(t')] \equiv e^{i\mathcal{H}_{Rt}/\hbar} \mathcal{H}_I[x(t')] e^{-i\mathcal{H}_{Rt}/\hbar}$. Inserting (6.33) into (6.30) we have

$$\mathcal{F}[x(t'), y(t')] = \mathrm{Tr}_R \left(\rho_R T \exp \frac{i}{\hbar} \int_0^t dt' \tilde{\mathcal{H}}_I[x(t')] \right.$$

$$\left. \times T \exp -\frac{i}{\hbar} \int_0^t dt'' \tilde{\mathcal{H}}_I[x(t'')] \right). \tag{6.34}$$

Assuming that the particle only disturbs its environment weakly and, consequently, expanding the chronological product up to second order in $\tilde{\mathcal{H}}_I$ (Hedegård and Caldeira, 1987), we have

$$T \exp -\frac{i}{\hbar} \int_0^t dt' \tilde{\mathcal{H}}_I[x(t')] \approx 1 - \frac{i}{\hbar} \int_0^t dt' \tilde{\mathcal{H}}_I[x(t')]$$

$$-\frac{1}{\hbar^2} \int_0^t dt' \int_0^{t'} dt'' \tilde{\mathcal{H}}_I[x(t')] \tilde{\mathcal{H}}_I[x(t'')], \tag{6.35}$$

which inserted in (6.34) yields (after performing the tracing operation indicated therein)

$$\mathcal{F}[x(t'), y(t')] \approx 1 - \frac{1}{\hbar^2} \int_0^t dt' \int_0^{t'} dt'' \left(\left\langle \tilde{\mathcal{H}}_I[x(t')] \tilde{\mathcal{H}}_I[x(t'')] \right\rangle \right.$$

$$+ \left\langle \tilde{\mathcal{H}}_I[y(t'')] \tilde{\mathcal{H}}_I[y(t')] \right\rangle - \left\langle \tilde{\mathcal{H}}_I[y(t')] \tilde{\mathcal{H}}_I[x(t'')] \right\rangle$$

$$\left. - \left\langle \tilde{\mathcal{H}}_I[y(t'')] \tilde{\mathcal{H}}_I[x(t')] \right\rangle \right), \tag{6.36}$$

where the brackets mean averages over the initial bath operator ρ_R.

The equation above can be re-exponentiated, yielding

$$\mathcal{F}[x(t'), y(t')] = \exp -\frac{1}{\hbar^2} \int_0^t dt' \int_0^{t'} dt'' \left(\left\langle \tilde{\mathcal{H}}_I[x(t')] \tilde{\mathcal{H}}_I[x(t'')] \right\rangle \right.$$

$$+ \left\langle \tilde{\mathcal{H}}_I[y(t'')] \tilde{\mathcal{H}}_I[y(t')] \right\rangle - \left\langle \tilde{\mathcal{H}}_I[y(t')] \tilde{\mathcal{H}}_I[x(t'')] \right\rangle$$

$$\left. - \left\langle \tilde{\mathcal{H}}_I[y(t'')] \tilde{\mathcal{H}}_I[x(t')] \right\rangle \right), \tag{6.37}$$

which is the main result of our approach. In this approximation we are obviously considering the bath within linear response theory, and consequently it is an exact expression for our non-linear coupling model where the environment coordinates appear linearly in the interaction Hamiltonian (5.40).

Considering explicitly the coupling (5.41) and using the fact that $\langle R_k(t')R_{k'}(t'')\rangle = 0$ unless $k' = -k$, we reach

$$\langle \tilde{\mathcal{H}}_I[x(t')]\tilde{\mathcal{H}}_I[x(t'')]\rangle = \frac{1}{2}\sum_k \{C_{-k}[x(t')]C_k[x(t'')]$$

$$+C_k[x(t')]C_{-k}[x(t'')]\}\,\alpha_k(t'-t''), \qquad (6.38)$$

where $\alpha_k(t'-t'') \equiv \langle R_k(t')R_{-k}(t'')\rangle$ can be related to the dynamical susceptibility of the kth mode of the bath through the fluctuation–dissipation theorem (see, for example, Forster (1990)):

$$\alpha_k(t'-t'') = \frac{\hbar}{\pi}\int\limits_{-\infty}^{\infty} d\omega\,\mathrm{Im}\,\chi_k(\omega)\frac{e^{-i\omega(t'-t'')}}{1-e^{-\omega\hbar\beta}}, \qquad (6.39)$$

where $\beta \equiv 1/k_B T$.

Inserting this result in (6.37), we can write its final form as

$$\mathcal{F}[x(t'), y(t')] = \exp -\frac{1}{\hbar^2}\int\limits_0^t dt'\int\limits_0^{t'} dt''$$

$$\times \left\{\sum_k \kappa_k\kappa_{-k}\alpha_k(t'-t'')\left[\cos k(x(t')-x(t''))-\cos k(y(t')-x(t''))\right]\right.$$

$$\left.+ \sum_k \kappa_k\kappa_{-k}\alpha_k^*(t'-t'')\left[\cos k(y(t')-y(t''))-\cos k(y(t'')-x(t'))\right]\right\}. \qquad (6.40)$$

Now, using the fact that $\mathrm{Im}\chi_k(-\omega) = -\mathrm{Im}\chi_k(\omega)$, we write the correlation function $\alpha_k(t'-t'')$ as

$$\alpha_k(t'-t'') = \alpha_k^{(R)}(t'-t'') + i\alpha_k^{(I)}(t'-t''), \qquad (6.41)$$

with the real and imaginary parts defined respectively as

$$\alpha_k^{(R)}(t'-t'') = \frac{\hbar}{\pi}\int\limits_0^{\infty} d\omega\,\mathrm{Im}\,\chi_k(\omega)\cos\omega(t'-t'')\coth(\hbar\beta\omega/2) \qquad (6.42)$$

and

$$\alpha_k^{(I)}(t' - t'') = -\frac{\hbar}{\pi} \int_0^\infty d\omega \, \mathrm{Im} \, \chi_k(\omega) \sin \omega(t' - t''). \qquad (6.43)$$

The functional (6.40) can also be written in terms of these real and imaginary parts. We should remember that it is not our intention to treat the full microscopic dynamics of the environment, but stick to the previous form of $\mathrm{Im}\chi_k(\omega)$ as modeled by (5.47) instead. Therefore, using the latter, we can evaluate the frequency integrals in the imaginary part (6.43) and end up with

$$\mathcal{F}[x(t'), y(t')] = \exp -\frac{1}{\hbar^2} \int_0^t dt' \int_0^{t'} dt'' \sum_k \kappa_k \kappa_{-k} \alpha_k^{(R)}(t' - t'')$$

$$\times \left[\cos k(x(t') - x(t'')) - \cos k(y(t') - x(t'')) + \cos k(y(t') - y(t'')) \right.$$
$$\left. - \cos k(y(t'') - x(t')) \right]$$

$$\times \exp \frac{i}{2\hbar} \sum_k \kappa_k \kappa_{-k} f(k) k \int_0^t dt' \sin k(y(t') - x(t')) \left(\dot{x}(t') + \dot{y}(t') \right). \quad (6.44)$$

This functional can be simplified further if we assume that the most important trajectories $(x(t'), y(t'))$ which contribute to the double path integral determining the superpropagator are confined to a region that is small compared with the length k_0^{-1} introduced in (5.53) through the function $g(k)$. For example, in fermionic environments this length is related to the Fermi wave number k_F, as mentioned in the previous chapter (see also Guinea (1984), Hedegård and Caldeira (1987)). Operationally, this implies that $k(y(t') - x(t')) \ll 1$ and the functional becomes

$$\mathcal{F}[x(t'), y(t')] = \exp \left\{ \frac{i\eta}{2\hbar} \int_0^t dt' (y(t') - x(t')) \left(\dot{x}(t') + \dot{y}(t') \right) \right.$$

$$- \frac{\eta}{\hbar\pi} \int_0^t dt' \int_0^{t'} dt'' (x(t') - y(t'))(x(t'') - y(t''))$$

$$\left. \times \int_0^\infty d\omega \, \omega \coth \left(\frac{\hbar\beta\omega}{2} \right) \cos \omega(t' - t'') \right\}, \qquad (6.45)$$

where $\eta = \sum_k k^2 \kappa_k \kappa_{-k} f(k)$ as in (5.48). Notice that the above equation reproduces (6.21) exactly, as it should. In other words, in the non-linear coupling model the particle behaves locally exactly as it does in the minimal model. Actually, we had already concluded the same when studying this model within the equation of motion method in the previous chapter, where we showed the Langevin equation (5.48) reproduced for high temperatures.

Now, we can extend this approach to deal with more than one external particle, as in Section 5.2. All we have to do, at least in one dimension, is consider the interaction of a couple of particles with the reservoir in the interaction term (5.50), which consequently doubles the number of paths $x(t) \rightarrow (x_1(t), x_2(t))$ and $y(t) \rightarrow (y_1(t), y_2(t))$ in the evaluation of the influence functional. The latter now accounts for the dynamics of the two quantum Brownian particles, with all the possible effects originated from the indirect coupling between them mediated by the environment as predicted in Section 5.2. Although the procedure we have just outlined here does not present any particular difficulty, the resulting expressions are quite lengthy and cumbersome (Duarte and Caldeira, 2009). Therefore, we shall not pursue the subject any further and content ourselves with what we have achieved so far.

6.1.3 The collision model case

Let us return to the discussion we have postponed from Section 5.3, where we stressed that, contrary to the other models presented so far, we were not able to find a simple way to treat the classical version of the model

$$\mathcal{H} = \frac{p^2}{2M} + \sum_{i=1}^{N} U(q - q_i) + \sum_{i=1}^{N} \frac{p_i^2}{2m}. \tag{6.46}$$

We have argued that besides representing the motion of a heavy particle coupled through $\sum_{i=1}^{N} U(q - q_i)$ to a set of non-interacting massive and identical particles, this model could also be applied to the motion of topological excitations in a non-linear material medium. Now we are going to adapt the strategy we have been developing in the last two sections to the present problem and show that, despite its complexity, we can deal with its quantum mechanical version. In so doing, we have to deviate a little from the standard steps taken so far.[2]

Our starting point is once again the object that gives us the dynamics of the particle of interest, namely its reduced density operator

$$\tilde{\rho}(x, y, t) = \int dq_1 dq_2 \dots dq_N \langle x, q_1, \dots, q_N | e^{i \frac{\mathcal{H}t}{\hbar}} \rho(0) e^{-i \frac{\mathcal{H}t}{\hbar}} | y, q_1, \dots, q_N \rangle, \tag{6.47}$$

where $\rho(0)$ is the initial density operator of the whole system.

---

[2] In the remainder of this section we shall be following closely part of Caldeira and Castro Neto (1995).

We can translate the reservoir coordinates by q through the unitary transformation $\mathcal{U}$ given by

$$\mathcal{U} = \exp{-\frac{i}{\hbar} \sum_{j=1}^{N} p_j q},$$ (6.48)

and use the identity $\mathcal{U}^{-1}\mathcal{U} = \mathbb{1}$ four times between any two operators in (6.47) to get

$$\tilde{\rho}(x, y, t) = \int dq_1 dq_2 \ldots dq_N \langle x, q_1, \ldots, q_N | \mathcal{U}^{-1} e^{i\frac{\mathcal{H}'t}{\hbar}} \rho'(0) e^{-i\frac{\mathcal{H}'t}{\hbar}} \mathcal{U} | y, q_1, \ldots, q_N \rangle$$

$$= \int dq_1 dq_2 \ldots dq_N \langle x, q_1, \ldots, q_N | \mathcal{U}^{-1} \rho'(t) \mathcal{U} | y, q_1, \ldots, q_N \rangle, \quad (6.49)$$

where $\mathcal{H}'$ is the transformed Hamiltonian (Hakim and Ambegaokar, 1985)

$$\mathcal{H}' = \mathcal{U}\mathcal{H}\mathcal{U}^{-1} = \frac{1}{2M}\left(p - \sum_{j=1}^{N} p_j\right)^2 + \sum_{j=1}^{N}\left(\frac{p_j^2}{2m} + U(q_j)\right)$$ (6.50)

and $\rho'(0) = \mathcal{U}\rho(0)\mathcal{U}^{-1}$ is the transformed initial condition.

The application of the unitary transformation $\mathcal{U}$ to the Hamiltonian $\mathcal{H}$ has changed the coordinate coupling via $\sum_{i=1}^{N} U(q - q_i)$ into a momentum–momentum coupling with a set of particles which are scattered by a potential $U(q)$. This transformation was also used in Castella and Zotos (1993) for the exact diagonalization of the problem via the Bethe ansatz.

Now, inserting

$$\int \ldots \int dx dq_1 dq_2 \ldots dq_N |x, q_1, \ldots, q_N \rangle \langle x, q_1, \ldots, q_N| = \mathbb{1}$$ (6.51)

twice between the unitary operators $\mathcal{U}$ and $\mathcal{U}^{-1}$ and the transformed density operator $\rho'(t)$ in (6.49) we have

$$\tilde{\rho}(x, y, t) = \int \ldots \int dq_1 dq_2 \ldots dq_N dr_1 dr_2 \ldots dr_N$$

$$\times \langle r_1, \ldots, r_N | e^{-\frac{i}{\hbar}\sum_{j=1}^{N} p_j(x-y)} | q_1, \ldots, q_N \rangle \langle x, q_1, \ldots, q_N | \rho'(t) | y, r_1, \ldots, r_N \rangle,$$ (6.52)

which relates the original reduced density operator of the system to a sort of reduced "$\mathcal{U}$-transform" of the full density operator in the new (and hopefully more convenient) representation.

Since we are mostly interested in the average values of observables such as $\langle x(t) \rangle$, $\langle p(t) \rangle$, $\langle x^2(t) \rangle$, $\langle p^2(t) \rangle$, we will show next that what we need are only the diagonal elements of (6.52) rather than its full form.

We start with observables which are only dependent on position. Then, because

$$\langle x^n(t) \rangle = \int dx\, x^n \tilde{\rho}(x, x, t), \qquad (6.53)$$

the first term on the r.h.s. of (6.52) becomes $\delta(q_1 - r_1) \ldots \delta(q_N - r_N)$, which can be integrated trivially to yield

$$\langle x^n(t) \rangle = \int \ldots \int dx\, dq_1 \ldots dq_N x^n \langle x, q_1, \ldots, q_N | \rho'(t) | x, q_1, \ldots, q_N \rangle. \qquad (6.54)$$

For momentum-dependent operators, what we need is

$$\langle p(t) \rangle = \lim_{x \to y} \frac{\hbar}{2i} \left(\frac{d}{dx} - \frac{d}{dy} \right) \tilde{\rho}(x, y, t), \qquad (6.55)$$

which can be applied to (6.52) resulting in

$$\langle p(t) \rangle = \frac{\hbar}{2i} \lim_{x \to y} \int \ldots \int dq_1 dq_2 \ldots dq_N dr_1 dr_2 \ldots dr_N$$

$$\times \langle r_1, \ldots, r_N | e^{-\frac{i}{\hbar} \sum_{j=1}^{N} p_j(x-y)} | q_1, \ldots, q_N \rangle$$

$$\times \left(\frac{d}{dx} - \frac{d}{dy} \right) \langle x, q_1, \ldots, q_N | \rho'(t) | y, r_1, \ldots, r_N \rangle. \qquad (6.56)$$

In the limit $x \to y$ the first term of the integrand above becomes once again $\delta(q_1 - r_1) \ldots \delta(q_N - r_N)$, which can be integrated to give

$$\langle p(t) \rangle = \frac{\hbar}{2i} \lim_{x \to y} \int \ldots \int dq_1 dq_2 \ldots dq_N$$

$$\times \left(\frac{d}{dx} - \frac{d}{dy} \right) \langle x, q_1, \ldots, q_N | \rho'(t) | y, q_1, \ldots, q_N \rangle, \qquad (6.57)$$

and so, what we really need in these cases is the reduced density operator of the particle in the new representation,

$$\tilde{\rho}'(x, y, t) = \int \ldots \int dq_1 dq_2 \ldots dq_N \langle x, q_1, \ldots, q_N | \rho'(t) | y, q_1, \ldots, q_N \rangle. \qquad (6.58)$$

However, it should be stressed here that this result is valid only if we want to compute the above-mentioned averages. Were we interested in the time evolution of operators, which would involve knowledge of the off-diagonal elements of $\tilde{\rho}(x, y, t)$, this statement would no longer be valid.

Another important point we should clarify at this stage refers to the choice we make for the initial condition $\rho'(0)$ in this section. As we have seen from (6.49), $\rho'(0)$ can be prepared in terms of the eigenstates of the transformed Hamiltonian

(6.50). Our particular choice is that the particle is in a pure state (for example, a wave packet centered at the origin) and the environment is in equilibrium in the presence of the particle. This means we can write $\rho'(0)$ as

$$\rho'(0) = \rho_s \rho'_{eq} = \rho_s e^{-\beta \mathcal{H}_R}, \tag{6.59}$$

where ρ_s refers only to the particle and $\mathcal{H}_R$ is given by

$$\mathcal{H}_R = \sum_{j=1}^{N} \left(\frac{p_j^2}{2m} + U(q_j) \right). \tag{6.60}$$

Once we have specified the initial state whose reduced time evolution we are going to study, we must now choose the representation in which we should do it. Following our previous choice for other environmental models, the Feynman path integral approach seems to be the most appropriate one in this new situation too. However, as our environment is now composed of indistinguishable particles (bosons or fermions), it is convenient to write the transformed Hamiltonian (6.50) in its second quantized version which reads (see, for example, Fetter and Walecka (2003))

$$\mathcal{H}' = \frac{1}{2M} \left(p - \sum_{i,j} \hbar g_{ij} a_i^\dagger a_j \right)^2 + \sum_i (\hbar \Omega_i - \mu) a_i^\dagger a_i, \tag{6.61}$$

where $g_{ij} = \frac{1}{\hbar} \langle i|p'|j \rangle$ is the matrix element of the momentum operator of a single environment particle between eigenstates $|i\rangle$ and $|j\rangle$ of $\mathcal{H}_R$ in (6.60), Ω_i are the eigenfrequencies of these states, and μ is the chemical potential. The operators a_i and $a_i^\dagger$ are the standard annihilation and creation operators for bosons or fermions.

Then we can use the initial conditions (6.59) in (6.58) to write the standard Feynman–Vernon expression (Feynman and Vernon Jr., 1963; Caldeira and Leggett, 1983b)

$$\tilde{\rho}'(x, y, t) = \int dx' \int dy' \mathcal{J}(x, y, t; x', y', 0) \rho_s(x', y', 0), \tag{6.62}$$

where

$$\mathcal{J}(x, y, t; x', y', 0) = \int \dots \int dq_1 \dots dq_N dr_1 \dots dr_N ds_1 \dots ds_N$$

$$\times \langle x, q_1, \dots, q_N | \exp -\frac{i}{\hbar} \mathcal{H}' t | x', r_1, \dots, r_N \rangle$$

$$\times \langle r_1, \dots, r_N | \exp -\beta \mathcal{H}_R | s_1, \dots, s_N \rangle$$

$$\times \langle y', s_1, \dots, s_N | \exp \frac{i}{\hbar} \mathcal{H}' t | y, q_1, \dots, q_N \rangle. \tag{6.63}$$

However, there is an important difference between the present model and the previous ones. As we have represented our Hamiltonian $\mathcal{H}'$ in its second quantized form, we must evaluate all the endpoint integrals in (6.63) in the coherent state representation[3] for the creation and annihilation operators (Negele and Orland, 1998) involved therein (see also Appendix C). Therefore, we can rewrite $\mathcal{J}$ in (6.63) as

$$\mathcal{J}(x, y, t; x', y', 0) = \int \dots \int d\mu(\boldsymbol{\alpha}) \, d\mu(\boldsymbol{\alpha}') \, d\mu(\boldsymbol{\gamma}') \, \langle x, \boldsymbol{\alpha}| \exp -\frac{i}{\hbar}\mathcal{H}'t|x', \boldsymbol{\alpha}' \rangle$$

$$\times \langle \boldsymbol{\alpha}'| \exp -\beta \mathcal{H}_e|\boldsymbol{\gamma}' \rangle \langle y', \boldsymbol{\gamma}'| \exp \frac{i}{\hbar}\mathcal{H}'t|y, \boldsymbol{\alpha} \rangle, \qquad (6.64)$$

where $\boldsymbol{\alpha}$ and $\boldsymbol{\gamma}$ are complex vectors with an infinite number of components and $d\mu(*)$ is an integration measure which depends on whether we are dealing with bosons or fermions (see Appendix C).

Equation (6.64) can be written suitably if we employ the functional integral representation for all the time evolution operators of its integrand. We can write

$$\mathcal{J}(x, y, t; x', y', 0) = \int_{x'}^{x} \int_{y'}^{y} \mathcal{D}x(t') \, \mathcal{D}y(t') \, \mathcal{F}[x, y] \, \exp \frac{i}{\hbar} \left(S_0[x] - S_0[y] \right),$$

$$(6.65)$$

where

$$S_0[x] = \int_0^t dt' \frac{1}{2} M \dot{x}^2(t') \qquad (6.66)$$

is the action of the particle if it were not coupled to the environment. $\mathcal{F}[x, y]$, the well-known influence functional, is now represented by

$$\mathcal{F}[x, y] = \int \dots \int d\mu(\boldsymbol{\alpha}) \, d\mu(\boldsymbol{\alpha}') \, d\mu(\boldsymbol{\gamma}') \, \rho_R(\boldsymbol{\alpha}'^*, \boldsymbol{\gamma}')$$

$$\times \int_{\boldsymbol{\alpha}'}^{\boldsymbol{\alpha}^*} \mathcal{D}\mu(\boldsymbol{\alpha}) \int_{\boldsymbol{\gamma}'^*}^{\boldsymbol{\alpha}} \mathcal{D}\mu(\boldsymbol{\gamma}) \exp \frac{i}{\hbar} \left(S_{RI}[x, \boldsymbol{\alpha}] - S_{RI}^*[y, \boldsymbol{\gamma}^*] \right), \quad (6.67)$$

with

$$S_{RI}[x, \boldsymbol{\alpha}] = \int_0^t dt' \left(\frac{i\hbar}{2} \sum_n (\alpha_n^* \dot{\alpha}_n - \alpha_n \dot{\alpha}_n^*) \right.$$

$$\left. + \dot{x} \sum_{m,n} \hbar g_{mn} \alpha_m^* \alpha_n - \sum_n (\hbar \Omega_n - \mu) \alpha_n^* \alpha_n \right), \qquad (6.68)$$

---

[3] Actually, this is not mandatory. We can still use other representations, but certainly the coherent state representation is the most economical in our case.

$$\rho_R(\boldsymbol{\alpha}'^*, \boldsymbol{\gamma}') = \frac{\exp\left\{e^{-\beta(\hbar\Omega_n - \mu)}\alpha_n'^* \gamma_n'\right\}}{Z}, \tag{6.69}$$

and

$$Z = \int d\mu(\boldsymbol{\alpha}) \exp\left\{e^{-\beta(\hbar\Omega_n - \mu)}|\alpha_n|^2\right\}. \tag{6.70}$$

Notice that above, in (6.67), since $S_{RI}[y, \gamma]$ is a complex functional of the complex function $\gamma(t)$, we have used the fact that $(f[z])^* = f^*[z^*]$. The functional integration measure $\mathcal{D}\mu(*)$ introduced in (6.67) is also defined in Appendix C.

The integral in (6.67) has been evaluated elsewhere in the context of the polaron dynamics (Castro Neto and Caldeira, 1992). The only difference is that here we must deal with the environmental particles, which are massive bosons or fermions (see Appendix C). The result of the integration, apart from a multiplicative time-dependent function which can be determined by normalization, reads

$$\mathcal{F}[x, y] = [\det(1 \mp \bar{N}\Gamma[x, y])]^{\mp 1}, \tag{6.71}$$

where the upper (lower) sign refers to bosons (fermions),

$$\bar{N}_{ij} = \delta_{ij}\bar{n}_i,$$

$$\bar{n}_i = \frac{1}{e^{\beta(\Omega_i - \mu)} \mp 1}, \tag{6.72}$$

and the functional $\Gamma[x, y]$ is such that

$$\Gamma_{nm}[x, y] = W_{nm}[x] + W_{nm}^*[y] + \sum_k W_{nk}^*[y]W_{km}[x], \tag{6.73}$$

with W_{nm} satisfying the integral equation

$$W_{nm}[x, \tau] = \int_0^\tau dt' \, W_{nm}^{(0)}(\dot{x}, t') + \sum_k \int_0^\tau dt' \int_0^{t'} dt'' \, W_{nk}^{(0)}(\dot{x}, t')W_{km}(\dot{x}, t''), \tag{6.74}$$

where

$$W_{nm}^{(0)}(t') = ig_{nm}\dot{x}(t')e^{i(\Omega_n - \Omega_m)t'}(1 - \delta_{nm}). \tag{6.75}$$

In order to evaluate the double functional integral in (6.65) we have to perform some approximations in the expression (6.71) for $\mathcal{F}[x, y]$, which was so far exact. From (6.74) and (6.75), we see that $\Gamma[x, y]$ depends on the velocities $\dot{x}$ and $\dot{y}$ and then, assuming that only slow paths will contribute to (6.65), we expand (6.71) up to quadratic terms in $\dot{x}$ and $\dot{y}$.

After some tedious algebraic steps we reach

$$\mathcal{F}[x, y] = \exp -\frac{i}{\hbar} \Phi_I[x, y] \exp -\frac{1}{\hbar} \Phi_R[x, y], \tag{6.76}$$

where

$$\Phi_I = \int_0^t dt' \int_0^{t'} dt'' (\dot{x}(t') - \dot{y}(t')) \hbar \Gamma_I(t' - t'')(\dot{x}(t'') + \dot{y}(t'')) \tag{6.77}$$

and

$$\Phi_R = \int_0^t dt' \int_0^{t'} dt'' (\dot{x}(t') - \dot{y}(t')) \hbar \Gamma_R(t' - t'')(\dot{x}(t'') - \dot{y}(t'')), \tag{6.78}$$

with

$$\Gamma_R(t) = \frac{1}{2} \sum_{i,j} |g_{ij}|^2 (\bar{n}_i + \bar{n}_j \pm 2\bar{n}_i\bar{n}_j) \cos(\Omega_i - \Omega_j)t, \tag{6.79}$$

where $+ (-)$ applies to bosons (fermions) and

$$\Gamma_I(t) = \frac{1}{2} \sum_{i,j} |g_{ij}|^2 (\bar{n}_i - \bar{n}_j) \sin(\Omega_i - \Omega_j)t. \tag{6.80}$$

Performing integrations by parts over t' in (6.77) and over t' and t'' in (6.78) we recover a superpropagator of the same form as (6.22), apart from transients that depend on the endpoints $x(t)$, $y(t)$, $x(0)$, and $y(0)$ (which we hope to have convinced the reader can be safely neglected). From that alternative form we can determine the damping function

$$\gamma(t) = -\frac{\hbar}{M} \frac{d\Gamma_I(t)}{dt}$$

$$= -\frac{\hbar}{2M} \int d\omega \int d\omega' S(\omega, \omega')[\bar{n}(\omega) - \bar{n}(\omega')](\omega - \omega') \cos(\omega - \omega')t$$

$$\tag{6.81}$$

and the diffusion function

$$D(t) = -\hbar \frac{d^2\Gamma_R}{dt^2}$$

$$= \frac{\hbar^2}{2} \int d\omega \int d\omega' S(\omega, \omega')[\bar{n}(\omega) + \bar{n}(\omega') \pm 2\bar{n}(\omega)\bar{n}(\omega')](\omega - \omega')^2 \cos(\omega - \omega')t.$$

$$\tag{6.82}$$

In the expressions for $\gamma(t)$ and $D(t)$ we have introduced the scattering function $S(\omega, \omega')$, which we define as

$$S(\omega, \omega') = \sum_{i,j} |g_{ij}|^2 \delta(\omega - \Omega_i)\delta(\omega' - \Omega_j). \qquad (6.83)$$

The function $S(\omega, \omega')$ introduced above plays a role very similar to $J(\omega)$ in the minimal model. We can either model it in such a way that given dissipative behavior can be recovered or extract its specific form from a more microscopic approach as in Caldeira and Castro Neto (1995), where the authors dealt with the problem of a heavy particle coupled through a repulsive delta potential to bosons or fermions in one dimension. The results obtained therein for fermions are in agreement with the so-called "piston model for friction" (Gross, 1975) and also with the linear response approach of d'Agliano *et al.* (1975). However, it fails to properly describe the more exact results of Castella and Zotos (1993), Castro Neto and Fisher (1996) for very low temperatures and we believe this is due to the Gaussian approximation we have employed to our influence functional (6.76), which is not appropriate for one-dimensional fermionic environments (see, for example, Caldeira and Castro Neto (1995), Castro Neto and Fisher (1996)).

6.2 The equilibrium reduced density operator

Despite the fact that we have developed an expression for the real-time evolution of $\tilde{\rho}(x, y, t)$, there are several problems for which only knowledge of the reduced density operator in equilibrium is necessary. It is also true that this operator can be obtained by taking the limit $t \to \infty$ of our previous time evolution, but this task might sometimes be fairly laborious. However, there is a shortcut we can take by evaluating its functional integral representation directly in terms of the system variables only. The latter will prove very useful since it can be handled by some very suitable techniques.

Defining $\beta \equiv 1/k_B T$, the density operator for the composite system in equilibrium reads

$$\langle x\mathbf{R}|e^{-\beta\mathcal{H}}|y\mathbf{Q}\rangle = \rho(x, \mathbf{R}; y, \mathbf{Q}, \beta), \qquad (6.84)$$

where $\mathcal{H}$ is the Hamiltonian of the complete system, $\mathcal{H} = \mathcal{H}_S + \mathcal{H}_I + \mathcal{H}_R + \mathcal{H}_{CT}$, as in (4.43).

We are going to treat the problem employing only the minimal model because, as we have already shown, the other two models with which we have been dealing reduce to the former under very general circumstances. Actually, they reproduce the same results as the minimal model for most of the cases in which we are interested. However, if we want or need to pursue this issue further we refer the reader to Valente and Caldeira (2010) for the case of more than one Brownian particle, or Caldeira and Castro Neto (1995) for the collision model.

The operator in (6.84) also admits a functional integral representation (Feynman and Hibbs, 1965) that reads

$$\rho(x, \mathbf{R}; y, \mathbf{Q}, \beta) = \int_y^x \int_{\mathbf{Q}}^{\mathbf{R}} \mathcal{D}q(\tau)\mathcal{D}\mathbf{R}(\tau)\exp -\frac{1}{\hbar}S_E[q(\tau), \mathbf{R}(\tau)], \quad (6.85)$$

where

$$S_E[q(\tau), \mathbf{R}(\tau)] = \int_0^{\hbar\beta} d\tau \left\{ \frac{1}{2}M\dot{q}^2 + V(q) \right.$$

$$\left. + \sum_k \left(C_k q R_k + \frac{1}{2}m_k\dot{R}_k^2 + \frac{1}{2}m_k\omega_k^2 R_k^2 + \frac{C_k^2 q^2}{2m_k\omega_k^2} \right) \right\} \quad (6.86)$$

is the so-called "Euclidean action" of the composite system. The reduced density operator of the particle is then

$$\tilde{\rho}(x, y, \beta) \equiv \int d\mathbf{R}\, \rho(x, \mathbf{R}; y, \mathbf{R}, \beta)$$

$$= \int_y^x \mathcal{D}q(\tau) \int d\mathbf{R} \int_{\mathbf{R}}^{\mathbf{R}} \mathcal{D}\mathbf{R}(\tau)\exp -\frac{1}{\hbar}S_E[q(\tau), \mathbf{R}(\tau)],$$

$$(6.87)$$

which involves a product of functional integrals of the form

$$A_k(\beta) \equiv \int d R_k \int_{R_k}^{R_k} \mathcal{D}R_k(\tau)$$

$$\times \exp -\frac{1}{\hbar}\left\{ \int_0^{\hbar\beta} d\tau \left[\frac{1}{2}m_k\dot{R}_k^2 + \frac{1}{2}m_k\omega_k^2 R_k^2 + C_k q R_k \right] \right\}. \quad (6.88)$$

These integrals are exactly the same as those we have dealt with in (6.5) and (6.6) if we replace t by $-i\hbar\beta$ and make $R_k = R_k'$. The final result is

$$A_k(\beta) = I_k(0)\exp \frac{\lambda_k}{\hbar} \int_0^{\hbar\beta} d\tau \int_0^{\hbar\beta} d\tau' \frac{q(\tau)q(\tau')\cosh\omega_k(|\tau - \tau'| - \hbar\beta/2)}{\sinh(\hbar\beta\omega_k/2)} \quad (6.89)$$

where

$$I_k(0) = \frac{1}{2}\operatorname{cosech}\left(\frac{\hbar\beta\omega_k}{2} \right) \quad \text{and} \quad \lambda_k = \frac{C_k^2}{4m_k\omega_k}. \quad (6.90)$$

Expression (6.89) acquires an extremely simple form if we define $q(\tau)$ outside the domain $0 \leq \tau < \hbar\beta$ by imposing that $q(\tau + \hbar\beta) = q(\tau)$ (Feynman, 1998). Then,

$$A_k(\beta) = I_k(0) \exp\left\{\frac{\lambda_k}{\hbar} \int_{-\infty}^{+\infty} d\tau' \int_0^{\hbar\beta} d\tau q(\tau)q(\tau') \exp -(\omega_k|\tau - \tau'|)\right\}. \quad (6.91)$$

Using this result in (6.87) we have

$$\tilde{\rho}(x, y, \beta) = \tilde{\rho}_0(\beta) \int_y^x \mathcal{D}q(\tau) \exp -\frac{1}{\hbar}\tilde{S}_E^{(0)}[q(\tau)] \exp\frac{\Lambda[q(\tau)]}{\hbar} \quad (6.92)$$

where $\tilde{\rho}_0(\beta) = \prod_k I_k(0)$,

$$\tilde{S}_E^{(0)}[q(\tau)] = S_E^{(0)}[q(\tau)] + S_{CT}[q(\tau)] = \int_0^{\hbar\beta} d\tau\left\{\frac{1}{2}M\dot{q}^2 + V(q)\right\} + S_{CT}[q(\tau)],$$
$$(6.93)$$

and

$$\Lambda[q(\tau)] = \sum_k \lambda_k \int_{-\infty}^{+\infty} d\tau' \int_0^{\hbar\beta} d\tau q(\tau)q(\tau')\exp -\omega_k|\tau - \tau'|. \quad (6.94)$$

Now, completing squares in the integrand of (6.94) we cancel the counter-term contribution from the Euclidean action and get

$$\tilde{\rho}(x, y, \beta) = \tilde{\rho}_0(\beta) \int_y^x \mathcal{D}q(\tau) \exp -\frac{1}{\hbar}S_{eff}[q(\tau)], \quad (6.95)$$

where

$$S_{eff}[q(\tau)] = \int_0^{\hbar\beta} d\tau\left\{\frac{1}{2}M\dot{q}^2 + V(q)\right\}$$
$$+ \frac{1}{2} \int_{-\infty}^{+\infty} d\tau' \int_0^{\hbar\beta} d\tau\, \alpha(\tau - \tau')\{q(\tau) - q(\tau')\}^2 \quad (6.96)$$

and

$$\alpha(\tau - \tau') \equiv \sum_k \lambda_k\exp -\omega_k|\tau - \tau'| = \frac{1}{2\pi} \int_0^{\infty} d\omega\, J(\omega) \exp -\omega_k|\tau - \tau'|. \quad (6.97)$$

In (6.97) we have transformed the sum into an integral with the help of (5.14), as usual. For the specific case of ohmic dissipation, this becomes

$$\alpha(\tau - \tau') = \frac{1}{2\pi} \int_0^\infty d\omega \, \eta\omega \exp - \omega_k |\tau - \tau'| = \frac{\eta}{2\pi} \frac{1}{(\tau - \tau')^2}. \qquad (6.98)$$

In conclusion, we have been able to write the reduced density operator (equilibrium or dynamical) for the particle of interest in the path integral representation. Now, it is our goal to try to solve those integrals for some particular cases.

7

The damped harmonic oscillator

In this section we will address two specific problems of the motion of wave packets in the classically accessible region, namely the time evolution of a single Gaussian packet and the destruction of quantum interference between two packets (*decoherence*), both in a harmonic potential.

7.1 Time evolution of a Gaussian wave packet

This first application is the simplest one. We want to follow the time evolution of a harmonic oscillator coupled to our standard environment. We thus initially prepare a particle of mass M subject to a potential $V(q) = M\omega_0^2 q^2/2$ in a pure state given by the wave packet

$$\psi(x') = \frac{1}{(2\pi\sigma^2)^{1/4}} \exp\frac{ip\,x'}{\hbar} \exp-\frac{x'^2}{4\sigma^2} \tag{7.1}$$

and allow it to interact with a bath of oscillators which is in equilibrium at temperature T. This wave packet is shown in Fig. 7.1 and has finite initial average momentum p. The first step of the calculation is to evaluate the specific form of the superpropagator $\mathcal{J}(x, y, t; x', y', 0)$ for the harmonic potential introduced above. Although the integrations might involve non-local kernels in time, they are all Gaussian and can easily be performed (Caldeira and Leggett, 1983b). For this particular case, (6.21) reads

$$\mathcal{J}(x, x', t; y, y', 0) = \int\limits_{x'}^{x} \int\limits_{y'}^{y} \mathcal{D}x(t')\mathcal{D}y(t') \, \exp\frac{i}{\hbar}\tilde{S}[x(\tau), y(\tau)]$$

$$\times \exp-\frac{1}{\hbar}\phi[x(\tau), y(\tau)] \tag{7.2}$$

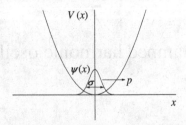

Figure 7.1 Gaussian wave packet

where

$$\tilde{S}[x(\tau), y(\tau)] = \int\limits_0^t \tilde{L}(x, \dot{x}, y, \dot{y})\, d\tau - M\gamma \int\limits_0^t (x\dot{x} - y\dot{y})dt', \qquad (7.3)$$

$$\tilde{L}(x, \dot{x}, y, \dot{y}) = \frac{1}{2}M\dot{x}^2 - \frac{1}{2}M\dot{y}^2 - \frac{1}{2}M\omega_0^2 x^2 + \frac{1}{2}M\omega_0^2 y^2 - M\gamma x\dot{y} + M\gamma y\dot{x}, \quad (7.4)$$

and

$$\phi[x(\tau), y(\tau)] = \exp{-\frac{2M\gamma}{\pi\hbar}} \int\limits_0^{\Omega} d\nu\, \nu\, \coth\frac{\hbar\nu}{2k_B T}$$

$$\times \int\limits_0^t \int\limits_0^{\tau} d\tau d\sigma\, [x(\tau) - y(\tau)] \cos \nu(\tau - \sigma)[x(\sigma) - y(\sigma)]\, d\tau\, d\sigma.$$

$$(7.5)$$

As the path integral above is Gaussian, it can be solved exactly. Let us initially expand its exponent about the paths $x_c(\tau)$ and $y_c(\tau)$ such that

$$\left.\frac{\delta\tilde{S}}{\delta x}\right|_{x=x_c, y=y_c} = \frac{d}{dt}\frac{\partial\tilde{L}}{\partial\dot{x}} - \frac{\partial\tilde{L}}{\partial x} = M\ddot{x}_c + 2M\gamma\dot{y}_c + M\omega_0^2 x_c = 0, \qquad (7.6)$$

$$\left.\frac{\delta\tilde{S}}{\delta y}\right|_{x=x_c, y=y_c} = \frac{d}{dt}\frac{\partial\tilde{L}}{\partial\dot{y}} - \frac{\partial\tilde{L}}{\partial y} = M\ddot{y}_c + 2M\gamma\dot{x}_c + M\omega_0^2 y_c = 0. \qquad (7.7)$$

Defining the paths

$$q(\tau) \equiv \frac{x(\tau) + y(\tau)}{2} \qquad \text{and} \qquad \xi(\tau) \equiv x(\tau) - y(\tau), \qquad (7.8)$$

we can write (7.6) and (7.7) as

$$\ddot{q}_c + 2\gamma\dot{q}_c + \omega_0^2 q_c = 0, \tag{7.9}$$

$$\ddot{\xi}_c - 2\gamma\dot{\xi}_c + \omega_0^2\xi_c = 0. \tag{7.10}$$

Given the initial conditions $q(0) = q'$, $q(t) = q$, $\xi(0) = \xi'$, and $\xi(t) = \xi$, the solutions of these equations are

$$q_c(\tau) = (\sin\omega t)^{-1}\{qe^{\gamma t}\sin\omega\tau + q'\sin\omega(t-\tau)\}e^{-\gamma\tau}, \tag{7.11}$$

$$\xi_c(\tau) = (\sin\omega t)^{-1}\{\xi e^{-\gamma t}\sin\omega\tau + \xi'\sin\omega(t-\tau)\}e^{\gamma\tau}, \tag{7.12}$$

where $\omega \equiv \sqrt{\omega_0^2 - \gamma^2}$ if $\omega_0 > \gamma$ and $\omega \equiv i\sqrt{\gamma^2 - \omega_0^2}$ if $\omega_0 < \gamma$. Notice that with these replacements all the trigonometric functions appearing here for the underdamped motion will become hyperbolic functions for the overdamped motion. Now we define new variables of integration, $\tilde{q}(\tau) \equiv q(\tau) - q_c(\tau)$ and $\tilde{\xi}(\tau) \equiv \xi(\tau) - \xi_c(\tau)$, and rewrite (7.2) as

$$\mathcal{J}(q, \xi, t; q', \xi', 0) = \exp\left\{\frac{i}{\hbar}\tilde{S}_c\right\}\exp-\frac{1}{\hbar}\left\{A(t)\xi^2 + B(t)\xi\xi' + C(t)\xi'^2\right\}$$
$$\times G(q, \xi, t; q', \xi', 0), \tag{7.13}$$

whose terms we analyze in what follows.

The first exponential in (7.13) involves the "classical action," $\tilde{S}_c$, given by

$$\tilde{S}_c = K(t)[q\xi + q'\xi'] - L(t)q'\xi - N(t)q\xi' - M\gamma[q\xi - q'\xi'], \tag{7.14}$$

where

$$K(t) = M\omega\cot\omega t, \quad L(t) = \frac{M\omega e^{-\gamma t}}{\sin\omega t}, \quad \text{and} \quad N(t) = \frac{M\omega e^{\gamma t}}{\sin\omega t}, \tag{7.15}$$

which originates from (7.3) evaluated at $q_c(\tau)$ and $\xi_c(\tau)$.

The second exponential in (7.13) comes from the integration (7.5) performed over the same extreme paths as above and the functions $A(t)$, $B(t)$, and $C(t)$ are all of the form

$$f(t) = \frac{M\gamma}{\pi}\int_0^\Omega dv\, v\, \coth\frac{\hbar v}{2k_B T}\, f_v(t), \tag{7.16}$$

where

$$A_v(t) = \frac{e^{-2\gamma t}}{\sin^2\omega t}\int_0^t\int_0^t \sin\omega\tau\cos v(\tau - \sigma)\sin\omega\sigma\, e^{\gamma(\tau+\sigma)}\, d\tau\, d\sigma, \tag{7.17}$$

$$B_\nu(t) = \frac{2e^{-2\gamma t}}{\sin^2 \omega t} \int_0^t \int_0^t \sin \omega \tau \cos \nu(\tau - \sigma) \sin \omega(t - \sigma) e^{\gamma(\tau+\sigma)} d\tau \, d\sigma, \quad (7.18)$$

$$C_\nu(t) = \frac{1}{\sin^2 \omega t} \int_0^t \int_0^t \sin \omega(t - \tau) \cos \nu(\tau - \sigma) \sin \omega(t - \sigma) e^{\gamma(\tau+\sigma)} d\tau \, d\sigma.$$

$$(7.19)$$

Finally, the expression for $G(q, \xi, t; q', \xi', 0)$ is given by

$$G(q, \xi, t; q', \xi', 0) = \int_0^0 \int_0^0 \mathcal{D}\tilde{q}(\tau)\mathcal{D}\tilde{\xi}(\tau) \exp\frac{i}{\hbar}\tilde{S}[\tilde{q}(\tau), \tilde{\xi}(\tau)]$$

$$\times \exp - \frac{1}{\hbar}\phi_T[\tilde{\xi}(\tau), \tilde{\xi}(\tau)] \exp\frac{2}{\hbar}\phi_T[\xi_c(\tau), \tilde{\xi}(\tau)], \quad (7.20)$$

where the functional ϕ_T is defined as

$$\phi_T[f(\tau), g(\tau)] = \frac{M\gamma}{\pi} \int_0^\Omega d\nu \, \nu \, \coth\frac{\hbar\nu}{2k_B T}$$

$$\times \int_0^t \int_0^t d\tau \, d\sigma \, f(\tau) \cos \nu(\tau - \sigma)g(\sigma) \, d\tau \, d\sigma. \quad (7.21)$$

If it were not for the functional integral (7.20), the expression for $\mathcal{J}(q, \xi, t; q', \xi', 0)$ would already be in the appropriate form for performing the time evolution of the wave packet in this potential. When we treat simple functional integrals, this term is a function of time only, which is not obvious in the case of (7.20) due to its functional dependence on $\xi_c(\tau)$. Nevertheless, we shall show next that this is also the case for that integral.

If we discretize (7.20) we can easily convince ourselves that its symbolic form is

$$N \int \delta U \exp -\frac{1}{2}U^T \mathcal{M} U \exp - \mathcal{A}U, \quad (7.22)$$

where $U^T = (\tilde{q}_1, \ldots, \tilde{q}_N, \tilde{\xi}_1, \ldots, \tilde{\xi}_N)$ is a row vector in $2N$ dimensions, $\delta U = d\tilde{q}_1 \ldots d\tilde{q}_N d\tilde{\xi}_1 \ldots d\tilde{\xi}_N$ is the volume element in this space, $\mathcal{A}$ is another row vector in $2N$ dimensions that comes from the discretization of $\xi_c(\tau)$, and $\mathcal{M}$ is a $2N \times 2N$ matrix. The interesting point about the specific form of (7.20) is that

$$\mathcal{A} = \begin{pmatrix} 0 \\ a \end{pmatrix} \quad (7.23)$$

and

$$M = \begin{pmatrix} 0 & p \\ p & r \end{pmatrix}, \tag{7.24}$$

where a is an N-dimensional vector and p and r are $N \times N$ matrices. The specific forms of $\mathcal{A}$ and $\mathcal{M}$ result from the fact that there is no product like $q(\tau)q(\sigma)$ in (7.20).

The integral (7.25) is a well-known example of multidimensional Gaussian integrals, and results in

$$N \int \delta U \exp -\frac{1}{2} U^T \mathcal{M} U \exp -\mathcal{A} U = \frac{1}{\sqrt{\det \mathcal{M}}} \exp \frac{1}{2} \mathcal{A} \mathcal{M}^{-1} \mathcal{A}. \tag{7.25}$$

But, owing to the fact that $\mathcal{M}$ has its upper left block null, its inverse has its lower right block null and, consequently, the product $\mathcal{A} \mathcal{M}^{-1} \mathcal{A} = 0$. As we know that the whole dependence of (7.20) on q, ξ, q', and ξ' is in $\mathcal{A}$, we conclude that $G(q, \xi, t; q', \xi', 0)$ is only a function of time. This function can be computed directly from the evaluation of $\det i\mathcal{M}$ or determined at the end of the calculation by the normalization condition of the reduced density operator. In order to simplify our computations we adopt the latter procedure. Then, we can rewrite (7.13) as

$$\mathcal{J}(q, \xi, t; q', \xi', 0) = G(t) \exp \frac{i}{\hbar} \Big\{ [K(t) - M\gamma]q\xi + [K(t) + M\gamma]q'\xi'$$

$$- L(t)q'\xi - N(t)q\xi' \Big\} \exp -\frac{1}{\hbar} \Big\{ A(t)\xi^2 + B(t)\xi\xi' + C(t)\xi'^2 \Big\}, \tag{7.26}$$

and determine $G(t)$ later on.

In terms of the new variables q, ξ, q', and ξ', the reduced density operator of the particle can be written as

$$\tilde{\rho}(q, \xi, t) = \int \int dq' d\xi' \, \mathcal{J}(q, \xi, t; q', \xi', 0) \, \tilde{\rho}(q', \xi', 0), \tag{7.27}$$

where $\tilde{\rho}(q', \xi', 0)$ is its initial state. In our specific example we can write it from our pure state (7.1) as $\psi(x')\psi^*(y')$ or

$$\tilde{\rho}(q', \xi', 0) = \frac{1}{(2\pi\sigma^2)^{1/2}} \exp \frac{ip\,\xi'}{\hbar} \exp -\frac{q'^2}{2\sigma^2} \exp -\frac{\xi'^2}{8\sigma^2}. \tag{7.28}$$

Inserting (7.28) in (7.27) and using (7.26), we get after some labor

$$\tilde{\rho}(q, \xi, t) = G(t) \sqrt{\frac{\pi \hbar^2}{N^2(t)\sigma^2(t)}} \exp -\frac{1}{2\sigma^2(t)}(q - x_0(t))^2$$

$$\times \exp -F(t)\xi^2 \exp \frac{i}{\hbar} D(q, p, t)\xi, \tag{7.29}$$

where

$$\sigma^2(t) = \frac{\sigma^2 K_1^2(t) + 2\hbar C_1^2(t)}{N^2(t)}, \qquad x_0(t) = \frac{p}{N(t)}, \qquad (7.30)$$

$$F(t) = \frac{A(t)}{\hbar} + \frac{\sigma^2 L^2(t)}{2\hbar^2} - \frac{(\sigma^2 K_1(t)L(t) - \hbar B(t))^2}{2\hbar^2\sigma^2(t)N^2(t)}, \qquad (7.31)$$

$$D(q, p, t) = K_2(t)q - \frac{(\sigma^2 K_1(t)L(t) - \hbar B(t))}{\sigma^2(t)N(t)}(q - x_0(t)), \qquad (7.32)$$

and

$$C_1(t) \equiv C(t) + \frac{\hbar}{8\sigma^2}, \qquad K_1(t) \equiv K(t) + M\gamma, \qquad K_2(t) \equiv K(t) - M\gamma. \quad (7.33)$$

The time evolution of the probability density associated with the particle is obtained by making $\xi = 0$ in (7.29), which means $\tilde{\rho}(q, 0, t) = \tilde{\rho}(x, x, t)$, the diagonal elements of $\tilde{\rho}(x, y, t)$. Its final expression after normalization must be

$$\tilde{\rho}(x, x, t) = \left(\frac{1}{2\pi\sigma^2(t)}\right)^{1/2} \exp -\frac{1}{2\sigma^2(t)}(x - x_0(t))^2, \qquad (7.34)$$

from which we can identify the time evolution of its center as $x_0(t)$, and width as $\sigma(t)$. The normalization of (7.34) has been used to determine

$$G(t) = \frac{N(t)}{\sqrt{2\pi\hbar}}. \qquad (7.35)$$

In the specific case of an underdamped oscillator ($\omega_0 > \gamma$), for example, it can be shown (Caldeira and Leggett, 1983b) that

$$x_0(t) = \frac{p}{M\omega} \sin \omega t \, e^{-\gamma t} \qquad \text{where} \qquad \omega = \sqrt{\omega_0^2 - \gamma^2}, \qquad (7.36)$$

which is exactly the classical trajectory of a harmonic particle with initial momentum p and position zero, which are clearly the initial conditions obtained from (7.1).

In the limit $t \to \infty$, the width of the packet (7.30) reduces to

$$\sigma^2(\infty) = \frac{\hbar}{\pi} \int_0^\infty dv \coth \frac{\hbar v}{2k_B T} \left(\frac{1}{M} \frac{2\gamma v}{\left(\omega_0^2 - v^2\right)^2 + 4\gamma^2 v^2}\right). \qquad (7.37)$$

Recognizing the term inside parentheses in the integrand as the imaginary part of the susceptibility of a damped oscillator, $\chi''(v)$, we rewrite the above expression as

$$\sigma^2(\infty) = \frac{\hbar}{\pi} \int_0^\infty dv \coth \frac{\hbar v}{2k_B T} \chi''(v), \qquad (7.38)$$

which is nothing but the celebrated *fluctuation–dissipation theorem* (see, for example, Forster (1990)). Now, just for general information, we see that the width of the probability density of the damped oscillator in equilibrium at $T = 0$ behaves as

$$\sigma^2(\infty) = \frac{\hbar}{2M\omega_0} f(\alpha) \qquad \left(\alpha \equiv \frac{\gamma}{\omega_0} \right), \qquad (7.39)$$

where

$$f(\alpha) = \begin{cases} \frac{1}{\sqrt{1-\alpha^2}} \left(1 - \frac{2}{\pi} \tan^{-1} \frac{\alpha}{\sqrt{1-\alpha^2}} \right) & \text{if } \alpha < 1 \\[3mm] \frac{1}{\sqrt{\alpha^2-1}} \frac{1}{\pi} \ln \left| \frac{\alpha+\sqrt{\alpha^2-1}}{\alpha-\sqrt{\alpha^2-1}} \right| & \text{if } \alpha > 1. \end{cases} \qquad (7.40)$$

This last expression shows us that the width of the packet always decreases with increasing dissipation, which is a signature of a more efficient localization of a particle in a dissipative environment. Consequently, we must have an increase in the uncertainty of the momentum distribution of the particle. This is characteristic of the kind of dissipation we have in the equation of motion of that particle. In the Langevin equation the dissipative term is such that $dE/dt \propto \dot{q}^2$, which originates from the coupling of the form $\sum_k C_k q_k q$ with a specific choice for the spectral function $J(\omega)$. When the dissipative term is such that $dE/dt \propto \dot{p}^2$, the effect on the uncertainties of position and momentum is opposite to the previous one. This is the so-called *anomalous dissipation* and was treated explicitly in Leggett (1984).

With (7.29) we have all the information needed to study the damped harmonic oscillator at any temperature. Moreover, we can also use the same expression to study the damped motion of a particle in other potentials. A particularly important one is the motion of a free Brownian particle ($V(q) = 0$), which we describe by taking the limit $\omega_0 \to 0$ and $\omega \to i\gamma$ in (7.29). Notice that in this case some care must be exercised when treating $\sigma(t)$ because it now spreads out, since the motion is no longer bounded to a fixed region in space.

Very weakly damped case. Although (7.29) indeed allows us to describe the solution of the damped quantum oscillator in very general circumstances, there are situations in which simpler expressions are more useful for our needs. This is the case, for example, of the extremely underdamped motion ($\omega_0 \gg \gamma$). In this case the path integral describing the superpropagator $\mathcal{J}(x, y, t; x', y', 0)$ can also be cast into an instantaneous form exactly as in the high-temperature limit (6.23) (Caldeira *et al.*, 1989). This can be seen if we evaluate one of the double time integrals in (7.17)–(7.19) and take the limit $\gamma \to 0$ of the resulting expressions. When this is done, we can show that $A_\nu(t)$, $B_\nu(t)$, and $C_\nu(t)$ all become proportional to $\delta(\nu - \omega_0)$ with multiplicative factors $g^{(\alpha)}(t)$ ($\alpha = A, B,$ or C) given by

$$g^{(A)}(t) = \frac{\pi}{\sin^2 \omega_0 t} \int_0^t d\tau \sin^2 \omega_0 \tau,$$

$$g^{(B)}(t) = \frac{2\pi}{\sin^2 \omega_0 t} \int_0^t d\tau \sin \omega_0 \tau \sin \omega_0 (t - \tau),$$

$$g^{(C)}(t) = \frac{\pi}{\sin^2 \omega_0 t} \int_0^t d\tau \sin^2 \omega_0 (t - \tau), \tag{7.41}$$

which if taken to (7.16) give

$$\left\{ \begin{array}{c} A(t) \\ B(t) \\ C(t) \end{array} \right\} = M\gamma \, \omega_0 \coth \frac{\hbar \omega_0}{2k_B T} \, g^{(\alpha)}(t), \tag{7.42}$$

where $\alpha = A$, B, or C. With this form for $A(t)$, $B(t)$, and $C(t)$, expression (7.26) is the same as we could have obtained had we started with (7.5) replaced by

$$\Phi[x(t'), y(t')] = \frac{1}{\hbar} M\gamma \, \omega_0 \coth \frac{\hbar \omega_0}{2k_B T} \int_0^t dt' [x(t') - y(t')]^2, \tag{7.43}$$

leading us to an instantaneous master equation (see Appendix B) with a diffusion constant given by $D(T) = M\gamma \hbar \omega_0 \coth(\hbar \omega_0 / 2k_B T)$ which, in the high-temperature limit ($\hbar \omega_0 \ll k_B T$), recovers $D(T) = 2M\gamma k_B T$. Therefore, we have reached the conclusion that for either $\hbar \omega_0 \ll k_B T$ (for any γ) or $\gamma \ll \omega_0$ (for any T) we can describe the quantum dynamics of a Brownian oscillator by an instantaneous quantum master equation as in (6.24) and (6.25).

In order to conclude this section we would like to make some additional comments. In the first place, we wish to stress that the model we have used for the environment allowed us to develop exactly the influence functional (6.4) of our particular system. However, the resulting functional integral (7.2) can only be evaluated analytically for potentials of the form $V(q) = Cq^n$, where $n = 0, 1, 2$. For these values of n, (7.2) is a Gaussian functional integral and, therefore, exactly soluble. The paths $q_c(\tau)$ and $\xi_c(\tau)$ were chosen only for convenience, because equations (7.3) and (7.10) are independent of that choice.

In the case of more complex potentials, we could employ approximate methods to solve that integral such as, for instance, the stationary phase (or saddle point) method, although this approximation is only a good one in the semi-classical limit ($\hbar \to 0$). The evaluation of integrals like (7.2) for more complicated potentials will appear in future specific applications.

7.2 Time evolution of two Gaussian packets: Decoherence

In this application we repeat exactly the same steps as before, with the difference that now the initial state of the oscillator is given by (see Fig. 7.2)

$$\psi(x) = \psi_1(x) + \psi_2(x) = \tilde{\mathcal{N}}\left[\exp-\frac{x^2}{4\sigma^2} + \exp-\frac{(x+x_0)^2}{4\sigma^2}\right], \qquad (7.44)$$

where $\tilde{\mathcal{N}}$ is a normalization constant. From this state we write the initial density operator of the particle as

$$\rho(x', y', 0) = \rho_1(x', y', 0) + \rho_2(x', y', 0) + \rho_{int}(x', y', 0), \qquad (7.45)$$

where $\rho_{int}(x', y', 0) = \psi_2(x')\psi_1^*(y') + \psi_1(x')\psi_2^*(y')$. The first (second) term on the r.h.s. of (7.45) represents the wave packet centered at the origin (position $-x_0$) whereas the last term represents the interference between these two packets. When $x_0 \gg \sigma$ the latter term is negligible.

It is our intention to study the time evolution of $\tilde{\rho}(x, x, t) \equiv \tilde{\rho}(x, t)$ of the dissipative system as before. As the time evolution is linear, we have

$$\tilde{\rho}(x, t) = \tilde{\rho}_1(x, t) + \tilde{\rho}_2(x, t) + \tilde{\rho}_{int}(x, t), \qquad (7.46)$$

where $\tilde{\rho}_1(x, t)$ and $\tilde{\rho}_2(x, t)$ are the probability densities corresponding to the time evolution of the initial wave packets ψ_1 and ψ_2, respectively, and $\tilde{\rho}_{int}(x, t)$ is obtained by

$$\tilde{\rho}_{int}(x, t) = \int dx' dy' \, \mathcal{J}(x, x, t; x', y', 0) \tilde{\rho}_{int}(x', y', 0). \qquad (7.47)$$

Evaluating (7.47) for the undamped case ($\gamma = 0$), we have

$$\rho_{int}(x, t) = 2\sqrt{\rho_1(x, t)\rho_2(x, t)} \cos \phi(x, t). \qquad (7.48)$$

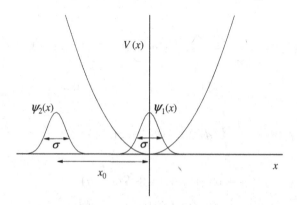

Figure 7.2 Delocalized initial state

This expression tells us that the interference term is more pronounced when the two packets overlap. In particular, when $t = n\pi/\omega_0 + \pi/2\omega_0$ we have

$$\rho_{int}(x, t = n\pi/\omega_0 + \pi/2\omega_0) = \cos\left(\frac{q_0}{\sigma^2}x\right)\exp-\frac{x^2}{\sigma^2}. \tag{7.49}$$

This expression, together with (7.45), reminds us of the interference pattern which results from a double-slit experiment where

$$I = I_1 + I_2 + 2\sqrt{I_1 I_2}\cos\phi(x, t), \tag{7.50}$$

with $I_1(I_2)$ being the intensity due to slit 1(2) and $\phi(x, t)$ a function of a linear distance on the screen. Now, let us return to the dissipative problem.

Evaluating (7.47) one reaches

$$\tilde{\rho}_{int}(x, t) = 2\sqrt{\tilde{\rho}_1(x, t)}\sqrt{\tilde{\rho}_2(x, t)}\cos\phi(x, t)\exp-f(t), \tag{7.51}$$

where $\phi(x, t)$ is a function that determines the interference pattern and $\exp -f(t)$ is an attenuation factor for the intensity of this pattern. An interesting question is to compare the time scale for the destruction of the interference term with the relaxation time of the system (γ^{-1}). Since the two wave packets relax to equilibrium within a time interval of the order of γ^{-1}, we would be tempted to think that this is also the time scale for the disappearance of the interference term. We will see next that this is not true.

In Caldeira and Leggett (1985) it is shown that in terms of the variables

$$\alpha \equiv \frac{\gamma}{\omega_0}, \qquad S \equiv \frac{\omega}{\omega_0}, \qquad \theta \equiv \omega_0 t, \qquad \kappa \equiv \frac{\hbar\omega_0}{2k_B T}, \qquad \lambda_c \equiv \frac{\Omega}{\omega_0}, \tag{7.52}$$

and the functions

$$q(\theta) = q_0\left[\frac{\alpha\sin S\theta}{S} + \cos S\theta\right]\exp-\alpha\theta,$$

$$C_R(\theta, \lambda) \equiv \frac{1}{\sin^2 S\theta}\int_0^\theta\int_0^\theta\sin\left[S(\theta - \theta_1)\right]\cos\left[\lambda(\theta_1 - \theta_2)\right]$$
$$\times\sin\left[S(\theta - \theta_2)\right]\exp\left[\alpha(\theta_1 + \theta_2)\right]d\theta_1\,d\theta_1,$$

$$I_R(\theta) = \frac{4}{\pi}\int_0^{\lambda_c}d\lambda\,\lambda\,C_R(\theta, \lambda)\coth(\kappa\lambda),$$

$$Q(\theta) \equiv 1 + \alpha I_R(\theta) + (\alpha + S\cot S\theta)^2,$$

$$\sigma^2(\theta) = \frac{\sigma^2 Q(\theta)\sin^2(S\theta)\exp-(2\alpha\theta)}{S^2}, \tag{7.53}$$

the attenuation factor becomes

$$f(t) = \frac{q_0^2 \alpha I_R(\theta)}{8\sigma^2 Q(\theta)}, \tag{7.54}$$

and it is the analysis of this term that we are aiming at.

Before we embark on this project it is worth noticing that those terms in (7.53) came from the integration (7.47) for the underdamped motion ($\gamma < \omega_0$ and $\omega = \sqrt{\omega_0^2 - \gamma^2}$). When $\gamma > \omega_0$ we must remember that $\omega = i\sqrt{\gamma^2 - \omega_0^2}$ and all the expressions therein must be modified accordingly.

For both the underdamped and overdamped motions the function $g(\theta) \equiv \alpha I_R(\theta)/Q(\theta)$ is such that $g(0) = 0$ and $g(\infty) = 1$. Consequently, the attenuation factor tends to $\exp -(N/2)$ when $t \to \infty$, where $N \equiv q_0^2/4\sigma^2$ (notice that N is the average number of quanta of energy $\hbar\omega_0$ present in the initial state of the system). As we are interested in situations such that $q_0 >> \sigma$, we have $N >> 1$ and therefore the attenuation factor vanishes at long times. As a matter of fact, the residual interference $\exp -(N/2)$, still present when $t \to \infty$, is due to the fact that we have disregarded the initial overlap of the packets in the normalization of $\tilde{\rho}(x', y', 0)$. If we take this correction into account, this anomalous residual contribution must disappear.

The analysis of $g(\theta)$ that follows involves a very careful examination of its behavior in several different limits of interest. In this section we shall be interested in studying the destruction of quantum interference, or *decoherence*, in the limits of low and high temperatures for both the underdamped and overdamped motions.

The time scale for decoherence can readily be obtained by linearizing $g(\theta)$, which yields

$$\exp -f(t) \approx \exp -\Gamma t, \tag{7.55}$$

where the general behavior of the so-called *decoherence rate* Γ is illustrated below:

$$\Gamma = \begin{cases} \text{high temperatures} \quad (\kappa \ll 1) \begin{cases} \frac{2Nk_BT}{\hbar\omega_0}\gamma & \text{if } \gamma \ll \omega_0 \\ \frac{2Nk_BT}{\hbar\omega_0}\frac{\omega_0^2}{2\gamma} & \text{if } \gamma \gg \omega_0 \end{cases} \\ \\ \text{low temperatures} \quad (\kappa \gg 1) \begin{cases} N\gamma & \text{if } \gamma \ll \omega_0 \\ N\frac{\omega_0^2}{2\gamma} & \text{if } \gamma \gg \omega_0. \end{cases} \end{cases} \tag{7.56}$$

So, for wave packets initially prepared quite far apart from one another, Γ is always given by the relaxation constant of the packet (γ if $\gamma << \omega_0$ or $\omega_0^2/2\gamma$ if $\gamma >> \omega_0$) multiplied by a function of the temperature (N if $k_BT << \hbar\omega_0$ or $2Nk_BT/\hbar\omega_0$ if $k_BT >> \hbar\omega_0$). As we are treating the case $N >> 1$, we see that the time for decoherence can be much shorter than the relaxation time of the packets. Depending on the initial preparation of the state of the oscillator, this time can be even shorter than the natural period of oscillation of the particle. If so, the two packets would behave as two moving classical probability distributions[1] in damped harmonic motion. The introduction of the dissipative medium turned the system "more classical".

This result can easily be understood through the following analysis. Suppose that the initial state of the composite system can be approximated by

$$|\phi_1\rangle \approx \{|\psi_0\rangle + |\psi_z\rangle\} \otimes |0\rangle, \tag{7.57}$$

where $|\psi_0\rangle + |\psi_z\rangle$ is the initial state of the system and $|0\rangle$ the ground state of the environment, which we assume to be initially at zero temperature. Moreover, since $|\psi_z\rangle$ is displaced from the origin, the initial state of the system contains an average number N of quanta of energy.

After a time interval τ (the relaxation time of the system), the composite system will be in the state

$$|\phi_f\rangle = |\psi_0\rangle \otimes |N\rangle, \tag{7.58}$$

where $|N\rangle$ is the state of the reservoir containing N quanta of energy $\hbar\omega_0$.

Now, let us investigate the state of the composite system once the oscillator has lost only one quantum of energy to the environment.

Since the emission of N quanta takes place in τ, the emission of a single quantum is expected to happen within τ/N. Besides, once the interaction Hamiltonian of our model involves coordinate–coordinate coupling, the state of the composite system at this instant can be approximated by

$$|\phi_1\rangle \approx |\tilde{\psi}_z\rangle \otimes |1\rangle + |\tilde{\psi}_0\rangle \otimes |0\rangle, \tag{7.59}$$

because states at different positions correlate differently with the environment. Therefore, using the orthogonality between $|0\rangle$ and $|1\rangle$, we can compute the reduced density operator of the system as

$$\tilde{\rho} \equiv \mathrm{tr}_R |\phi_1\rangle\langle\phi_1| = |\tilde{\psi}_z\rangle\langle\tilde{\psi}_z| + |\tilde{\psi}_0\rangle\langle\tilde{\psi}_0|, \tag{7.60}$$

which is a statistical mixture. So, since the off-diagonal terms have been washed out within such a short time scale, the overlap of the wave packets will present no interference pattern any more.

The finite-temperature effect can also be accounted for with a very similar reasoning (Caldeira and Leggett, 1985). The only difference is that now we have to consider the transitions induced by the coupling to the bath.

As we have seen before, the average loss of a quantum of energy is a sufficient condition for the destruction of the interference between the two packets. In

---

[1] This is not entirely true because the packets themselves can still be approximately pure states. We shall soon return to this point.

contrast, we know that $|\tilde{\psi}_z\rangle$, which evolves from $|\psi_z\rangle$, and $|\psi_z\rangle$ itself are both centered at positions $\tilde{z}$ and z, respectively, and therefore are not energy eigenstates of the system. Nevertheless, as we are only trying to understand the mechanism that leads us to (7.56), let us assume that in the expansion of $|\psi_z\rangle$ and $|\tilde{\psi}_z\rangle$ in energy eigenstates only those components close to E_1 and E_2, such that $E_1 - E_2 \approx \hbar\omega_0$, are dominant. The time evolution of the occupation number of these two states is given by the system (Murray *et al.*, 1978)

$$\dot{n}_1 = -An_1 - An(\omega_0)(n_1 - n_2),$$
$$\dot{n}_2 = An_1 + An(\omega_0)(n_1 - n_2), \qquad (7.61)$$

where we have neglected the influence of energy eigenstates other than the two we are interested in. The factor A is of the order of $(\tau/N)^{-1}$, which is the spontaneous emission rate of one quantum of energy at $T = 0$, and

$$n(\omega_0) = \frac{1}{\exp\frac{\hbar\omega_0}{k_B T} - 1}. \qquad (7.62)$$

The solution of (7.61) is simple and reads

$$n_1 - n_2 \propto \exp -[2n(\omega_0) + 1]At, \qquad (7.63)$$

and, consequently, the decoherence time reduces to

$$\tau_d \propto \frac{1}{2n(\omega_0) + 1}\frac{\tau}{N}, \qquad (7.64)$$

which, for high temperatures, can be approximated by

$$\tau_d \approx \frac{\hbar\omega_0}{2Nk_B T}\tau, \qquad (7.65)$$

in accordance with the result we obtained in (7.56).

It is obvious that our latest approximations and hypotheses were by no means rigorous. Our intention was only to understand this specific case of decoherence by some physical reasoning.

Pointer basis. Another point which is worth emphasizing here is the agreement of our back-of-the-envelope arguments with the more systematic approach of the *pointer basis* advocated by Zurek (2003) regarding the *quantum theory of measurement*, which we briefly outline below.

The idea behind it is that a measuring apparatus might be composed of a macroscopic observable (*the pointer P*) described by a set of states $|P_i\rangle$ coupled to a huge number of microscopic – and passive – components each of which is by itself described by $|E_i^{(n)}\rangle$, where n refers to one of those specific subsystems. This latter system is referred to as the environment R.

Now, suppose that there is a microscopic system S of which a given observable $\hat{O}$ with eigenstates O_i must be measured. We assume the system is initially prepared in a state $|\psi\rangle = \sum_i a_i |\varphi_i\rangle$ such that $\hat{O}|\varphi_i\rangle = O_i|\varphi_i\rangle$. Moreover, let us assume that the initial apparatus state $|A_0\rangle$ is given by the direct product $|A_0\rangle = |P_0\rangle \otimes |E_0\rangle$, where $|P_0\rangle$ is a reference pointer state and $|E_0\rangle$ is the initial environment state ($|E_0\rangle = |E_0^{(1)}\rangle \otimes |E_0^{(2)}\rangle \otimes \ldots$). The state of the whole universe (system-plus-pointer-plus-environment) is then

$$|\Psi\rangle = |\psi\rangle \otimes |P_0\rangle \otimes |0\rangle = \sum_i a_i(|\varphi_i\rangle \otimes |P_0\rangle \otimes |0\rangle), \qquad (7.66)$$

which after some time interval t is assumed to have evolved to

$$|\Psi(t)\rangle = \sum_i a_i(|\varphi_i\rangle \otimes |P_i\rangle \otimes |E_i\rangle). \qquad (7.67)$$

In order for this to happen, the interaction Hamiltonians $\mathcal{H}_{SP}$ (between S and P) and $\mathcal{H}_{PR}$ (between P and R) must be such that they induce specific correlations between the states $|\varphi_i\rangle$, $|P_i\rangle$, and $|E_i\rangle$. In the particular case of measurement theory they must also commute with the free evolution Hamiltonians $\mathcal{H}_S$, $\mathcal{H}_P$, and $\mathcal{H}_R$ (Zurek, 2003). If the properties of the environment are such that $\langle E_i|E_j\rangle = \delta_{ij}$ within a very short t, it is a simple exercise to show that the reduced density matrix of the system–pointer subsystem becomes

$$\rho_{SP} \equiv \text{tr}_R \rho = \sum_i |a_i|^2 |\varphi_i\rangle\langle\varphi_i| \otimes |P_i\rangle\langle P_i|, \qquad (7.68)$$

a statistical mixture of the states of the composite system SP which carries information about the states $|\varphi_i\rangle$ to be measured. The measurement itself is performed by observing the pointer state correlated with the latter.

Although the model for dissipation we have applied to describing the dynamics of the two wave packets does not exactly fit this scheme, the requirement that the interaction Hamiltonian correlates certain states of the system of interest with orthogonal ones from the environment is exactly what was shown in (7.58) and (7.59). The difference is that since the interaction Hamiltonian commutes with neither the system nor the environment Hamiltonians, we do not have the correlation being established between fixed states. The reason behind it is that the operators showing up in the interaction Hamiltonian are the position operators of the particle and environment oscillators and Gaussian wave packets are only approximate eigenstates of the position of the particle. Nevertheless, even for these kind of decaying states the transition from a pure state to a mixed one still applies.

We should bear in mind that the mixed state which results within a very short time scale can only be considered as such if referred to the block subspaces labeled by the center of the packets. These two indeed behave as classical probability densities when their interference terms are taken into account. However, if we use any measure of purity for the full state the result is that it is not a maximally mixed state. The reason for this remaining purity is the "internal" coherence of each Gaussian packet, which will decay in a different time scale.

Internal decoherence. In order to analyze this phenomenon we could appeal to the qualitative approach employed in Zurek (1991), where the author uses the high-temperature master equation (6.24) to estimate the time scale for decay of the off-diagonal elements of the reduced density operator of a free particle initially prepared in a Gaussian wave packet of width Δx. Although this gives a plausible estimate of the internal decoherence time of the packet,

$$\tau_D \approx \gamma^{-1} \left(\frac{\lambda_T}{\Delta x} \right)^2, \tag{7.69}$$

where $\lambda_T = \hbar/\sqrt{2Mk_BT}$ is the *thermal de Broglie wavelength* of the particle, the most reliable maneuver would be to study the time evolution of (7.29) directly, from which we could extract the exact time dependence of the decay of each of its off-diagonal elements.

Actually, this has been done for a very weakly damped harmonic oscillator (Santos, 2013) through the linearization of (7.31), which together with (7.29) gives us

$$\tau_D = \frac{1}{\gamma} \left(\frac{\sigma}{\xi} \right)^2 \left(\exp \frac{\hbar \omega_0}{k_B T} - 1 \right)$$

$$= \begin{cases} \frac{1}{\gamma} \frac{\hbar \omega_0}{k_B T} \left(\frac{\sigma}{\xi} \right)^2 & \text{if } k_B T \gg \hbar \omega_0 \\ \frac{1}{\gamma} \left(\frac{\sigma}{\xi} \right)^2 \exp \frac{\hbar \omega_0}{k_B T} & \text{if } k_B T \ll \hbar \omega_0. \end{cases} \tag{7.70}$$

Once again we see that the further the off-diagonal elements of $\tilde{\rho}(x, y, t)$ are from the diagonal terms ($x = y$), the faster they decay and the increase in temperature also tends to enhance the destruction of coherence. However, although the off-diagonal elements present this peculiar behavior, the purity of the state is independent of its preparation details. This can be seen if we use (7.29) to compute $\text{Tr}\,\tilde{\rho}^2(x, y, t)$, which for short times reads

$$\text{Tr}\,\tilde{\rho}^2(x, y, t) \approx 1 - 4 \left(\exp \frac{\hbar \omega_0}{k_B T} - 1 \right)^{-1} \gamma t. \tag{7.71}$$

Therefore, we see that it ceases to be a pure state within a time interval $\tau_c = \left(\exp\frac{\hbar\omega_0}{k_B T} - 1\right)/4\gamma$, which is independent of σ. Another curious feature of τ_c is that, for $T \to 0$, it diverges, meaning that this damped state never becomes a mixture at $T = 0$. Despite being strange, this characteristic is typical of a coherent state of a harmonic oscillator coupled to a bath within the rotating wave approximation (see Section 5.4), whose behavior can be reproduced by the very weak damping approximation of the minimal model (Rosenau da Costa *et al.*, 2000). Obviously, this picture changes as higher-order terms in γ are taken into account, as it should.

This procedure has also been adopted in Venugopalan (1994) for the case of a free damped particle in the high-temperature limit, which in this case means $\hbar\gamma/2k_B T \ll 1$. The author found that contrary to the expectation that the off-diagonal elements of the reduced density operator in the position representation should be quickly washed out, this takes place much faster in the momentum representation. It is shown in the above-mentioned reference that the reduced density operator of the particle does not even become fully diagonal in the position representation, at variance with what happens in the momentum representation where it becomes completely diagonal with its off-diagonal elements decaying exponentially within a time scale $t_d = M^2\gamma^2/DQ^2$. In the latter, $Q \equiv p_x - p_y$ where the momenta p_x and p_y are defined through the Fourier transform

$$\tilde{\rho}(p_x, p_y) = \int\limits_{-\infty}^{+\infty} \int\limits_{-\infty}^{+\infty} dx\, dy\, \tilde{\rho}(x, y) \exp\frac{-i(p_x x - p_y y)}{\hbar}. \tag{7.72}$$

Although the approach presented in Venugopalan (1994) shows us a somewhat unexpected result, a more refined treatment of the free particle problem can be implemented if we use the exact diagonalization of Hakim and Ambegaokar (1985), from which the relevance of position representation to the decoherence problem can be more properly addressed.

8

Dissipative quantum tunneling

This is one of the examples where all that is needed can be extracted from the reduced density operator of the system in equilibrium. As we have seen in the introductory chapters, there are many realistic examples where the decay of a metastable configuration of a given system can take place by quantum tunneling. Those metastable configurations were described either by a single coordinate or a dynamical field. In this chapter we will describe how to deal with these problems.

8.1 Point particles

We are interested in studying the decay of a particle out of a metastable potential well. The potential we are going to employ in our exercise is given by (see Fig. 8.1)

$$V(q) = \frac{1}{2} M \omega_0^2 q^2 - \frac{1}{6} M \lambda q^3 \qquad (\lambda > 0). \qquad (8.1)$$

Evidently, we need to define precisely the problem we are going to attack because of the many questions that can be raised regarding this same problem. We are going to try to answer the following questions:

(i) Considering that the Brownian particle is in equilibrium with its reservoir at temperature T about the metastable position $q = 0$, what is the probability of it leaving this position by quantum tunneling?

(ii) Is it possible to describe this tunneling rate as a function of the phenomenological damping parameter and temperature only?

In order to establish a criterion for comparison of our future expressions with those already known for the undamped problem, let us start by briefly reviewing the latter.

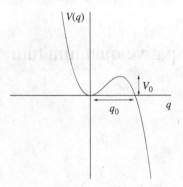

Figure 8.1 Metastable potential

8.1.1 The zero-temperature case

The central point of the calculation is the expression for the tunneling rate within the WKB approximation, which is obtained from the evaluation of the imaginary part of the energy, $E_0^{(0)}$, of the particle at the metastable energy minimum (see Appendix D) (Callan and Coleman, 1977; Coleman, 1988) and reads

$$\Gamma_0 = \frac{2\,|\mathrm{Im}E_0^{(0)}|}{\hbar} = \left(\frac{B_0}{2\pi\hbar M}\right)^{1/2}\left|\frac{\det(-M\partial_t^2 + M\omega_0^2)}{\det'(-M\partial_t^2 + V''(q_c^{(0)}))}\right|^{1/2}\exp-\frac{B_0}{\hbar},$$

(8.2)

where det′ means that the zero eigenvalue must be omitted from the evaluation of the determinant and B_0 is the Euclidean action of the *bounce* $q_c^{(0)}(\tau)$. The latter is a solution of the imaginary time equation of motion

$$\left.\frac{\delta S_E}{\delta q}\right|_{q_c} = -M\ddot{q}_c + V'(q_c) = 0,$$

(8.3)

where the Euclidean action for an isolated particle is

$$S_E[q(\tau')] = \int_{-\infty}^{\infty} d\tau'\left\{\frac{1}{2}M\left(\frac{dq}{d\tau'}\right)^2 + V(q)\right\}.$$

(8.4)

Notice that (8.3) is an equation of motion of a fictitious particle in a potential $-V(q)$ (see Fig. 8.2), which can easily be integrated with the boundary condition $q_c^{(0)}(-\infty) = q_c^{(0)}(\infty) = 0$. On top of the trivial solution $q_c = 0$, there is also a solution that spends most of the time close to $q = 0$ at the remote past ($\tau \to -\infty$), makes a rapid excursion to $q = \bar{q} = q_0$ at $\tau = 0$, and slowly returns to $q = 0$ at $\tau \to \infty$. This is the *bounce* solution we have mentioned above, which is analyzed in Appendix D.

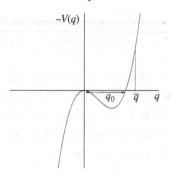

Figure 8.2 Inverted potential

This solution (Callan and Coleman, 1977; Caldeira and Leggett, 1983a; Coleman, 1988) is

$$q_c^{(0)}(\tau) = \frac{3\omega_0^2}{\lambda} \operatorname{sech}^2 \frac{\omega_0 \tau}{2},$$ (8.5)

and therefore

$$B_0 = \int_{-\infty}^{\infty} \left[\frac{1}{2} M \dot{q}_c^2 + V(q_c) \right] d\tau = \frac{36}{5} \frac{V_0}{\omega_0},$$ (8.6)

where V_0 is the height of the potential barrier, which can be written in terms of the remaining parameters as

$$V_0 = \frac{2}{3} \frac{M \omega_0^6}{\lambda^2}.$$ (8.7)

Proceeding with the evaluation of the other terms in (8.2), we reach

$$\Gamma_0 = A_0 \exp -\frac{36}{5} \frac{V_0}{\hbar \omega_0} \quad \text{with} \quad A_0 = 6\omega_0 \sqrt{\frac{6}{\pi} \frac{V_0}{\hbar \omega_0}},$$ (8.8)

which is the well-known WKB expression for the decay rate by quantum tunneling, as expected.

Now, suppose we want to study the problem of the tunneling of this particle if it is coupled to the standard environment we have been using in our applications. So, what we have now is a system in an $(N + 1)$-dimensional space that, besides the metastable minimum of energy at $(q, \mathbf{R}) = (0, \mathbf{0})$, has a saddle point at $(q, \mathbf{R}) = (2\omega_0^2/\lambda, \ldots, 2 C_k \omega_0^2/m_k \omega_k^2, \ldots)$. As we are treating the composite system at $T = 0$, the only reasonable interpretation for $\mathrm{Im} E_0$ (E_0 is the metastable energy of the composite system) is that it is proportional to the tunneling rate of this $(N + 1)$-dimensional configuration out of the metastable well about the position $(0, \mathbf{0})$. Besides, since the particle of interest is the only one able to tunnel, we are going to associate $\mathrm{Im}E_0$ with its tunneling rate subject to the influence of the remaining degrees of freedom.

Therefore, when $\eta \neq 0$ the procedure to obtain the new rate Γ is exactly the same as for the undamped case with the difference that we must now evaluate $\mathrm{Im}E_0$ for the composite system instead of $\mathrm{Im}E_0^{(0)}$ for the isolated particle of interest.

Let us start by writing the density operator of the composite system in equilibrium at inverse temperature β, which reads

$$\rho(x, \mathbf{R}; y, \mathbf{Q}, \beta) = \sum_n \psi_n(x, \mathbf{R})\psi_n^*(y, \mathbf{Q}) \exp{-\beta E_n}. \qquad (8.9)$$

The reduced density operator of the particle in equilibrium is then

$$\tilde{\rho}(x, y, \beta) = \int d\mathbf{R}\rho(x\mathbf{R}; y\mathbf{R}) = \int d\mathbf{R}\sum_n \psi_n(x, \mathbf{R})\psi_n^*(y, \mathbf{R}) \exp{-\beta E_n}, \qquad (8.10)$$

or, using the functional integral representation,

$$\tilde{\rho}(x, y, \beta) = \int d\mathbf{R}\langle x\mathbf{R}|e^{-\beta\mathcal{H}}|y\mathbf{R}\rangle$$

$$= \int d\mathbf{R}\int_{y,\mathbf{R}}^{x,\mathbf{R}} \mathcal{D}q(\tau')\mathcal{D}\mathbf{R}(\tau') \exp{-\frac{S_E[q(\tau'), \mathbf{R}(\tau')]}{\hbar}}, \qquad (8.11)$$

where $\mathcal{H}$ is the Hamiltonian of the composite system and S_E its corresponding Euclidean action. The energy of the metastable state of the composite system can be computed if we take the limit $\beta \to \infty$ in (8.10) when $x = y = 0$. Defining $\tau \equiv \hbar\beta$, we have

$$\int d\mathbf{R}|\psi_0(0, \mathbf{R})|^2 e^{-\tau E_0/\hbar} \approx \int d\mathbf{R}\int_{0,\mathbf{R}}^{0,\mathbf{R}} \mathcal{D}q(\tau')\mathcal{D}\mathbf{R}(\tau') \exp{-\frac{S_E[q(\tau'), \mathbf{R}(\tau')]}{\hbar}}. \qquad (8.12)$$

The trace of the functional integral over $\mathbf{R}(\tau')$ in (8.12) has already been performed and the result is, as in (6.95) and (6.96),

$$\tilde{\rho}(0, 0, \beta) = \tilde{\rho}_0(\beta)\int_0^0 \mathcal{D}q(\tau') \exp{-\frac{S_{eff}[q(\tau')]}{\hbar}} \qquad (8.13)$$

and

$$S_{eff}[q(\tau')] = \int_0^\tau \left\{ \frac{1}{2} M \dot{q}^2 + V(q) \right\} d\tau'$$

$$+ \frac{\eta}{4\pi} \int_{-\infty}^\infty d\tau'' \int_0^\tau d\tau' \frac{\{q(\tau') - q(\tau'')\}^2}{(\tau' - \tau'')^2}. \tag{8.14}$$

Now we have to evaluate the path integral (8.13) in order to identify the imaginary part of the energy of the metastable state of the composite system through (8.12). We shall associate this quantity with the effective decay rate (Γ) of the metastable state of the system of interest influenced by the presence of the environment, as explained before.

This computation is not simple. It involves the application of the saddle point method to the functional integral (8.13) in the same way as we did for the undamped case in Appendix D. We shall reproduce here some steps of the calculation which can be found in Caldeira and Leggett (1983a) for the case of zero temperature and dedicate the next section to the finite-temperature case.

Our first step is the determination of $q_c(\tau)$. When $\tau \to \infty$ (or $T \to 0$), the functional derivative of (8.14) at $q_c(\tau')$ reads

$$\left. \frac{\delta S_{eff}}{\delta q} \right|_{q_c} = M \ddot{q}_c - \frac{\partial V}{\partial q_c} - \frac{\eta}{\pi} \int_{-\infty}^\infty d\tau'' \frac{[q_c(\tau') - q_c(\tau'')]}{(\tau' - \tau'')^2} = 0, \tag{8.15}$$

where the last equality follows from the stationary phase approximation we have been employing. This equation also represents the motion of a fictitious particle in imaginary time. Likewise, the $\eta = 0$ case is invariant under time reversal $(\tau' \to -\tau')$ and by translation of the time origin $(\tau' \to \tau' + \tau_0)$, which allows us to show the existence of a bounce solution when $\tau \to \infty$. Multiplying (8.15) by $\dot{q}_c$, we have

$$\frac{d}{d\tau'} \left[\frac{1}{2} M \dot{q}_c^2 - V(q_c) \right] = \dot{q}_c(\tau') \frac{\eta}{\pi} \int_{-\infty}^\infty d\tau'' \frac{[q_c(\tau') - q_c(\tau'')]}{(\tau' - \tau'')^2}, \tag{8.16}$$

which is an explicit expression of the variation of the "Euclidean energy" with imaginary time. In this way, if we take $q_c(-\infty) = 0^+$ (local maximum of the potential of Fig. 8.2), the "total energy" of the solution $q_c(\tau')$ starts to increase as $\dot{q}_c(\tau')$ increases until it reaches the turning point $\bar{q}$ at $\tau' = 0$. From then onwards, $\dot{q}_c(\tau') < 0$ and all the energy gained by the fictitious particle on its way to $\bar{q}$ will be lost at the same rate on its way back to $q_c(\infty) = 0^+$. Therefore, when $\tau \to \infty$, (8.15) admits a bounce-like solution.

Once the reader has been convinced of this argument, what we have to do next is study the stability of the bounce solution by investigating the eigenvalue problem originated from the second functional derivative of (8.14), namely

$$\hat{D}q(\tau') = \lambda q(\tau'),\tag{8.17}$$

where the differential operator $\hat{D}$ acting on $q(\tau')$ reads

$$\hat{D}q(\tau') = -M\frac{d^2q(\tau')}{d\tau'^2} + V''(q_c)q(\tau') + \frac{\eta}{\pi}\int\limits_{-\infty}^{\infty} d\tau'' \frac{[q(\tau') - q(\tau'')]}{(\tau' - \tau'')^2}.$$

$$\tag{8.18}$$

Deriving (8.15) with respect to τ', we can show that $\dot{q}_c$ is a solution of (8.18) with zero eigenvalue. As $q_c(\tau')$ is an even function of its argument, $\dot{q}_c(0) = 0$, and, therefore, the solution with $\lambda = 0$ has a node for $\tau' \neq \pm\infty$. This implies that there is an eigenfunction $q_0(\tau')$ with $\lambda < 0$.

Since in (8.13) we have to integrate over all possible directions in function space, this particular one will cause us problems because it gives rise to an increasing Gaussian integral. The way to avoid this problem is to perform an appropriate analytical extension of the variable along this dangerous direction as we have done in Appendix D (Callan and Coleman, 1977; Caldeira and Leggett, 1983a; Coleman, 1988), which gives us an imaginary part for the energy E_0.

Because of this analysis, we can write the result corresponding to the tunneling of the Brownian particle at $T = 0$ directly as

$$\Gamma = A \exp-\frac{B}{\hbar},\tag{8.19}$$

where

$$A = \sqrt{\frac{||\dot{q}_c||^2}{2\pi\hbar}} \sqrt{\left|\frac{\det \hat{D}_0}{\det' \hat{D}}\right|} \quad \text{with} \quad ||\dot{q}_c||^2 \equiv \int\limits_{-\infty}^{\infty} d\tau' \, \dot{q}_c^2(\tau') \tag{8.20}$$

and

$$B \equiv S_{eff}[q_c] = \int\limits_{-\infty}^{\infty} \left\{\frac{1}{2}M\dot{q}_c^2 + V(q_c)\right\} d\tau'$$

$$+ \frac{\eta}{4\pi}\int\limits_{-\infty}^{\infty} d\tau' \int\limits_{-\infty}^{\infty} d\tau'' \frac{\{q_c(\tau') - q_c(\tau'')\}^2}{(\tau' - \tau'')^2}.\tag{8.21}$$

In (8.20), $\hat{D}_0$ is the same as $\hat{D}$ (see (8.18)) with $V''(q_c)$ replaced by $M\omega_0^2$ and det' means that the determinant must be computed omitting the zero eigenvalue (see

Appendix D). Notice that, in general, $||\dot{q}_c||^2 \neq B/M$. The equality only holds when the "Euclidean energy" is conserved or, in other words, $\eta = 0$.

The prefactor A represents the quantum fluctuations about this extremum solution and its evaluation involves the subtle computation of the ratio of determinants present therein. In the remainder of this chapter we will eventually perform the computation of the latter in some specific case. Actually, we have already done that for the prefactor of the real-time propagator in Appendix A, although the method employed there is not applicable to any situation. A very reliable approach to the problem is the employment of the Fredholm scattering theory (Gottfried, 1966), as done in Chang and Chakravarty (1984).

Although A might increase with dissipation, there is no doubt that B does (see below). Now, since only the latter appears in the exponent of the rate expression it is evident that the overall influence of dissipation on quantum tunneling will be to reduce it. Therefore, we shall concentrate our efforts on understanding the possible modifications induced by the damping mainly on the effective action, although some estimates of the prefactor will also be touched upon whenever possible.

This increase of the action is seen trivially, because the term that depends on η in (8.21) is a positive definite integral. Therefore, $B > B_0 \Rightarrow \Gamma < \Gamma_0$ and we can conclude that dissipation tends to inhibit quantum tunneling of a Brownian particle. At this point it is worth emphasizing an important fact; dissipation or, still better, the coupling to an environment does not always tend to inhibit tunneling. The particular form (8.21) is dependent on our choice for the particle–bath interaction. As a matter of fact, there are other forms of coupling that tend to speed the tunneling process up, even if the effect on the classical equation of motion comes about as a dissipative term. This is the case we have already mentioned, of anomalous dissipation (Leggett, 1984). Although extremely interesting, these examples are not our main goal in this section, where we wish to address the specific case of a Brownian particle (ohmic dissipation).

Weakly damped systems

Let us study first the correction to B_0 as a function of η. Despite being very complex, this problem presents two particularly interesting limits in which it can be solved analytically, namely $\eta \to 0$ and $\eta \to \infty$. When $\eta \to 0$ (or $\gamma << \omega_0$), we assume that the solution $q_c(\tau')$ has the form

$$q_c(\tau') = q_c^{(0)}(\tau') + \eta q_c^{(1)}(\tau'), \tag{8.22}$$

where $q_c^{(0)}(\tau')$ is the bounce when $\eta = 0$. Now, inserting (8.22) in (8.21) we see that in order to get corrections $\mathcal{O}(\eta)$ to the action B we should only evaluate the double integral in (8.21) with $q_c(\tau') = q_c^{(0)}(\tau')$. The other terms will be $\mathcal{O}(\eta^2)$. This integral can be solved in two steps. Firstly, we define new variables of integration $u = \tau' - \tau''$ and $v = (\tau' + \tau'')/2$, which allows us to reduce it to

$$\Delta B \equiv B - B_0 = \frac{2\eta}{3\pi}\bar{q}^2 \int\limits_{-\infty}^{\infty} \frac{dx}{x^2}\left[\frac{\sinh^2 x - 3x \coth x + 3}{\sinh^2 x}\right], \tag{8.23}$$

which can be solved by residues, yielding

$$\Delta B = \frac{45}{\pi^3}\zeta(3)\,\alpha B_0 = \frac{12}{\pi^3}\zeta(3)\,\eta\,\bar{q}^2 \tag{8.24}$$

where $\alpha \equiv \eta/2M\omega_0 = \gamma/\omega_0$ and $\zeta(3) = \sum\limits_{n=1}^{\infty} 1/n^3$.

Prefactor. Following what we have just done, all we need now is to evaluate the dissipative correction to lowest order in η due to its effect on the undamped bounce solution $q_c^{(0)}(\tau')$. Consequently, if we keep the dissipative corrections to the same order in the prefactor, they appear in a term proportional to the action itself (see (D.36) and (8.20)) and through the ratio of determinants. Therefore, if we write $B = B_0 + \Delta B$ where $\Delta B \ll B_0$ and expand both the ratio of determinants and the square-root contributions up to first order in ΔB, the prefactor reads

$$A \approx A_0(1 + c\,\alpha) \tag{8.25}$$

where $c \approx 2.8$ (Freidkin *et al.*, 1986; Ankerhold, 2007). Then, expanding the exponent up to the same order in ΔB one has

$$\Gamma \approx \Gamma_0(1 + c\,\alpha)\left(1 - \frac{\Delta B}{\hbar}\right), \tag{8.26}$$

where by (8.24) $\Delta B/\hbar \approx 1.7\alpha B_0/\hbar$. Consequently, as for the validity of the semi-classical approximation $B_0 \gg \hbar$, we see that the decay rate, up to this order in η, is still dominated by the exponential factor and then diminishes with increasing damping as we had foreseen. Notice that to reach this conclusion we have assumed that $B_0 \gg \hbar$, which is not necessarily the case in practical applications. However, the same conclusion can be reached for values of $B_0 \gtrsim \hbar$ for which the semi-classical approach is still a good approximation.

Strongly damped systems

In the case of strongly damped systems ($\gamma \gg \omega_0$) we had best write the equation of motion (8.15) in terms of the Fourier transform $q(\omega)$ of $q(\tau')$. Using the fact that

$$q(\tau') = \frac{1}{2\pi} \int\limits_{-\infty}^{\infty} d\omega\, q(\omega)e^{-i\omega\tau'} \tag{8.27}$$

we have

$$\{M\omega^2 + \eta|\omega| + M\omega_0^2\}q(\omega) - \frac{M\lambda}{4\pi} \int\limits_{-\infty}^{\infty} d\omega' \, q(\omega - \omega') \, q(\omega') = 0. \quad (8.28)$$

As we are interested in the overdamped solutions we hope the main contribution to $q(\omega)$ comes from the low-frequency part of its spectrum and, therefore, we should neglect the term in ω^2 in (8.28). Then, assuming that $q_c(\omega)$ is of the form

$$q_c(\omega) = A \, e^{-\kappa|\omega|}, \quad (8.29)$$

we have

$$\eta|\omega|A \, e^{-\kappa|\omega|} + M\omega_0^2 A \, e^{-\kappa|\omega|} - \frac{3M\lambda A^2}{4\pi}\left|\frac{1}{\kappa} + |\omega|\right|e^{-\kappa|\omega|} = 0 \quad (8.30)$$

which implies that

$$A = \frac{4\pi\eta}{3M\lambda} \quad \text{and} \quad \kappa = \frac{2\alpha}{\omega_0}. \quad (8.31)$$

The expression for κ in (8.31) really confirms that the main contribution to $q_c(\omega)$ in (8.29) comes from frequencies such that $\omega << \omega_0$, once $\alpha >> 1$. This justifies our self-consistent approach to ignore the $M\omega^2$ term in (8.28).

Inserting (8.29) in (8.27), using (8.31), and performing the resulting integral, one gets

$$q_c(\tau') = \frac{4}{3} \frac{q_0}{1 + \left(\frac{\omega_0^2 \tau'}{2\gamma}\right)^2}. \quad (8.32)$$

The action corresponding to this solution can readily be evaluated as

$$\begin{aligned} B = \frac{1}{2\pi} \int\limits_{-\infty}^{\infty} &\left\{ \frac{1}{2}(M\omega^2 + \eta|\omega| + M\omega_0^2) \, q_c^2(\omega) \right. \\ &\left. - \frac{M\lambda}{12\pi} \int\limits_{-\infty}^{\infty} d\omega' \, q_c(\omega) \, q_c(\omega - \omega') \, q_c(\omega') \right\} d\omega \\ &= \frac{2\pi}{9}\eta q_0^2 + \mathcal{O}\left(\frac{\omega_0}{\gamma}\right). \end{aligned} \quad (8.33)$$

Once again we conclude that dissipation increases the action of the bounce, which implies a decrease of Γ. Moreover, the correction that shows up due to coupling to the environment can be written solely in terms of the phenomenological dissipation constant η. In spite of still having to analyze the prefactor of Γ, we can safely say that there is an overall decrease of this decay rate since it is the exponential term which dominates its behavior.

For intermediate values of η, it is not possible to write an analytical solution to (8.28). However, variational arguments can be used in order to establish general bounds for $B(\eta)$. Although we are not going to present them in this book, we would like to recommend Caldeira and Leggett (1983a) for details about the referred procedure. A numerical approach to the problem is available in Chang and Chakravarty (1984).

Prefactor. In order to analyze the behavior of the prefactor in the overdamped case (actually the subsequent general analysis can also be applied to underdamped situations), it is convenient to introduce the dimensionless variable $t \equiv \omega_0 \tau / 2\alpha$ with ω denoting the frequency conjugate to t. Then, defining the dimensionless classical bounce trajectory measured in units of q_0 by $\tilde{q}_c(t)$, we can rewrite the prefactor expression (8.20) as

$$A = \omega_0 \left(\frac{\|\dot{\tilde{q}}_c\|^2}{2\pi\hbar} \right)^{1/2} K^{1/2} \tag{8.34}$$

where the dimensionless factor K is given by

$$K \equiv \left| \frac{\det \hat{H}_0}{\det'(\hat{H}_0 + \hat{V})} \right|. \tag{8.35}$$

In the representation $|\omega\rangle$ such that $\langle t|\omega\rangle = (2\pi)^{-1/2} \exp{-i\omega t}$, for any wave function $\psi_n(\omega)$, the operators $\hat{H}_0$ and $\hat{V}$ are defined by the relations

$$\hat{H}_0 \psi_n(\omega) \equiv \left(1 + |\omega| + \frac{\omega^2}{4\alpha^2} \right) \psi_n(\omega) = \epsilon_0(\omega)\psi_n(\omega),$$

$$\hat{V}\psi_n(\omega) \equiv \int_{-\infty}^{+\infty} V(\omega - \omega' : \alpha)\psi_n(\omega')d\omega', \tag{8.36}$$

where

$$V(\omega) \equiv -\frac{3}{2\pi} \int_{-\infty}^{+\infty} dt\, \tilde{q}_c(t) \exp{-i\omega t}.$$

Now, analyzing the spectrum of $\hat{H}_0 + \hat{V}$, we realize that it admits a negative eigenvalue λ_0, associated with the eigenfunction $\psi_0(\omega)$, and an eigenvalue $\lambda_1 = 0$, associated with the eigenfunction $\psi_1(\omega) \propto \omega V(\omega)$,[1] due to the translation invariance of the bounce along the time axis. It may also have other discrete eigenvalues (bound states) for $0 < \lambda < 1$, but definitely has a quasi-continuous

---

[1] As this function has a node it implies the existence of a nodeless eigenfunction with negative eigenvalue.

spectrum starting at $\lambda = 1$. If we assume periodic boundary conditions with period $\Omega \equiv \omega_0 \bar{t}/2\alpha$ ($\bar{t} \to \infty$) to the eigenstates of $\hat{H}_0 + \hat{V}$, the level spacing in the quasi-continuous spectrum vanishes as Ω^{-1}. Consequently, the effect of $\hat{V}$ on the states of the quasi-continuous spectrum is of the order Ω^{-1} relative to that of $\hat{H}_0$.

The next step in proceeding with our analysis of K is to use the well-known equality $\ln(\det \hat{O}) = \mathrm{tr}(\ln \hat{O})$, which when applied to $\ln K$ reads

$$\ln K = \sum_{n=0}^{b} \langle \psi_n | \ln \hat{H}_0 | \psi_n \rangle + \sum_{n>b} \langle \psi_n | \ln \hat{H}_0 | \psi_n \rangle$$

$$- \sum_{n=0, \neq 1}^{b} \ln |\lambda_n| - \sum_{n>b} \langle \psi_n | \ln(\hat{H}_0 + \hat{V}) | \psi_n \rangle, \quad (8.37)$$

where $|\psi_n\rangle$ is an eigenstate of $\hat{H}_0 + \hat{V}$ with eigenvalue λ_n and b is the number of bound states for $0 < \lambda < 1$. Therefore, writing the eigenvalue equation $(\hat{H}_0 + \hat{V})|\psi_n\rangle = \lambda_n |\psi_n\rangle$ in the ω-representation, we have

$$\epsilon_0(\omega)\psi_n(\omega) + \int_{-\infty}^{+\infty} V(\omega, \omega')\,\psi_n(\omega')\,d\omega' = \lambda_n \psi_n(\omega), \quad (8.38)$$

which, when multiplied on the left by $\psi_n^*(\omega)$ and integrated over ω, yields

$$\int_{-\infty}^{+\infty} \epsilon_0(\omega)|\psi_n(\omega)|^2 + \int_{-\infty}^{+\infty} \int_{-\infty}^{+\infty} \psi_n^*(\omega) V(\omega, \omega')\,\psi_n(\omega')\,d\omega'\,d\omega = \lambda_n. \quad (8.39)$$

For short, we write $\lambda_n = I_1^{(n)} + I_2^{(n)}$, where $I_k^{(n)}$ represents the kth integral in (8.39), to express the last term on the r.h.s. of (8.37) as

$$\sum_{n>b} \ln \left(I_1^{(n)} + I_2^{(n)} \right) = \sum_{n>b} \ln I_1^{(n)} + \sum_{n>b} \ln \left(1 + \frac{I_2^{(n)}}{I_1^{(n)}} \right)$$

$$\approx \sum_{n>b} \ln I_1^{(n)} + \sum_{n>b} \left(\frac{I_2^{(n)}}{I_1^{(n)}} \right) \quad (8.40)$$

since $I_2^{(n)} \ll I_1^{(n)}$, as we have argued before.

In this way we can rewrite (8.37) as

$$\ln K = \sum_{n=0}^{b} \langle \psi_n | \ln \hat{H}_0 | \psi_n \rangle - \sum_{n=0, \neq 1}^{b} \ln |\lambda_n| + \sum_{n=0}^{b} \langle \psi_n | \hat{H}_0^{-1} \hat{V} | \psi_n \rangle$$

$$+ \sum_{n>b} \left[\int_{-\infty}^{+\infty} \ln \epsilon_0(\omega)\,|\psi_n(\omega)|^2\,d\omega - \ln \left(\int_{-\infty}^{+\infty} \epsilon_0(\omega)\,|\psi_n(\omega)|^2\,d\omega \right) \right] - \mathrm{tr}(\hat{H}_0^{-1} \hat{V}),$$

$$(8.41)$$

where now, due to the presence of the third term on the r.h.s. of (8.41), the trace of $\hat{H}_0^{-1}\hat{V}$ is unrestricted and can consequently be evaluated in any representation. Choosing the ω-representation for simplicity we have

$$\mathrm{tr}\hat{H}_0^{-1}\hat{V} = V(0) \int\limits_{-\infty}^{+\infty} \frac{d\omega}{1 + |\omega| + \omega^2/4\alpha^2} = -\frac{3}{\pi} f(\alpha) \int\limits_{-\infty}^{+\infty} \tilde{q}_c(t')\, dt', \quad (8.42)$$

where $f(\alpha)$ is given by (7.40) for $\alpha \lesssim 1$.

Notice that the approach we have used to evaluate the desired ratio of determinants is valid for any value of α. Rewriting (8.41) as $\ln K = D_0(\alpha) - \mathrm{tr}\hat{H}_0^{-1}\hat{V}$, all we need to do is compute $D_0(\alpha)$. In order to achieve that, we need to solve the eigenvalue problem (8.38) which can be, in general, a hard task. However, there are limits where this program can be carried out explicitly, for example, the case of very high damping.

In the limit $\alpha \gg 1$ we can use the appropriate form of $f(\alpha)$ from (7.40) and the bounce solution (8.32) in (8.42) to reduce the latter to $8 \ln 2\alpha$. Moreover, as in this limit the term depending on ω^2 can be dropped from the eigenvalue equation, it becomes evident that $D_0(\alpha)$ is actually a constant, independent of α. It has been shown explicitly in Caldeira and Leggett (1983a) that for values of α for which the eigenvalue problem admits only three bound states in the spectrum (which seems to be the case up to $\alpha \lesssim 10$ (Chang and Chakravarty, 1984)), their contribution to D_0 is about -1.5. Since the contribution of the quasi-continuous part of the spectrum to D_0 was overlooked in Caldeira and Leggett (1983a), we shall ignore its specific numerical value in our final expressions because it will have no influence on the already established α-dependence of the prefactor. Once again we see the reduction of the tunneling rate as a function of the damping parameter.

Using all these results in (8.35) we get

$$A = c\, A_0\, \alpha^{7/2}, \quad (8.43)$$

where A_0 is the undamped prefactor and c is a constant which was evaluated numerically in Chang and Chakravarty (1984) and shown analytically (Larkin and Ovchinnikov, 1984; Grabert *et al.*, 1987) to be equal to $4\sqrt{\pi}/3$ to leading order in α.

8.1.2 The finite-temperature case

Once we have learnt about the zero-temperature tunneling of a particle out of a metastable potential well, it is time to try to generalize it to finite temperatures. We say try because, as we shall see shortly, this generalization is far from being a completely settled issue. To the best of our knowledge there are three basic approaches used to generalize the $T = 0$ case.

The first one is very similar to what we have done so far. The only difference is that now we interpret, at least for low enough temperatures (see below), the imaginary part of the *free energy*, F, as the decay rate. This is given by

$$\Gamma = \frac{2 \operatorname{Im} F}{\hbar} = A(T) \exp - \frac{B(T)}{\hbar}, \tag{8.44}$$

where F is computed exactly like what has been done for E_0, but with finite limits of integration instead (Larkin and Ovchinnikov, 1983, 1984; Grabert and Weiss, 1984; Weiss, 1999). Although there is no rigorous proof of the validity of (8.44), many results originated from the application of that expression to specific problems reproduce appropriate limits of the predicted (or even measured) decay rate. In order to keep pace with the formulation we have chosen to follow in this book, this will be the method we briefly resume in this section. We call it the *free energy method*.

The second approach (Waxman and Leggett, 1985) is a generalization for finite temperatures of an alternative way to deal with the $T = 0$ case. In the latter, the authors redefine the decay rate in terms of WKB wave functions of the metastable potential in order to avoid any sort of analytical continuation procedure. The results obtained for the argument of the exponential term in (8.44) still coincide with those one obtained by the previous method, whereas for the prefactor only the low-temperature and vanishing dissipation limit match up in the two methods.

The third approach is an adaptation of the method of Feynman and Vernon (Ueda, 1996) aimed at studying the quantum transmission through a dissipative potential barrier. Although the method envisages a different sort of problem, it can easily be adapted to our present needs. Nevertheless, we shall stick to the formulation we have chosen for dealing with the decay problem and not discuss this method here.

In any of these three cases analytical solutions are, in general, not available except in the limit of very low temperatures, where a correction proportional to T^2 can be found, or very high T, where the thermal activation results (Kramers, 1940) are recovered.

Undamped systems

Let us start by briefly showing how, within the free energy method, the problem is approached for undamped systems and then generalize it to finite damping. In the first part of this program we will be following closely Affleck (1981) and Weiss (1999).

Consider, once again, a particle in thermal equilibrium with its reservoir about $q = 0$ in Fig. 8.2, but this time let us assume that the temperature $T = (k_B \beta)^{-1} \neq 0$. Now, assuming that each approximate eigenstate of the particle in the well has a very tiny imaginary part, we can make the replacement $E_n^{(0)} \rightarrow E_n^{(0)} + i\hbar\Gamma_n^{(0)}/2$ and then, using the fact that

$$F = -\frac{1}{\beta} \ln Z = -\frac{1}{\beta} \ln \left[\sum_{n=0}^{\infty} \exp -\beta \left(E_n^{(0)} + i\frac{\hbar\Gamma_n^{(0)}}{2} \right) \right], \tag{8.45}$$

where F is the Helmholtz *free energy* of the metastable configuration and Z the corresponding partition function, we write

$$F = -\frac{1}{\beta} \ln Z_R + i\frac{1}{Z_R} \sum_{n=0}^{\infty} \frac{\hbar\Gamma_n^{(0)}}{2} \exp -\beta E_n^{(0)}, \tag{8.46}$$

where $Z_R = \sum_{n=0}^{\infty} \exp -\beta E_n^{(0)}$ is the real part of the partition function. Consequently, we can now say that

$$\frac{2 \operatorname{Im} F}{\hbar} = \langle \Gamma_n^{(0)} \rangle_\beta \equiv \Gamma_\beta^{(0)}, \tag{8.47}$$

which is a reasonable choice for the undamped decay rate at finite temperatures at least as long as we consider that only few energy levels are thermally populated or, in other words, we are working at finite but low temperatures.

Having expressed F in terms of the trace of $\exp -\beta H$, all we have to do now is evaluate

$$Z(\beta) = \int_{-\infty}^{+\infty} dx \, \langle x | \exp -\beta H | x \rangle = \int_{-\infty}^{+\infty} dx \int_x^x \mathcal{D}q(\tau') \exp -\frac{S_E[q(\tau')]}{\hbar}$$

$$\tag{8.48}$$

where

$$S_E[q(\tau')] = \int_{-\hbar\beta/2}^{+\hbar\beta/2} d\tau' \left\{ \frac{1}{2} M \left(\frac{dq}{d\tau'} \right)^2 + V(q) \right\}, \tag{8.49}$$

and, for convenience, we have used a symmetric imaginary time interval. From the resulting expression we can find $\operatorname{Im} F$ and finally establish its relation to the well-known limits of the decay rate computed by other methods.

There are two differences between the present case and that solved in Appendix D in the zero-temperature limit. The first is that the path integral in (8.48) must now be evaluated for paths within a region of finite imaginary time (finite β) such that $q(\hbar\beta/2) = q(-\hbar\beta/2)$. The second has to do with the bounce structure, which is no longer appropriate for exponentiation. Let us therefore see how we must handle this new situation.

The Euclidean action in (8.48) has two kinds of extrema, which obey (D.5) with $q(\hbar\beta/2) = q(-\hbar\beta/2)$ for inverted potentials like the one in Fig. 8.2. One kind is composed of non-periodic solutions which are similar to those of an inverted oscillator. They start at a given position, say $q(-\hbar\beta/2) = x$, move to the harmonic-like barrier centered at $q = 0$, and bounce back at $t = 0$ to $q(\hbar\beta/2) = x$, but with $\dot{q}(\hbar\beta/2) \neq \dot{q}(-\hbar\beta/2)$. It is a good approximation to consider these solutions as those of the inverted harmonic potential, in particular for positions x not far from the origin. We call these solutions type I and consider only the case where they have negative Euclidean energy:

$$E = -|E| = \frac{1}{2}M\dot{q}^2 - V(q). \tag{8.50}$$

The type I solutions of positive energy would bounce off the cubic part of the potential at $q > q_0$ (see Fig. 8.2) and will not be relevant for our purposes.

The other kind of solutions – which we will call type II – are periodic solutions with period $\hbar\beta$ which live within the metastable well of the inverted cubic potential of Fig. 8.2 for $0 < x < q_0$. These remind us of a bounce-like solution with the difference that they now have a finite period, unlike the genuine bounce of the zero-temperature case of Appendix D. As the Euclidean energy in (8.50) approaches $-V_0$, the period of these solutions tends to $2\pi/\omega_b$ where $\omega_b^2 \equiv V''(q_b)/M$ with q_b given by the solution of $V'(q_b) = 0$ ($V(q_b) = V_0$). Notice that in this case, only for $\hbar\beta = 2n\pi/\omega_b$ are there periodic solutions of the equation of motion (D.5). We shall return to this point shortly.

Now, with these two kinds of solution for the Euclidean equation of motion, we can solve (8.48) within the stationary phase approximation (SPA). As type I and II solutions are well separated in function space, we can consider their contribution to $Z(\beta)$ to be additive. Therefore, we can write it as

$$Z(\beta) \approx \int_{-\infty}^{+\infty} dx\, \Delta_I(\beta) \exp{-\frac{S_E^{(I)}(x, \beta)}{\hbar}} + Z_1(\beta), \tag{8.51}$$

where $S_E^{(I)}(x, \beta)$ and $\Delta_I(\beta)$ are respectively the Euclidean action (8.49) evaluated at $q_c(\tau) = q_c^{(I)}(\tau)$, which for small x turns out to be Gaussian, and the prefactor $N_0(\beta)\left\{\det\left[-M\partial_\tau^2 + M\omega_0^2\right]\right\}^{-1/2}$, with $N_0(\beta)$ being a normalization factor. The latter is chosen in such a way that the first integral in (8.51) reduces to the harmonic oscillator partition function $Z_0(\beta) = [2\sinh(\hbar\omega_0\beta/2)]^{-1}$ since, as the main contribution to the x integral comes from $x \approx 0$, it can be extended from minus to plus infinity. The contribution $Z_1(\beta)$ results from the SPA about $q_c(\tau) = q_c^{(II)}(\tau)$, which we analyze next.

For temperatures such that $\beta^{-1} < \beta_0^{-1} \equiv \hbar\omega_b/2\pi$ ($T_0 \equiv \hbar\omega_b/2\pi k_B$ is the so-called *crossover temperature* and its role will become clearer in the sequel), the eigenvalue problem coming from the second functional derivative of the action at the periodic solutions $q_c^{(II)}(\tau)$ still preserves the structure of the $T = 0$ case and admits a negative and a zero eigenvalue (translation mode). Therefore, it can be solved in a similar fashion to what we have done before for a single bounce. The result is (Affleck, 1981; Ankerhold, 2007)

$$Z_1(\beta) \approx -\frac{i\hbar\beta}{2}\frac{1}{\sqrt{2\pi\hbar|\tau'(E_\beta)|}} \exp-\left[\frac{2W(E_\beta)}{\hbar} + \beta E_\beta\right], \qquad (8.52)$$

where $\tau'(E) = d\tau(E)/dE$ and E_β is a solution of $2dW(E)/dE = -\hbar\beta$ with

$$2W(E) = \oint \sqrt{2M(V(q) - E)}dq. \qquad (8.53)$$

The expression for $Z_1(\beta)$ above is just another way (see, for example, Rajaraman (1987)) to write

$$Z_1(\beta) \approx -\frac{i\hbar\beta}{2}\left(\frac{\|\dot{q}_c\|^2}{2\pi\hbar}\right)^{1/2}\frac{1}{\sqrt{\det'\left[-M\partial_\tau^2 + V''(q_c(\tau))\right]}}$$

$$\times \exp-\frac{S_E\left[q_c(\tau)\right]}{\hbar}, \qquad (8.54)$$

where $q_c(\tau) = q_c^{(II)}(\tau)$. In (8.52) the complex i appears because of the deformation of the integration contour into the complex plane and the factor $1/2$ is due to the fact that the integration is performed only along half of the steepest descent through the saddle point (see Appendix D). There is also ambiguity in the sign of (8.54) (Langer, 1967), but we have chosen the minus sign for convenience. The multiplicative factor $\hbar\beta$, in contrast, appears because of the imaginary time translation invariance of the periodic solutions and comes out naturally from the integration over x.

For temperatures $\beta^{-1} > \beta_0^{-1}$ the zero mode disappears but there is still a negative eigenvalue corresponding to the constant solution q_b which preserves the imaginary contribution to $Z_1(\beta)$. The disappearance of the translation mode is a direct consequence of the fact that for such short imaginary times, there is no way we can accommodate periodic orbits satisfying $q_c(\hbar\beta/2) = q_c(-\hbar\beta/2)$ and, therefore, in the neighborhood of q_b we once again have to appeal for the non-periodic type I solutions. Another way to see the same result is by simply analyzing the eigenvalues of the second functional derivative of the action at q_b. If $\beta^{-1} > \beta_0^{-1}$ the lowest positive eigenvalue $\omega_1 - \omega_b > 0$, where $\omega_1 \equiv 2\pi/\hbar\beta$. In this limit, applying the SPA to the path integral in $Z_1(\beta)$ and performing the integration over x yields

$$Z_1(\beta) \approx -\frac{i}{2}N_b(\beta)\frac{1}{\sqrt{|\det[-M\partial_\tau^2 - M\omega_b^2]|}}\exp{-\beta V_0}$$

$$= \frac{-i}{4\sin(\beta\hbar\omega_b/2)}\exp{-\beta V_0}, \tag{8.55}$$

where $N_b(\beta)$ is the result of the x integration (appropriately continued into the complex plane) divided by the normalization constant $\mathcal{N}_R$ defined in Appendix A. The last step of (8.55) follows from the results of Appendices A and D. Finally, using the fact that $Z_0 \gg |Z_1|$ we can write

$$F = -\frac{1}{\beta}\ln Z(\beta) \approx -\frac{1}{\beta}\ln Z_0(\beta) + \frac{i}{\beta}\frac{|Z_1(\beta)|}{Z_0(\beta)}. \tag{8.56}$$

Now, if we establish the following relations:

$$\Gamma_\beta^{(0)} = \begin{cases} (2/\hbar)\,\mathrm{Im}\,F & \text{if } T \leq T_0 \\[2ex] (\omega_b\beta/\pi)\,\mathrm{Im}\,F & \text{if } T \geq T_0 \end{cases} \tag{8.57}$$

we get

$$\Gamma_\beta^{(0)} \approx \frac{2\sinh(\hbar\omega_0\beta/2)}{\sqrt{2\pi\hbar|\tau'(E_\beta)|}}\exp{-\left[\frac{2W(E_\beta)}{\hbar} + \beta E_\beta\right]}, \tag{8.58}$$

if $T < T_0$, and

$$\Gamma_\beta^{(0)} \approx \frac{\omega_b\sinh(\beta\hbar\omega_0/2)}{2\pi\sin(\beta\hbar\omega_b/2)}\exp{-\beta V_0}, \tag{8.59}$$

if $T > T_0$. Notice that (8.59) diverges as $T \to T_0$, which establishes the range of validity of that expression.

The motivation behind writing relations (8.57) is to reach the desired connection between $\mathrm{Im}\,F$ and the decay rate Γ which can, at least, be tested in some limiting cases. Actually, they have also been motivated by the application of WKB methods to the tunneling problem in the energy domain (Affleck, 1981; Weiss, 1999; Ankerhold, 2007), but these will not be reviewed here.

The first check is obviously the zero-temperature case. With the expressions we have defined in this section it is not hard to show that when $\beta \to \infty$ the first of (8.57) reproduces the zero-temperature expression (8.8), whereas when $\beta \to 0$ ($T \gg T_0$) the second of (8.57) gives us

$$\Gamma_\beta^{(0)} \approx \frac{\omega_0}{2\pi}\exp{-\beta V_0}, \tag{8.60}$$

which is the well-known *Arrhenius law* for the escape rate, a result that can be obtained by the classical *transition state theory* (TST) (see, for example, Weiss (1999) and references cited therein). Notice that these two relations coincide at $T = T_0$.

Here we should mention that by a somewhat different application of WKB methods combined with a semi-classical evaluation of the equilibrium density operator of the particle in a metastable well, it was shown (Waxman and Leggett, 1985) that

$$
\Gamma_\beta^{(0)} \approx \frac{1}{\sqrt{2\pi \langle q^2 \rangle}} \frac{||\dot{q}_c||^2}{2M|\ddot{q}_c(\hbar\beta/2)|} \left| \frac{\det(-M\partial_\tau^2 + M\omega_0^2)}{\det''(-M\partial_\tau^2 + V''(q_c(\tau)))} \right|^{1/2}
$$
$$
\times \exp -\frac{S_E[q_c(\tau)]}{\hbar}, \tag{8.61}
$$

where now $\det''$ means that both the zero and negative eigenvalues must be omitted from the evaluation of the ratio of determinants and $\langle * \rangle$ is the thermal average of the variable being considered. This alternative expression has exactly the same exponential behavior as those two forms in (8.57), which is a good hint that the exponential term has indeed the correct dependence at all temperatures.

Contrary to the exponent, the prefactor in (8.61) does not provide us with a meaningful result for $T \lesssim T_0$. When we reach this range of temperatures the prefactor vanishes because the term $||\dot{q}_c||^2/|\ddot{q}_c(\hbar\beta/2)| \to 0$, at variance with the expected high-temperature behavior of (8.60). Probably this takes place because the method used by Waxman and Leggett (1985) only takes into account spatially decaying WKB wave functions under the potential barrier of the metastable cubic potential.

In contrast, if we apply similar WKB methods to the same problem as in Affleck (1981), it becomes clear that the high-temperature contribution comes about because scattering states just above the top of the barrier are considered. Therefore, we can say that the free energy method presented here is a quantum version of the TST (Weiss, 1999), with which the method presented in Waxman and Leggett (1985) coincides at low temperatures. In other words, viewed in this way, the free energy method, applied to very low temperatures, has generalized the Arrhenius law to include quantum tunneling. Nevertheless, this sort of generalization is not yet free from criticism.

Classical activation rate

As is already known from the classical theory of the activation rate (Kramers, 1940), the question of the determination of the prefactor is a very subtle point that deserves special attention. We only highlight here a few results that exemplify the complexity of the problem (for a more thorough discussion on the subject, see also Weiss (1999)).

The classical TST assumes implicitly that, even for the undamped case, there is a thermal distribution of particles in the metastable well at finite temperatures. If this is indeed the case, there must be an exponential tail of the energy distribution extending past over V_0 and particles above this value can move freely across the potential barrier. So, after a very short (transient) time, all particles with energies above V_0 will have left their initial positions and it is very unlikely that the initial distribution function will keep its form since we are entirely neglecting the coupling to an external environment at temperature T. In order for the remaining particles to rearrange themselves into a new equilibrium distribution, it is crucial that they either interact among themselves or are acted on by an external agent. Therefore, apart from a fast transient during which an exponentially small number of particles leave the well, the decay rate becomes zero.

Once we introduce the interaction with an external environment and dissipation appears explicitly in the determination of the equilibrium distribution, it can be shown (Weiss, 1999) that the decay rate is given by

$$\Gamma_\beta \approx A_{cl}(\beta) \exp{-\beta V_0}, \tag{8.62}$$

where $A_{cl}(\beta) \propto \gamma \beta \omega_0 / 2\pi$, which clearly vanishes for zero damping as we have anticipated above.

For moderate-to-strong damping (Kramers, 1940), the decay factor can be computed from a Fokker–Planck-like equation for the probability density and its final form is exactly like that in (8.62), with $A_{cl}(\beta)$ given by (Weiss, 1999)[2]

$$A_{cl}(\beta) = f(\gamma)\frac{\omega_0}{2\pi} \tag{8.63}$$

where

$$f(\gamma) = \sqrt{1 + \left(\frac{\gamma}{\omega_b}\right)^2} - \frac{\gamma}{\omega_b}. \tag{8.64}$$

Notice that although this expression reproduces (8.60) when $\gamma \to 0$ we should bear in mind that it is not valid in this regime, for which one must apply (8.62).

For overdamped systems (8.63) still holds and in the particular case of very strongly damped systems ($\gamma \gg \omega_b$) it reduces to

$$A_{cl}(\beta) = \frac{\omega_0 \omega_b}{4\pi \gamma}, \tag{8.65}$$

which, once again, is much smaller than the prefactor of (8.60). The reason for this reduction is no longer the same as for the undamped case. Now, despite the existence of the interaction with the environment which ensures a fast recovery of the thermal equilibrium distribution in the well, the particles move sluggishly in

---

[2] The definition of γ in this reference and also in Ankerhold (2007) is twice what we use in this book.

space and therefore, the rate at which they leave the well is also small in this regime. In other words, for very weakly damped systems the activation rate is dominated by the process of energy diffusion whereas for very strongly damped systems it is dominated by spatial diffusion.

Damped systems

Once we have performed the former analysis, let us include dissipation in the present finite-temperature generalization of the undamped quantum tunneling problem. Fortunately, we have chosen to work with the free energy method which, albeit not free from criticism, provides us with a framework that can incorporate both finite temperature and damping effects straightforwardly. All one has to do is modify expression (8.48), replacing the finite-temperature action (8.49) by its damped version (8.14), and proceed as we have done for the finite T (or β) undamped case. Actually, using the symmetric imaginary time interval $-\hbar\beta/2 < \tau < \hbar\beta/2$ and, unlike what we have done to reach (6.91), not extending $q(\tau)$ outside this domain, we rewrite (8.14) in a more symmetric form as

$$
S_{eff}[q(\tau')] = \int_{-\hbar\beta/2}^{\hbar\beta/2} \left\{ \frac{1}{2}M\dot{q}^2 + V(q) \right\} d\tau'
$$

$$
+ \frac{\eta\pi}{4\hbar^2\beta^2} \int_{-\hbar\beta/2}^{\hbar\beta/2} d\tau'' \int_{-\hbar\beta/2}^{\hbar\beta/2} d\tau' \frac{\{q(\tau') - q(\tau'')\}^2}{\sin^2(\pi(\tau'-\tau'')/\hbar\beta)}, \quad (8.66)
$$

from which the following equation of motion and eigenvalue problem arise:

$$
\left. \frac{\delta S_{eff}}{\delta q} \right|_{q_c} = M\ddot{q}_c - \frac{\partial V}{\partial q_c} - \frac{\eta\pi}{\hbar^2\beta^2} \int_{-\hbar\beta/2}^{\hbar\beta/2} d\tau'' \frac{[q_c(\tau') - q_c(\tau'')]}{\sin^2(\pi(\tau'-\tau'')/\hbar\beta)} = 0
$$

$$(8.67)$$

and

$$
\hat{D}_\beta q(\tau') = -M\frac{d^2q(\tau')}{d\tau'^2} + V''(q_c)q(\tau') + \hat{O}_\beta q(\tau') = kq(\tau'), \quad (8.68)
$$

where the differential operator $\hat{O}_\beta$ acting on $q(\tau')$ reads

$$
\hat{O}_\beta q(\tau') = \frac{\eta\pi}{\hbar^2\beta^2} \int_{-\hbar\beta/2}^{\hbar\beta/2} d\tau'' \frac{[q(\tau') - q(\tau'')]}{\sin^2(\pi(\tau'-\tau'')/\hbar\beta)}. \quad (8.69)
$$

Now we proceed as before and evaluate the partition function of the system using SPA. The same sort of analysis for low or high temperatures is also possible in the present case, although there are some important differences the reader should be warned of.

On top of having to deal with periodic paths in finite imaginary time intervals, the concept of Euclidean "energy" is no longer very useful here.[3] For example, the reasoning that led us to (8.52) cannot be applied in the present case. However, a simple qualitative analysis of the problem makes us conclude that we must still have the same sort of structure of stationary paths for $S_{eff}[q_c(\tau)]$ as before, but this time for the deformed solutions of the dissipative equations of motion (8.67). Therefore, we are going to use a generalization of the finite damping expression (8.19) to finite temperatures,

$$
\Gamma_\beta \approx \sqrt{\frac{||\dot{q}_c||^2}{2\pi\hbar}} \left| \frac{\det D_\beta^{(0)}}{\det' D_\beta} \right|^{1/2} \exp -\frac{S_{eff}[q_c(\tau)]}{\hbar}, \tag{8.70}
$$

where $\det D_\beta^{(0)}$ is evaluated for $V''(q_c(\tau)) = M\omega_0^2$ in (8.68). Notice that the form (8.70) already accounts for the existence of a zero mode in our problem, which is not generally the case. It is only valid for temperatures lower than the modified crossover temperature for damped systems T_R, which we shall define below. For temperatures $T > T_R$ the zero mode disappears and the resulting expression is no longer applicable.

Although the problem is basically the same as we have been dealing with so far, the solution of the equation of motion (8.67) and the eigenvalue problem (8.68) is by no means an easy task now.

The method widely adoped in the literature to tackle these issues (see, for example, Waxman and Leggett (1985), Grabert *et al.* (1987), Weiss (1999), Ankerhold (2007)) makes use of the Fourier representation of the periodic paths $q(\tau)$ in which we decompose

$$
q(\tau) = \sum_{n=-\infty}^{\infty} q_n \exp -i\omega_n\tau, \tag{8.71}
$$

where $\omega_n = 2n\pi/\hbar\beta$ are the well-known *Matsubara frequencies*

$$
(M\omega_n^2 + 2M\gamma|\omega_n| + M\omega_0^2)q_n - \frac{3M\omega_0^2}{2q_0} \sum_{m=-\infty}^{\infty} q_m q_{m-n} = 0, \tag{8.72}
$$

where, for convenience, we have used the fact that $\lambda = 3M\omega_0^2/q_0$. Note that this equation must be complemented with the boundary condition $q(\hbar\beta/2) = q(-\hbar\beta/2) = x$, which implies that

---

[3] Although not constant, it is still symmetric within the finite imaginary time interval.

$$x = \sum_{n=-\infty}^{\infty} (-1)^n q_n. \tag{8.73}$$

This constraint must be included in the variational problem that generates (8.72) through the introduction of the Lagrange multiplier κ, which gives us

$$(M\omega_n^2 + 2M\gamma|\omega_n| + M\omega_0^2)q_n - \frac{3M\omega_0^2}{2q_0} \sum_{m=-\infty}^{\infty} q_m q_{m-n} = \kappa(-1)^n. \tag{8.74}$$

This allows us to determine q_n as a function of κ, which in turn can be taken to (8.73) to finally determine $\kappa = \kappa(x, \omega_n)$ and, consequently, $q_n = q_n(x, \omega_n)$.

The action can also be written in terms of the Fourier amplitudes and reads

$$S_{eff}(z_n) = \hbar\omega_0\beta B_0 \left[\sum_{n=-\infty}^{\infty} (\nu_n^2 + 2\alpha|\nu_n| + 1)z_n z_{-n} \right.$$
$$\left. - \sum_{n=-\infty}^{\infty} \sum_{m=-\infty}^{\infty} z_m z_{m-n} z_n \right], \tag{8.75}$$

where $\nu_n = \omega_n/\omega_0$, $z_n = q_n/q_0$, and $B_0 = M\omega_0 q_0^2/2$. It is a simple matter to rewrite the equation of motion (8.72) in terms of these dimensionless variables as

$$(\nu_n^2 + 2\alpha|\nu_n| + 1)z_n - \frac{3}{2} \sum_{m=-\infty}^{\infty} z_m z_{m-n} = 0, \tag{8.76}$$

whose variation allows us to write the eigenvalue equation $D_\beta f_n = k_n f_n$ as

$$(\nu_n^2 + 2\alpha|\nu_n| + 1)f_n - 3 \sum_{m=-\infty}^{\infty} z_{n-m}^{(c)} f_m = k_n f_n, \tag{8.77}$$

where $z_n^{(c)}$ is a solution of (8.76) (or the dimensionless version of (8.74)) and k_n is measured in units of $\hbar\omega_0\beta B_0$.

Although our problem looks much simpler in its present form, in general we have to appeal to numerical methods in order to find the complete behavior of the decay rate as a function of damping and temperature (Chang and Chakravarty, 1984; Waxman and Leggett, 1985; Grabert *et al.*, 1987), which allows us to draw very general conclusions about the semi-classical approach for the decay rate. However, there are few situations where analytical solutions are available and these can help us draw those general conclusions without recourse to numerical methods. Therefore, let us briefly highlight them in what follows (Ankerhold, 2007).

The low-temperature regime. The first analytical results available are in the very low-temperature domain, either for very weak or very strong damping. We have already obtained the zero-temperature results in these two latter limits in (8.23), (8.24), (8.33), and (8.43). What we have to do now is consider the solution of the equation of motion (8.67) for very large β as approximately the zero-temperature bounce (8.5) ((8.32)) for the very weakly (strongly) damped case and write the action at finite temperatures as the zero-temperature action minus $2 \int_{\hbar\beta/2}^{\infty} S_E[q_c(\tau)] d\tau$ to adjust it to a finite but still very large imaginary time interval. The small correction we have just mentioned gives rise to an asymptotic expansion in β^{-1} whose leading corrections are shown below in the final expression of the decay rate.

Writing it in the form $\Gamma_\beta = A_\beta \exp - B_\beta/\hbar$ we find that for very low damping

$$B_\beta \approx B - \frac{5B_0}{2\pi} \left(\frac{2\pi}{\hbar\omega_0\beta}\right)^2 = B - 10\pi B_0 \left(\frac{k_B T}{\hbar\omega_0}\right)^2, \tag{8.78}$$

where B already contains the first-order correction due to α as in (8.23) and B_0 is the undamped action of the bounce. Notice that to lowest order in α and T (or β^{-1}) the corrections are independent of each other. A combined contribution only shows up to order αT^4.

For high damping we have

$$B_\beta \approx B - \frac{4\pi B}{3} \left(\frac{2\alpha}{\hbar\omega_0\beta}\right)^2, \tag{8.79}$$

where now B is given by (8.33). Notice that in this case the correction to lowest order in temperature is also α dependent.

Therefore, we conclude that for both weakly as well as strongly damped cases, the lowest-order corrections to the decay rate due to finite-temperature effects behave as $a(\alpha)T^2$, where $a(\alpha)$ is always negative and independent of (strongly dependent on) α for very weakly (strongly) damped systems. This means that as the temperature is raised, the tendency of the system to occupy higher energy states within the metastable well clearly increases the decay rate as these states see lower and narrower energy barriers. Note that for low-temperature corrections we have ignored its influence coming from the prefactor A_β and this is because, as we have seen before, the exponential contribution dominates the behavior of the decay rate strongly even for $T = 0$. Another important remark about these results is that the T^2 law is dependent on the sort of damping the system is subject to. In other words, it depends on the low-frequency behavior of the spectral function $J(\omega)$ as in (5.30). In general, the temperature dependence of the decay rate is $a_s T^{s+1}$.

High-temperature regime. Now, for high temperatures (or small β) the main contributions to the path integral in the partition function come from the stationary points of the action (8.75) at $q(\tau) = 0$ and $q(\tau) = q_b$, about which we expand the fluctuations, respectively, as $x(\tau) = \sum_n X_n \exp{-i\omega_n \tau}$ and $y(\tau) = q_b + \sum_n Y_n \exp{-i\omega_n \tau}$. The action about each of these read

$$S(X_n) = \frac{M\hbar\beta}{2} \sum_{n=-\infty}^{\infty} \Lambda_n^{(0)} X_n X_{-n} \tag{8.80}$$

and

$$S(Y_n) = \hbar\beta V_0 + \frac{M\hbar\beta}{2} \sum_{n=-\infty}^{\infty} \Lambda_n^{(b)} Y_n Y_{-n}, \tag{8.81}$$

whose second functional derivative eigenvalues are $\Lambda_n^{(0)} = \omega_n^2 + \omega_0^2 + 2|\omega_n|\gamma$ and $\Lambda_n^{(b)} = \omega_n^2 - \omega_b^2 + 2|\omega_n|\gamma$.

For times $\hbar\beta \to 0$, similarly to what happened in the undamped case, there is a crossover temperature, $T_R \equiv \hbar\omega_R/2\pi k_B$, above which the zero eigenvalue at $q(\tau) = q_b$ vanishes. The value of ω_R is then determined in such a way that this condition is met, which means that it is the largest positive root of

$$\Lambda_R^{(b)} = \omega_R^2 - \omega_b^2 + 2\omega_R \gamma = 0. \tag{8.82}$$

It is a simple matter to show that when $\gamma \ll \omega_b$, $T_R = T_0 = \hbar\omega_b/2\pi k_B$, whereas it is given by $T_R = \hbar\omega_b^2/4\pi\gamma k_B \ll T_0$ if $\gamma \gg \omega_b$.

Now, evaluating (8.56) with the help of (8.80) and (8.81) and employing the high-temperature form in (8.57), it can be shown (Ankerhold, 2007) that

$$\Gamma = \frac{\omega_0}{2\pi} \frac{\omega_R}{\omega_b} f_q \exp{-\beta V_0}, \tag{8.83}$$

where

$$f_q = \prod_{n=1}^{\infty} \frac{\omega_n^2 + \omega_0^2 + 2\gamma|\omega_n|}{\omega_n^2 - \omega_b^2 + 2\gamma|\omega_n|}. \tag{8.84}$$

Expression (8.83) has been analyzed in Ankerhold (2007). It is shown that for very weakly damped systems at temperatures such that $\hbar\omega_0\beta \ll 1$ we have

$$f_q \approx \exp\left[\frac{\hbar^2\beta^2}{24}(\omega_0^2 + \omega_b^2)\right], \tag{8.85}$$

which represents a quantum mechanical correction to the classical Arrhenius factor (8.60). It now reads

$$\Gamma_\beta = \frac{\omega_0}{2\pi} \exp{-\beta\left(V_0 - \frac{\hbar^2\beta}{24}(\omega_0^2 + \omega_b^2)\right)}, \tag{8.86}$$

meaning that the potential barrier is reduced slightly by quantum fluctuations. On the contrary, for strongly overdamped systems ($\gamma \gg \omega_0$), we still have f_q given by (8.85) if $\hbar\gamma\beta \ll 1$. However, this time the corrected exponential term is multiplied by the prefactor $A_{cl}(\beta)$ given by (8.65).

Finally, for $T_R \ll T \ll \hbar\gamma/k_B$, the resulting form of f_q (see Ankerhold (2007)) yields

$$\Gamma_\beta = \frac{\omega_0 \omega_b}{4\pi\gamma} \exp -\beta \left(V_0 - \frac{\hbar(\omega_0^2 + \omega_b^2)}{4\pi\gamma} \ln \frac{\hbar\gamma\beta}{\pi} \right), \qquad (8.87)$$

which shows that, in this temperature range, overdamped quantum fluctuations reduce the potential barrier more efficiently, consequently enhancing the decay rate.

Now that we have analyzed all possible situations we can draw general conclusions about the problem of the decay rate of a damped particle out of a metastable well.

Figure 8.3 presents a qualitative picture of our findings of the behavior of the decay rate as a function of temperature for two different values of damping. Initially we see that, for very low temperatures, quantum tunneling is the dominant feature in the process of decay. Moreover, the tunneling rate depends on the damping constant and is reduced for increasing dissipation. At least this seems to be the case for most damped systems, in particular, systems subject to ohmic dissipation which has been the example treated explicitly in this chapter.

As we raise the temperature, the rate starts to increase. Even though damping tends to reduce tunneling, the increase in temperature reduces the action of bounce-like paths under the potential barrier, which tends to increase the decay rate. As we have seen, this reduces the action by a term proportional to T^2 for ohmic damping, and the proportionality constant increases with damping. This is seen clearly in Fig. 8.3, where the sensitivity of the decay rate to temperature is much more evident in the stronger-damped case.

As the temperature is increased further, higher-order powers of T will come into play, correcting the zero-temperature action until the crossover temperature T_R is reached. Above it, thermal activation dominates the decay rate, which is seen by the straight-line asymptote in Fig. 8.3 at high temperatures. Notice that this crossover takes place at much lower temperatures for highly damped systems as $T_R^{(1)} > T_R^{(2)}$, although quantum corrections to the classical decay formula are also much stronger in this case. In other words, the crossover region is clearly broader for highly damped systems.

Experimental evidence of the validity of this theory can be found in Clarke *et al.* (1988), where the probability of generation of a finite-voltage state between the terminals of a current-biased Josephson junction was measured. They performed

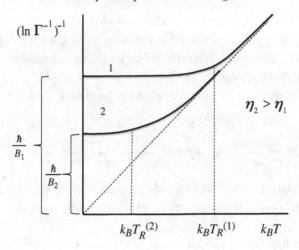

Figure 8.3 Escape rate as a function of temperature for two different values of damping, $\eta_1 \approx 0$ and η_2

these measurements for currents lower than the critical current at ultra-low temperatures (down to ≈ 5 mK). For these values of the parameters we would expect the generation of finite voltages through the junction only by quantum tunneling of the phase of the superconducting wave function (quantum phase slip). There is excellent agreement with the theoretical predictions as far as the exponential contribution to Γ is concerned. The contribution of the prefactor is still an open question.

8.2 Field theories

Having addressed the problem of the decay of a given configuration described by a single dynamical variable (point particle) out of a metastable state, it is now time to study the same situation when the physical system is described by a continuous dynamical variable. This is indeed the case for many examples we have given in our introductory chapters, such as nucleation and depinning in magnetic and superconducting systems.

Although there are many different realistic situations we could deal with, we exemplify the relevant physics of quantum decay in field theory through a specific model we believe contains the most important features of all those realistic situations mentioned in the introductory chapters. We concentrate on the so-called *double sine-Gordon* model (Furuya and Caldeira, 1989), which appeared naturally in the homogeneous nucleation problem in magnetic systems (see Section 2.3.2). Besides containing all the ingredients we need, this potential is bounded below and much more suitable for the analytical continuation procedure we often need in this kind of problem.

The double sine-Gordon equation is given by

$$\frac{1}{c^2}\frac{\partial^2\varphi}{\partial t^2} - \nabla_D^2\varphi + g\sin 2\varphi + \lambda\sin\varphi = 0, \tag{8.88}$$

where $g > 0$, $\varphi = \varphi(\mathbf{r}, t)$, $\mathbf{r} \in \mathbb{R}^D$, and ∇_D^2 is a D-dimensional Laplacian. Notice that this equation is the generalization of (2.76) for a scalar dynamical variable in D-dimensional space subject to the potential energy density

$$U(\varphi) = g\sin^2\varphi + \lambda(1 - \cos\varphi). \tag{8.89}$$

Therefore, the previous analyses of Sections 2.3.1 and 2.3.2 apply here also for the appropriate values of λ. Metastability takes place when $\lambda_c \equiv -2g < \lambda < 0$, as can be seen from Figs 2.5, 2.6, and 2.7. Once again we want to emphasize that although we are concentrating on the double sine-Gordon potential, our subsequent results can easily be reproduced for the asymmetric φ^4 model, the quadratic-plus-cubic potential energy density, or any other model presenting metastable configurations due to the modification of control parameters.

8.2.1 The undamped zero-temperature case

In order to extend our previous results for the point particle to a continuum model, we had best use the path integral formulation of the problem since the resulting expressions (8.2) and (8.19) for the undamped and damped particles, respectively, have a very direct generalization to field theory.

Let us start with the undamped case. Instead of going, for example, through the details of the evaluation of the partition function (8.48) for a field theoretical model

$$Z(\beta) = \text{Tr}\exp -\beta H = \int \mathcal{D}\bar{\varphi}(\mathbf{r}') \int_{\bar{\varphi}(\mathbf{r}')}^{\bar{\varphi}(\mathbf{r}')} \mathcal{D}\varphi(\mathbf{r}', \tau')\exp -\frac{S_E[\varphi(\mathbf{r}', \tau')]}{\hbar}, \tag{8.90}$$

where $\bar{\varphi}(\mathbf{r}')$ is an arbitrary field configuration and

$$S_E[\varphi(\mathbf{r}', \tau')] = \varphi_{0i}^2 \int_{-\hbar\beta/2}^{+\hbar\beta/2} \int_{V_D} d\tau' \, d^D x \left\{\frac{1}{2c^2}\left(\frac{\partial\varphi}{\partial\tau}\right)^2 + \frac{1}{2}|\nabla_D\varphi|^2 + U(\varphi)\right\}, \tag{8.91}$$

when $\hbar\beta \to \infty$, we rely on the direct extension of the resulting expression (8.2) for our present model. In the expression above, φ_0 is a parameter such that

$$\left[\frac{\varphi_0^2}{c}\right]L^{D-1} = [\hbar], \tag{8.92}$$

and φ is a dimensionless field.

The extremal solution of the action (8.91) with appropriate boundary conditions (the bounce) together with the stability analysis of its second functional derivative at that configuration is still very similar to the single-variable case (see below). Apart from the obvious increase in complexity due to the change in dimensionality of the problem, one simple modification must be stressed here. Like the zero-dimensional ($D = 0$) point particle problem, the double sine-Gordon equation is still time translation invariant. Moreover, it can easily be observed that it is also invariant by the translation of the origin of coordinates. Therefore, when evaluating the partition function path integral, on top of the multiplicative factor proportional to the imaginary time interval $\hbar\beta \to \infty$ coming from the integration of the "zero mode," another term appears proportional to the D-dimensional hypervolume, $V_D \to \infty$, due to the D-dimensional position integration of the Lagrangian density. Consequently, what is evaluated in the generalization of (8.2) is

$$\frac{\Gamma_0}{V_D} = A_0 \exp -\frac{B_0}{\hbar}, \tag{8.93}$$

or the decay rate per unit of the D-dimensional hypervolume.

In (8.93) we have $B_0 = S_E[\varphi_c]$, where the potential energy density in (8.91) is given by $U(\varphi_c) = g \sin^2 \varphi_c - |\lambda|(1 - \cos \varphi_c)$ because we are already assuming that $\lambda < 0$. In other words, $\varphi = 2n\pi$ are now metastable configurations. The bounce $\varphi_c(\mathbf{r}, \tau)$ is a solution of $\delta S_E/\delta\varphi = 0$ with "spherical" symmetry and obeys

$$\frac{d^2\varphi}{d\rho^2} + \frac{D}{\rho}\frac{d\varphi}{d\rho} - g \sin 2\varphi - |\lambda| \sin \varphi = 0, \tag{8.94}$$

where $\rho = \sqrt{c^2\tau^2 + r^2}$ and the boundary conditions are

$$\varphi_c(\rho = \infty) = 0 \qquad \text{and} \qquad \frac{d\varphi_c}{d\rho}\bigg|_{\rho=0} = 0. \tag{8.95}$$

Using this particular symmetry of the bounce, we can write its action as

$$B_0 = \frac{\varphi_0^2}{c} \int_{-\infty}^{+\infty} c\,d\tau \int_0^\infty r^{D-1}\,dr \int_0^{2\pi} d\theta_1 \int_0^\pi d\theta_2 \ldots \int_0^\pi d\theta_{D-1} \sin^{D-2}\theta_{D-1} \ldots \sin\theta_2$$

$$\times \left\{ \frac{1}{2c^2}\left(\frac{\partial\varphi_c}{\partial\tau}\right)^2 + \frac{1}{2}|\nabla_D\varphi_c|^2 + U(\varphi_c) \right\}, \tag{8.96}$$

which after the evaluation of the angular integrations can be written as

$$B_0 = \frac{\varphi_0^2}{c} N(D) \int_0^\infty d\rho\, \rho^D \left\{ \frac{1}{2}\left(\frac{\partial\varphi_c}{\partial\rho}\right)^2 + g\sin^2\varphi_c - |\lambda|(1 - \cos\varphi_c) \right\}, \tag{8.97}$$

where $N(D) = 2\pi^{(D+1)/2}/\Gamma((D+1)/2)$ is a numerical factor which depends on the dimensionality of our system. $\Gamma(*)$ above is the well-known Gamma function. Notice that the expression (8.96) is only defined for $D \geq 2$. When $D = 1$ it is still valid with no angular integration and replacing the spatial integration

$$\int_0^\infty dr \rightarrow \int_{-\infty}^\infty dx.$$

For $D = 0$ (point particle) only the time integration remains.

As we have mentioned above, the stability analysis of the second functional derivative of the action still follows closely what we have done for the point particle case, which leads us to the field theoretical prefactor

$$A_0 = \left(\frac{B_0}{2\pi\hbar}\right)^{(D+1)/2} c \sqrt{\frac{\det(-\nabla^2_{(D+1)} + U''(0))}{\det'(-\nabla^2_{(D+1)} + U''(\varphi_c))}}, \tag{8.98}$$

where the exponent $(D + 1)/2$ results from the omission of $D + 1$ translation-invariant modes from the denominator of the ratio of determinants and each contributes to the prefactor with a multiplicative factor $B_0/2\pi\hbar$ (Callan and Coleman, 1977).

The classical instability of $U(\varphi)$ takes place when the metastable configurations at $2n\pi$ become maxima of the potential energy density, which happens when $\lambda = -2g$. Therefore, in order to observe quantum tunneling within a reasonable time interval, we should work close to this instability which provides us with small energy barriers. Let us then define the parameter ϵ such that $|\lambda| \equiv 2g(1 - \epsilon)$, which measures how far we are from the classical instability. In terms of the latter, the potential energy density $U(\varphi)$ reads

$$U(\varphi) = -4g \sin^4 \frac{\varphi}{2} + 4g\epsilon \sin^2 \frac{\varphi}{2}. \tag{8.99}$$

Now, expressing the field φ in terms of a dimensionless variable $x \equiv \sqrt{2g}\rho$, we can rewrite the action B as

$$B = \frac{N(D)\varphi_0^2}{(2g)^{(D-1)/2}c} \int_0^\infty dx \, x^D \left\{ \frac{1}{2}\left(\frac{\partial\varphi_c}{\partial x}\right)^2 - 2\sin^4 \frac{\varphi_c}{2} + 2\epsilon \sin^2 \frac{\varphi_c}{2} \right\} \tag{8.100}$$

and the equation of motion (8.94) as

$$\frac{d^2\varphi_c}{dx^2} + \frac{D}{x}\frac{d\varphi_c}{dx} = \sin\varphi_c(\cos\varphi_c - 1) + \epsilon \sin\varphi_c, \tag{8.101}$$

with boundary conditions $\varphi_c(x = \infty) = 0$ and $d\varphi_c/dx|_{x=0} = 0$. Integrating (8.100) by parts and using (8.101), we end up with

$$B = P_D I_D(\epsilon) \qquad \text{where} \qquad P_D = \frac{2N(D)\varphi_0^2}{(2g)^{(D-1)/2}c} \qquad \text{and}$$

$$I_D(\epsilon) = \frac{1}{2}\int\limits_0^\infty dx\, x^D \left\{ 2\varphi_c \sin^3 \frac{\varphi_c}{2} \cos \frac{\varphi_c}{2} - 2\sin^4 \frac{\varphi_c}{2} \right.$$

$$\left. + \epsilon \left(2\sin^2 \frac{\varphi_c}{2} - \frac{\varphi_c}{2} \sin \varphi_c \right) \right\}. \tag{8.102}$$

In order to evaluate the bounce action (8.102), we need to find the bounce itself by solving (8.101). The solution of the latter is only available by numerical methods, but its qualitative behavior can easily be analyzed from a mechanical analogue. As we can see, (8.101) is an equation of motion of a particle of coordinate φ_c subject to the potential $-U(\varphi_c)$ (see Fig. 8.4) and a viscous damping inversely proportional to the "time" parameter x. A solution obeying the appropriate boundary conditions is obtained when we find an initial position $\bar{\varphi}$ from which the particle is released at rest at "time" $x = 0$ and reaches $\varphi = 0$ when $x \to \infty$. Notice that this choice is unique, since for $\varphi(0) < \bar{\varphi}$ the particle ends up at the nearest minimum of $-U(\varphi)$ whereas for $\varphi(0) > \bar{\varphi}$ it overshoots. Therefore, sweeping different values of $\varphi(0)$ enables us to identify the desired value $\bar{\varphi}$.

We think this is a suitable place to make the connection between the purely mathematical bounce solution in imaginary time and the physical droplet profile shown in Fig. 2.10. As a matter of fact, the bounce is exactly the solution we drew in the latter figure if we define $\rho \equiv \sqrt{c^2\tau^2 + r^2}$. Now, making a *Wick rotation*, which means replacing $\tau \to it$, we easily see that an expanding droplet $\varphi(\sqrt{-c^2t^2 + r^2})$ is found. Physically, this means that at $t = \tau = 0$ a deformation of the metastable configuration in the form of a droplet is materialized in real space and afterwards it expands, leaving the whole system excited about a more stable configuration. The

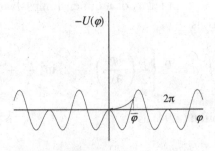

Figure 8.4 Inverted potential

relaxation to the stable configuration requires the presence of dissipation, which we have not yet introduced in this example.

After having obtained the bounce solution it can be taken to (8.102), from which we finally get the bounce action $B(\epsilon)$. The numerical evaluation of the latter was performed in Furuya and Caldeira (1989) for $D = 1, 2,$ and 3. Here, we should note that although we have computed the action of the bounce for any $0 < \epsilon < 1$, the limit $\epsilon \to 0$ cannot be taken at will. We must always keep in mind that due to the approximation we have been using, namely the WKB approximation, we need parameters c, g, and φ_0 such that $B(\epsilon)\hbar^{-1} \gtrsim 1$. For values of ϵ below which this relation is not obeyed, our approximation is no longer reliable.

An interesting feature of the dependence of the bounce action with the parameter ϵ is that it can be obtained analytically, at least for low dimensions ($D = 1$ or 2), from an approximate form of the potential energy density. Expanding $U(\varphi)$ about $\varphi = 0$ up to quartic terms in φ we have

$$U(\varphi) \approx 2g\epsilon \frac{\varphi^2}{2} - 2g \frac{\varphi^4}{8} - 2g\epsilon \frac{\varphi^4}{24}, \tag{8.103}$$

which, in terms of a dimensionless variable $\tilde{x} \equiv \sqrt{\epsilon}x$ and a dimensionless field $y(\tilde{x}) \equiv \varphi(\tilde{x})/\sqrt{2\epsilon}$, simplifies to

$$B(\epsilon) \approx P_D \epsilon^{(3-D)/2} \tilde{I}_D \quad \text{with} \quad \tilde{I}_D = \int_0^\infty d\tilde{x}\, \tilde{x}^D \left\{ \frac{1}{2}\left(\frac{dy_c}{d\tilde{x}}\right)^2 + \frac{y_c^2}{2} - \frac{y_c^4}{4} \right\}, \tag{8.104}$$

where P_D has been defined previously in (8.102) and $y_c(\tilde{x})$ is a solution of

$$\frac{d^2 y_c}{d\tilde{x}^2} + \frac{D}{\tilde{x}} \frac{dy_c}{d\tilde{x}} = y_c - y_c^3, \tag{8.105}$$

which can be solved as (8.101). Integrating $\tilde{I}_D$ above by parts and using (8.105), we have

$$\tilde{I}_D = \int_0^\infty d\tilde{x}\, \tilde{x}^D \frac{y_c^4}{4}, \tag{8.106}$$

which like (8.105) is independent of ϵ.

So, as can easily be seen, the ϵ dependence has been scaled out completely from (8.102) and it turns out that it matches exactly the behavior of the numerical solutions for $B(\epsilon)$ obtained with the exact form of $U(\varphi)$ when $\epsilon \to 0$ for $D = 1$ and 2 (see Furuya and Caldeira (1989)). Actually, it has been shown explicitly in the latter reference that there is no solution of (8.105) with the imposed boundary condition for $D = 3$ and, consequently, only the exact form of the potential energy density can account for the small ϵ dependence in this case. Fortunately, most of the examples we have presented in the introductory chapters of this book involve only low-dimensional ($D = 1$ or 2) dynamical variables, which allows us to use the simple analytical dependence on ϵ in (8.104) to treat the low-barrier phenomenon.

Now let us present a brief analysis of the prefactor A_0. As we have seen in previous sections, the evaluation of the ratio of determinants is not a simple task, even for the undamped point particle tunneling problem. The complexity increases naturally with dimensionality, as shown in Langer (1967) and Callan and Coleman (1977), which makes this evaluation a mathematical problem certainly beyond our aim in this book. Nevertheless, there is some useful information we can get without solving the problem explicitly. For example, let us concentrate on $D = 1$ or 2, from which we could extract the analytical dependence on ϵ close to the classical instability.

The eigenvalue problems we have to deal with in order to compute the ratio of determinants that come from the second variation of the action (8.91) are of the form

$$-\frac{d^2\psi}{d\rho^2} - \frac{D}{\rho}\frac{d\psi}{d\rho} + \frac{k_{(D)}}{\rho^2}\psi + U''(\varphi_c)\psi = E_{(D)}\psi, \qquad (8.107)$$

where $k_{(1)} = m^2$ with $m = 0, \pm 1, \pm 2, \ldots$, $k_{(2)} = \ell(\ell + 1)$ with $\ell = 0, 1, \ldots$, and $E_{(1)} = E_{nm}$ and $E_{(2)} = E_{n\ell}$ are the corresponding eigenvalues. Therefore, writing the potential energy density in the approximate form (8.103) and rescaling all variables to $\tilde{x}$ and $y(\tilde{x})$, we have

$$-\frac{d^2\psi}{d\tilde{x}^2} - \frac{D}{\tilde{x}}\frac{d\psi}{d\tilde{x}} + \frac{k_{(D)}}{\tilde{x}^2}\psi + U''(y_c)\psi = \frac{E_{(D)}}{2g\epsilon}, \qquad (8.108)$$

implying that all the eigenvalues we are interested in scale as $2g\epsilon \tilde{E}_D$, where the eigenvalues $\tilde{E}_D$ do not depend on ϵ. This means that the determinants of the two Schrödinger-like operators involved in the evaluation of the prefactor contain an infinite product of terms of that sort and if not for the omission of the zero eigenvalues from the ratio of determinants would become independent of ϵ, at least when $\epsilon \to 0$. However, remembering that we have $D + 1$ translation modes which must be omitted from the denominator of the ratio of determinants in (8.98), simple dimensional arguments (Callan and Coleman, 1977) lead us to

$$\sqrt{\frac{\det(-\nabla^2_{(D+1)} + 2g\epsilon)}{\det'(-\nabla^2_{(D+1)} + U''(\varphi_c))}} = (2g\epsilon)^{(D+1)/2} R_D, \qquad (8.109)$$

where R_D is the ratio of the determinants of operators of the form (8.108) with "potentials" $U''(0)$ and $U''(y_c)$ and, consequently, independent of ϵ.

Putting all our findings together in (8.98) and (8.93) we finally have

$$\frac{\Gamma}{V_D} \approx \left(\frac{g\epsilon B(\epsilon)}{\pi\hbar}\right)^{(D+1)/2} c\, R_D \exp-\frac{B(\epsilon)}{\hbar}, \qquad (8.110)$$

where $B(\epsilon)$ is given by (8.104). So, at least for $D = 1$ or 2, we have been able to compute analytically the dependence of Γ/V_D on ϵ.

Note that in order to write the final expression (8.110) we have assumed that there is only one negative eigenvalue of the Schrödinger-like operator in (8.107). Although this seems a very plausible assumption, it has only been proved (at least for $D = 3$) in the limit $\epsilon \to 1$ (Callan and Coleman, 1977), which is exactly the opposite to what we have been exploring here (see the thin-wall approximation below). If it happens that other negative eigenvalues exist, then the whole procedure for computing the decay rate must be modified.

Another important remark about the approach to the classical instability, apart from that referring to the validity of the WKB approximation, has to do with zero-point energy fluctuations about a metastable configuration. To simplify matters, let us consider a point particle in a metastable potential energy minimum. The idea is that even if we rely on the WKB approximation for energy barriers not so high, it does not make sense to reach values of ϵ such that $U_0(\epsilon) \approx \hbar\omega_0$. If we reduce its value further, the treatment we have dealt with here is no longer applicable since zero-point energy fluctuations will be large enough to drive the system unstable. The extension of this reasoning to field theoretical cases involves the evaluation of effective potentials through the so-called *one-loop approximation*, which is quite well known to field theorists and certainly beyond the scope of this book. The application of these ideas to the double sine-Gordon system was carried out in Furuya and Caldeira (1989) for $D = 1$ and 2.

Thin-wall approximation. Although quantum mechanical tunneling is much more likely to take place close to the classical instability due to the presence of small potential barriers we will now present, for the sake of completeness, the *thin-wall approximation* (TWA) which is actually suitable for the other extreme situation when the parameter λ in (8.89) is very small. In other words, the different minima of $U(\varphi)$ respectively at $\varphi = 2n\pi$ and $(2n + 1)\pi$ are almost degenerate. Despite the fact that tunneling is very unlikely in this situation, unless the barriers $U(\varphi \approx (2n + 1)\pi/2) \gtrsim \hbar\omega_0$ (a very weakly corrugated potential which, however, is not appropriate for the WKB approximation), the importance of the TWA is twofold. Firstly, it provides us with an approximate analytical solution to the tunneling problem in general continuum models and secondly, it might be an appropriate way to deal with the important *false vacuum problem* in quantum field theory and cosmology (Frampton, 1976; Stone, 1976; Callan and Coleman, 1977; Coleman, 1977; Guth, 1981).

The main difference between this present case and the one we have treated previously is the way we approximate the solution of (8.94) when $\lambda = -|\lambda| \to 0$. Now, the initial field configuration $\bar{\varphi}$ of Fig. 8.4 lies very close to one of the maxima of $-U(\varphi)$ at $\varphi = \pm 2\pi$, which is almost degenerate to the maximum at $\varphi = 0$. If we appeal to the interpretation of (8.94) as an equation of motion of a particle with coordinate φ at time ρ, it is evident that this fictitious motion presents a damping term[4] inversely proportional to time. As we are assuming that $\varphi = 0$ and $\varphi = \pm\pi$ are almost degenerate, the fictitious particle must stay at $\bar{\varphi} \approx \pi$ (we have chosen the plus sign only for convevience) for as long as it is necessary to neglect the "dissipative" term in (8.94) (see Coleman (1977)). After this very long time, say $\rho \approx R$, the particle leaves $\bar{\varphi}$ and moves swiftly to $\varphi = 0$, where it remains to satisfy the boundary condition $\varphi(\infty) = 0$. Therefore, it is a reasonable hypothesis to assume that $\varphi(\rho)$ has the form of the soliton solution (2.51) as a function of ρ. Moreover, this soliton is centered at the "instant" R, about which the quick excursion to the maximum at $\varphi = 0$ takes place. In other words, it behaves as the solution drawn in Fig. 2.11 as a function of $\rho = \sqrt{c^2\tau^2 + r^2}$, or $\varphi_c(\rho - R)$. The actual solution in real time is easily obtained by the analytic continuation through the replacement $\tau \to it$.

Physically this means that at $t = 0$ a very large droplet with $\varphi = \pi$ and radius R nucleates by quantum mechanical tunneling and subsequently expands as a function of time as $\varphi_c(\sqrt{r^2 - c^2t^2} - R)$, which implies that the radius of the droplet obeys $r_p^2(t) = R^2 + c^2t^2$.

It still remains to evaluate the tunneling rate per unit "volume." This can easily be accomplished if we take into account that since $\varphi_c(\rho - R)$ has a functional form approximately equal to the soliton-like solution (2.51) centered at $\rho \approx R \gg \zeta \equiv (2g)^{-1/2}$, instead of solving (8.101) we can assume that its solution is given by $\varphi_c(\rho - R)$ and consider R as a variational parameter. The latter can therefore be determined if we insert the functional form of $\varphi_c(\rho - R)$ into (8.100) resulting in $S_E(R)$, from which we obtain the bounce action as

$$B = S_E(R_c) \qquad \text{where} \qquad \left.\frac{dS_E(R)}{dR}\right|_{R=R_c} = 0. \qquad (8.111)$$

---

[4] Notice that this term represents only a formal damping and has nothing to do with the dynamical variable relaxation.

Proceeding with the replacement we have just mentioned, we write

$$S_E(R) = \frac{\varphi_0^2}{c} N(D) \int_0^\infty d\rho \, \rho^D \left\{ \frac{1}{2}\left(\frac{\partial\varphi_c}{\partial\rho}\right)^2 + g\sin^2\varphi_c - |\lambda|(1-\cos\varphi_c) \right\}$$

$$\approx -\frac{2\varphi_0^2}{(D+1)c} |\lambda| N(D) R^{D+1} + \frac{\varphi_0^2}{2c} N(D) S_1 R^D, \tag{8.112}$$

where we have used the fact that

$$\varphi_c(\rho - R) = \begin{cases} \pi & \text{if } \rho < R \\ 0 & \text{if } \rho > R \end{cases} \tag{8.113}$$

with a fast variation within $\zeta = (2g)^{-1/2}$ about $\rho \approx R$. Therefore, we have the first term above resulting from the integration of a constant function within a hypervolume in D dimensions and the second one due to the contribution of a very thin hyperspherical wall of thickness ζ, where we define

$$S_1 \equiv \int_0^\infty d\rho \left(\frac{\partial\varphi_c}{\partial\rho}\right)^2 \tag{8.114}$$

which must clearly be proportional to ζ^{-1}.

The value R_c which extremizes $S_E(R)$ is such that

$$\left.\frac{dS_E(R)}{dR}\right|_{R=R_c} = 0 \Rightarrow R_c = \frac{DS_1^{(D)}}{4|\lambda|} \quad \text{and} \quad B = \frac{\tilde{N}(D)\varphi_0^2}{c} \frac{S_1^{D+1}}{|\lambda|^D}, \tag{8.115}$$

where $\tilde{N}(D) \equiv 2N(D)D^D/[(D+1)4^{D+1}]$.

From (8.115) we see that R_c and $B \to \infty$ as $|\lambda| \to 0$, which means that tunneling cannot take place through the formation of a finite-size droplet. The only way it can happen is if the whole system collectively tunnels from one configuration to another, which would involve a tunneling probability proportional to $\exp -L^D$, where L is the typical linear size of the system. This means that in the thermodynamic limit, it vanishes. A simple energetic argument can be used to understand what happens in this case. When the two configurations are almost degenerate, the volumetric energy reduction by formation of the droplet can only compensate the positive distortion energy stored within the droplet wall when its size grows very large. In the case of perfect degeneracy, any droplet formed (in the thermodynamic limit) will shrink, no matter what its size is.

To conclude this section let us briefly say a few words about the prefactor evaluation in this limit. It is not hard to convince ourselves that the strategy developed
previously can be extended straightforwardly to the present case by repeating the
same steps taken before when we worked in the limit $\lambda \to 0$ (or $\epsilon \to 1$). Actually,
it is even easier in the TWA because we now have the analytic form of the bounce
$\varphi_c(\rho)$, which allows us to evaluate the explicit form of the "potential energy"
$U''(\varphi_c)$ of the Schrödinger-like equation $(-\nabla^2_{(D+1)} + U''(\varphi_c))$. Moreover, as we
have mentioned earlier in this section, it is in this limit that we can prove (Callan
and Coleman, 1977) that there is only one negative eigenvalue of the latter equation
which guarantees the validity of that approach to the determination of the prefactor.

8.2.2 *The damped case at finite temperatures*

Having developed the strategy to deal with a damped point particle tunneling out
of a metastable state at finite temperatures and the undamped zero-temperature
field theoretical decay of the false vacuum in Sections 8.1.2 and 8.2.1, respectively, we can generalize the latter to include, at the same time, damping and
finite-temperature effects in two alternative ways. Both of them make use of the
extension of (8.91) to

$$
S_{eff}[\varphi(\mathbf{r}', \tau')] = \varphi_0^2 \int\limits_{-\hbar\beta/2}^{+\hbar\beta/2} d\tau' \int\limits_{V_D} d^D x \left\{ \frac{1}{2c^2} \left(\frac{\partial\varphi}{\partial\tau} \right)^2 + \frac{|\nabla_D\varphi|^2}{2} + U(\varphi) \right\}
$$

$$
+ \frac{2\gamma\pi}{4\hbar^2\beta^2} \frac{\varphi_0^2}{c^2} \int\limits_{V_D} d^D x \int\limits_{-\hbar\beta/2}^{\hbar\beta/2} d\tau'' \int\limits_{-\hbar\beta/2}^{\hbar\beta/2} d\tau' \frac{\{\varphi(x, \tau') - \varphi(x, \tau'')\}^2}{\sin^2(\pi(\tau' - \tau'')/\hbar\beta)}, \quad (8.116)
$$

which is the field theoretical equivalent of (8.66) with $x = (x_1, \ldots, x_D)$. Notice
that (8.116) has been obtained from the latter by replacing $q(\tau) \to \varphi(x, \tau)$, M and
$V(q)$ by their respective densities in D dimensions, introducing an elastic term
proportional to $\nabla_D\varphi$, and finally integrating over $d^D x$. We should also be warned
that new dimensional variables (for example, φ_0 and c) should be defined in terms
of the previous ones in order to bring (8.66) to its new form.

From expression (8.116) we can obtain, as usual, the periodic bounce as the
solution of

$$
\frac{\delta S_{eff}}{\delta\varphi}\bigg|_{\varphi_c} = \frac{1}{c^2} \frac{\partial^2\varphi_c}{\partial\tau^2} + \nabla^2_D\varphi_c - \frac{\partial U}{\partial\varphi_c}
$$

$$
- \frac{2\gamma\pi}{\hbar^2\beta^2 c^2} \int\limits_{-\hbar\beta/2}^{\hbar\beta/2} d\tau'' \frac{[\varphi_c(x, \tau') - \varphi_c(x, \tau'')]}{\sin^2(\pi(\tau' - \tau'')/\hbar\beta)} = 0, \quad (8.117)
$$

with boundary conditions $\varphi_c(x, \hbar\beta/2) = \varphi_c(x, -\hbar\beta/2)$, whereas its stability can be studied by the eigenvalue problem

$$\hat{D}_\beta \varphi(x, \tau') = -\frac{1}{c^2}\frac{\partial^2 \varphi(x, \tau')}{\partial \tau'^2} - \nabla_D^2 \varphi(x, \tau') + U''(\varphi_c)\varphi(x, \tau')$$
$$+ \hat{O}_\beta \varphi(x, \tau') = kq(\tau'), \tag{8.118}$$

where the differential operator $\hat{O}_\beta$ acting on $\varphi(x, \tau')$ now reads

$$\hat{O}_\beta \varphi(x, \tau') = \frac{2\gamma\pi}{\hbar^2\beta^2 c^2} \int_{-\hbar\beta/2}^{\hbar\beta/2} d\tau'' \frac{[\varphi(x, \tau') - \varphi(x, \tau'')]}{\sin^2(\pi(\tau' - \tau'')/\hbar\beta)}. \tag{8.119}$$

In the first method mentioned above, we can create an ansatz for the functional form of the solution of (8.117) and allow for the existence of one or more free (variational) parameters that can be determined through the extremization of the resulting action, exactly as we have done in the zero-temperature TWA (Hida and Eckern, 1984). Notice that even for the latter case, our bounce solution is no longer "spherically" symmetric in $D+1$ dimensions because the non-local term resulting from the existence of relaxation processes prevents it from being so. This method is also appropriate if we wish to compute perturbatively finite-temperature and damping contributions to the undamped zero-temperature solution.

The second method is a direct extension of the approach described in (8.75)–(8.77) to a field variable $\varphi(x, \tau)$ (Smith *et al.*, 1996, 1997). Taking the Fourier transform of the latter for the Matsubara frequencies, we end up with

$$S_{eff}(\varphi_n(x)) = \hbar\omega_0\beta B_0 \int_{V_D} d^D x \sum_{n=-\infty}^{\infty} \left[(v_n^2 + 2\alpha|v_n|)\varphi_n(x)\varphi_{-n}(x)\right.$$
$$\left. + \nabla_D\varphi_n(x)\cdot\nabla_D\varphi_{-n}(x) - [U(\varphi(x, \tau))]_n\right], \tag{8.120}$$

where x is now a dimensionless coordinate. In the expression above we have introduced a characteristic frequency and action of the system as ω_0 and B_0, respectively, and $[U(\varphi(x, \tau))]_n$ is the nth Fourier component of the potential energy density. α is, as usual, the dimensionless damping γ/ω_0.

From (8.120) we can immediately find the field theoretical extensions of (8.76) and (8.77), which must be solved in order to find the desired decay rate. Although some regimes can be treated by approximate analytical solutions, the structure of the above equations in terms of Fourier components is very suitable for the application of numerical methods to the problem.

At this point we must notice that all we have been doing in this section is to adapt the machinery developed previously for dealing with tunneling of a point particle with dissipation at finite temperatures to the decay of a false vacuum in field theory. Apart from the technicalities inherent in this extension, the physics we have explored before is still present here. The quantum nucleation rate of the critical droplet depends on damping and temperature in a form very similar to that displayed in Fig. 8.3. Actually, it is the evaluation of the form of the critical droplet which turns out to be the main issue here.

As proceeding with this generalization will not add much to our knowledge of dissipative tunneling and, on top of that, there are very few realistic examples where dissipative quantum nucleation takes place (in particular the homogeneous case), we shall leave the subject and embark on the more important issue of coherent tunneling.

9

Dissipative coherent tunneling

Another important problem that appeared when we studied the dynamics of the flux in a SQUID ring was that of a Brownian particle in a quartic (bistable) potential (see Fig. 9.1),

$$V(q) = \frac{M\omega_0^2 q_0^2}{32} \left[\left(\frac{q}{q_0/2} \right)^2 - 1 \right]^2.$$

(9.1)

Actually this example is present in many areas of physics, ranging from chemical to high-energy physics. In the absence of dissipation, the particle tunnels coherently from one well to the other and this is the effect we want to study when we switch on the interaction with the environment.

9.1 The spin–boson Hamiltonian

Let us start by reviewing the quantum mechanical motion of a particle in the bistable potential. Since the tunneling rate between the two minima is finite it splits the otherwise degenerate lowest-energy states on each side of the barrier. This splitting is usually exponentially smaller than the energy scale $\hbar\omega_0$ and then, under appropriate conditions (for example, low temperatures), we can restrict ourselves to the two-dimensional Hilbert space formed by these lowest-energy states of the problem, the well-known even and odd parity states

$$\psi_E = \frac{1}{\sqrt{2}}(\psi_R + \psi_L) \quad \text{and} \quad \psi_O = \frac{1}{\sqrt{2}}(\psi_R - \psi_L),$$

(9.2)

where ψ_R (ψ_L) corresponds to the approximate ground state in the right (left) well. The splitting $\hbar\Delta_0 = E_O - E_E$ is proportional to $\langle \psi_R | \mathcal{H} | \psi_L \rangle = \langle \psi_L | \mathcal{H} | \psi_R \rangle$ and, within the WKB approximation, can be computed as (see Appendix D and Coleman (1988))

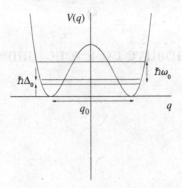

Figure 9.1 Bistable potential

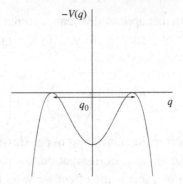

Figure 9.2 Inverted bistable potential

$$\Delta_0 = A_0 \exp -\frac{B_0}{\hbar} \qquad \text{where}$$

$$A_0 = 2\left(\frac{B_0}{2\pi M \hbar}\right)^{1/2} \left|\frac{\det(-M\partial_t^2 + M\omega_0^2)}{\det'(-M\partial_t^2 + V''(q_c^{(0)}))}\right|^{1/2}, \tag{9.3}$$

and B_0 is now the Euclidean action of the *instanton* $q_c^{(0)}(\tau)$. The latter is a solution of the imaginary time equation of motion (8.3) for the bistable potential (9.1) and B_0 is the action (8.4) evaluated at this solution.

Here, (8.3) is once again an equation of motion of a fictitious particle in a potential $-V(q)$ (see Fig. 9.2) which can easily be integrated with the boundary condition $q_c^{(0)}(-\infty) = -q_0/2$ and $q_c^{(0)}(\infty) = q_0/2$. Notice that we are only addressing the $T = 0$ case with these boundary conditions. This is the *instanton* solution mentioned above (Coleman, 1988).

The explicit form of this solution is

$$q_c^{(0)}(\tau) = \frac{q_0}{2} \tanh\frac{\omega_0\tau}{2}, \tag{9.4}$$

and therefore

$$B_0 = \int\limits_{-\infty}^{\infty} \left[\frac{1}{2} M \dot{q_c}^2 + V(q_c) \right] d\tau = \frac{4V_0}{\omega_0}, \tag{9.5}$$

where V_0 is the height of the potential barrier, which can be identified trivially in (9.1) as

$$V_0 = \frac{M \omega_0^2 q_0^2}{32}. \tag{9.6}$$

The next step is to compute the prefactor A_0 in (9.3), for which we refer the reader to Appendix D and Coleman (1988).

Now we generalize the procedure that led us to (9.3) to the damped case. In order to truncate the Hamiltonian of the composite system, we have to evaluate its partition function restricted to a given region of the parameter space. This can be done by integrating (6.95) for $x = y = q$ which yields

$$Z(\beta) = \text{const.} \int dq \int\limits_{q(0)=q}^{q(\hbar\beta)=q} \mathcal{D}q(\tau) \exp - \frac{S_{eff}[q(\tau)]}{\hbar}, \tag{9.7}$$

where S_{eff} is given by (6.96) and (6.98) for the bistable potential. For the undamped case ($\eta = 0$) at very low temperatures ($k_B T << \hbar\omega_0$), we have already seen that the path integral in (9.7) is finite. Moreover, the final result of the integration can be shown to be

$$Z_0(\beta) = \cosh(\beta\hbar\Delta_0/2), \tag{9.8}$$

which is nothing but the partition function of a two-state system with splitting Δ_0 as discussed above.

Therefore, it is now our goal to perform the same calculation for the system of interest when it is coupled to the ohmic bath of oscillators. We hope the resulting partition function will enable us to map our initial Hamiltonian onto a simpler one. The procedure has been worked out in great detail in Leggett *et al.* (1987), and we shall summarize its main steps in what follows.

The first difficulty appears when we naively compute the path integral (9.7) in the WKB approximation because the effective action of the instanton solution is infrared divergent for the ohmic spectral function (5.22). This problem has its origin in the fact that the environmental oscillators play quite a different role in the truncation process depending on their frequencies.

In order to understand what we mean by that, we should start by defining the characteristic frequency scales of the problem. Two obvious scales are the attempt frequency at each minimum of the bistable potential, ω_0, and the "thermal frequency," $k_B T / \hbar$. Moreover, if the problem is to be mapped onto a sort of effective two-state system, we should expect a new splitting frequency, Δ, which would be determined self-consistently. There can also be a "bias frequency" given by $\epsilon / \hbar$, which would originate from the presence of an external forcing term that tilts the bistable potential and consequently breaks the two-site degeneracy energy. Although we will not be interested in the latter, we shall sometimes mention its effects for the sake of completeness.

The approximation used for the undamped system leads us to expect that Δ, $k_B T / \hbar$, $\epsilon / \hbar << \omega_0$ if the mapping onto a two-state system is at all possible. Moreover, since we have been applying the WKB approximation to this problem, it is clear that we should take $\hbar \omega_0 << V_0$. Therefore, we can separate the environmental oscillators into two classes; the fast ones, with frequencies $\geq \omega_0$, which affect the tunneling transition and renormalize Δ and the slow ones, with frequencies $\leq \Delta$, which detune the two wells in a random fashion causing the destruction of the phase coherence of the motion. Before we go on with this analysis, let us replace the attempt frequency ω_0 by a more general ω_b, by which we can represent not only the previous quantity but also the slower frequency $M \omega_0^2 / \eta$ when we deal with very overdamped situations. We shall also use ω_b to denote the instanton width, which is of the order of one of the two above-mentioned frequencies in the appropriate dissipative regime. Let us start by remembering that the path integral in (9.7) contains the effect of the whole lot of environmental oscillators. So, if we want to take them into account carefully, we should first split their spectral function in the following way:

$$J_0(\omega) = J(\omega) + J'(\omega), \tag{9.9}$$

where $J_0(\omega)$ is the previously defined spectral function (5.22) and

$$J(\omega) \equiv J_0(\omega) \, e^{-\omega / \omega_c},$$
$$J'(\omega) \equiv J_0(\omega)(1 - e^{-\omega / \omega_c}). \tag{9.10}$$

We have introduced here a new frequency cutoff ω_c, which separates the slow from the fast oscillators. By our preceding argument we should write Δ, $k_B T / \hbar$, $\epsilon / \hbar << \omega_c << \omega_b$. Here we must warn the reader that this new cutoff frequency has nothing to do with the one appearing in equation (5.22), Ω. The latter refers to some characteristic frequency of the many-body environment where the original Brownian particle is immersed. It can be, for instance, the Debye frequency for a phonon bath or the Fermi frequency ($\epsilon_F / \hbar$) for a fermionic environment. Usually, this microscopic frequency is much higher than ω_b, although this is not required for our present purpose.

What we do next is perform the path integral (9.7) considering only the effect of the fast oscillators or, in other words, write the effective action in (9.7) as (6.96) with $\alpha(\tau - \tau')$ replaced by

$$\alpha'(\tau - \tau') = \frac{1}{2\pi} \int_0^\infty d\omega \, J'(\omega) e^{-\omega|\tau-\tau'|}. \qquad (9.11)$$

This integral yields

$$Z(\beta) = \cosh(\beta\hbar\Delta/2), \qquad (9.12)$$

once again a partition function of a two-state system whose level splitting is, within the so-called *dilute-instanton* approximation (Leggett *et al.*, 1987), given by

$$\Delta \equiv 2A \exp -S'_{eff}[q_c(\tau)]/\hbar. \qquad (9.13)$$

S'_{eff} is obviously (6.96) evaluated with (9.11) and q_c is the instanton solution for the equation of motion resulting from this action. It should be emphasized that within the above-mentioned approximation, we are considering a gas of non-interacting instantons and, therefore, the inverse instanton width, namely ω_b, plays the role of a high-frequency cutoff of our integrals.

It can be shown (Dorsey *et al.*, 1986) that within the dilute-instanton approximation and for ohmic dissipation, we have

$$\Delta \sim \Delta_0 \exp\left[-\frac{q_0^2}{2\pi\hbar} \int_{\omega_c}^{\omega_b} d\omega \frac{J_0(\omega)}{\omega^2} \right]$$

$$\sim \Delta_0 \left(\frac{\omega_c}{\omega_b} \right)^\alpha, \qquad (9.14)$$

where $\alpha \equiv \eta q_0^2/2\pi\hbar$ is a dimensionless damping constant. Later on we will see that all the measurable quantities of the problem depend on a very specific combination of parameters, which cancels the dependence of Δ on this unphysical cutoff.

Now, having convinced ourselves that the high-frequency sector of our problem is really that of an effective two-level system, we describe its Hilbert space, at least to zeroth order in Δ/ω_c, as

$$\Psi_\pm(q, \{x_j\}) = \frac{1}{\sqrt{2}} [\Psi_R(q, \{x_j\}) \pm \Psi_L(q, \{x_j\})], \qquad (9.15)$$

where the set $\{x_j\}$ stands for the coordinates of those high-frequency oscillators we have been considering in our calculations. The wave functions Ψ_R and Ψ_L are then, to this same order, the wave functions of damped harmonic oscillators centered at $q_0/2$ and $-q_0/2$, respectively. Finally, in order to proceed with

taking the other oscillators into account, all we have to do is project the original Hamiltonian of a Brownian particle in a bistable potential onto the subspace
spanned by Ψ_R and Ψ_L.

This leads us to

$$\mathcal{H} = -\frac{1}{2}\hbar\Delta\sigma_x + \frac{1}{2}\epsilon\,\sigma_z + \frac{1}{2}\,q_0\,\sigma_z \sum_k C_k\,q_k + \sum_k \frac{p_k^2}{2m_k} + \sum_k \frac{1}{2}\,m_k\,\omega_k^2\,q_k^2, \quad (9.16)$$

which is known in the literature as the *spin–boson Hamiltonian*. In (9.16), $\hbar\Delta$ is
the previously introduced effective energy splitting of the two lowest-energy states
in the presence of dissipation, whereas ϵ represents the possible external bias that
may break the degeneracy of the two minima. Finally, we should remember that
the remaining oscillators of (9.16) are distributed in accordance with the spectral
function $J(\omega)$ defined in (9.10), which implies that ω_c is now the high-frequency
cutoff for the oscillator bath.

Although (9.16) has been obtained from the truncation of the Hamiltonian of a
damped particle in a quartic potential, we should emphasize here that its form is
actually general enough to describe the dynamics of non-isolated truly two-level
systems such as, for example, a genuine spin coupled to a heat bath. In this case
Δ and ϵ represent the bare external field components along $\hat{\mathbf{x}}$ and $\hat{\mathbf{z}}$, respectively,
and ω_c turns out to be the microscopic cutoff Ω. In other words we can bypass the
whole former discussion on truncation and jump directly onto that particular model
in order to furnish a semi-phenomenological description for a given spin dynamics
(see the discussion on the Bloch equation below).

Another possible form of spin Hamiltonian in a harmonic environment is
provided by

$$\mathcal{H} = -\frac{1}{2}\hbar\Delta\sigma_z + \frac{1}{2}\epsilon\,\sigma_z + \frac{1}{2}\,q_0\,\sigma_z \sum_k C_k\,q_k + \sum_k \frac{p_k^2}{2m_k} + \sum_k \frac{1}{2}\,m_k\,\omega_k^2\,q_k^2, \quad (9.17)$$

where all we have done is replace σ_x by σ_z in (9.16). Notice that as $[\sigma_z, \mathcal{H}] = 0$
in (9.17), we have $\langle\sigma_z(t)\rangle = $ constant, contrary to what happens in the previous
spin–boson model where $[\sigma_z, \mathcal{H}] \neq 0$. Actually, as we will see shortly, in the latter case $\langle\sigma_z(t)\rangle$ relaxes with time and therefore (9.16) is a kind of the so-called
amplitude damping model. On the contrary, (9.17) is a *phase damping* model since
relaxation only takes place in $\langle\sigma_x(t)\rangle$ and $\langle\sigma_y(t)\rangle$ with $\langle\sigma_z(t)\rangle$ being preserved.
Studying the time evolution of the reduced density operator of the spin variable,
we can show that decoherence takes place in both models; in the first case due
to the full spin relaxation to the final equilibrium configuration and in the second case due to a pure dephasing process taking place in its off-diagonal matrix
elements.

9.2 The spin–boson dynamics

We can now think about the application of our previously developed real-time expression (6.21) to describe the dynamics of this effective two-state system coupled to the environment. However, there is still one problem regarding (9.16) that we would like to address beforehand, and which becomes quite evident from a simple equation of motion analysis of the problem.

9.2.1 Weak damping limit

Applying the Heisenberg equation of motion for all the operators involved in (9.16), taking their Laplace transforms, and proceeding exactly as done in Section 5.1, we can after some approximations (see below) reproduce a set of Bloch equations (see (2.61)) for the average "spin" coordinates $S_i = \hbar \langle \sigma_i \rangle / 2$ which read, for the particular case $\epsilon = 0$,

$$\frac{dS_x}{dt} = -\frac{S_x - S_x^{(eq)}}{T_1}, \tag{9.18}$$

$$\frac{dS_y}{dt} = \Delta S_z - \frac{S_y}{T_2}, \tag{9.19}$$

$$\frac{dS_z}{dt} = -\Delta S_y. \tag{9.20}$$

In these expressions $S_x^{(eq)} = \hbar \tanh(\hbar \beta \Delta)/2$ is the equilibrium value of the S_x spin component and

$$\frac{1}{T_1} = \frac{1}{T_2} = \frac{q_0^2}{2\hbar} J(\Delta) \coth \frac{\beta \hbar \Delta}{2} \tag{9.21}$$

are frequencies playing the same role as the inverse longitudinal and transverse relaxation times already introduced in (2.61).

However, in order to reach these equations, we have:

 (i) assumed that the coupling between the two-level system and the environment is very weak and can be treated perturbatively, and
(ii) ignored some terms that result from the interaction of the fictitious spin with high-frequency oscillators ($\omega_k > \Delta$) (see, for example, Harris and Silbey (1983) and Waxman (1985)), which is exactly the issue we will treat next but not before elaborating further on the set of phenomenological Bloch equations obtained above.

As we should already have noticed, the Bloch equations (9.18)–(9.20) represent the dynamics of a non-isolated spin coupled to an external field $\mathbf{B} = -\Delta \sigma_x \hat{\mathbf{x}} + \epsilon \sigma_z \hat{\mathbf{z}}$, which is exactly the situation found in magnetic resonance experiments where, conventionally, the x and y directions are interchanged as in (2.61). Here, the constant field component points along $\hat{\mathbf{x}}$ whereas the pumping (or transverse) field is nothing but $\epsilon(t)\sigma_z\hat{\mathbf{z}}$. When $\epsilon \neq \epsilon(t)$ this field component only breaks the degeneracy of the eigenstates $|\uparrow\rangle$ and $|\downarrow\rangle$ of σ_z.

The longitudinal and transverse relaxation times, T_1 and T_2, respectively, usually have quite distinct origins. In genuine magnetic resonance experiments, for example, they can be explained exactly by the same arguments employed just below (2.61), namely the process of inelastic scattering (and also spontaneous emission or absorption) of phonons by the magnetic particle in the case of T_1, and the random precession of the spin about the constant external field in the case of T_2. The latter effect clearly induces a relaxation behavior on the y and z components of the spin without any energy exchange between it and its reservoir. Therefore, we see that $S_x(t)$ relaxes to its equilibrium value $S_x^{(eq)}$ within T_1 independently of $S_y(t)$ or $S_z(t)$ which, in contrast, relaxes to zero according to T_2. Moreover, as we have already mentioned, depending on the specific system with which we are dealing, the relation between the relaxation times may vary from $T_1 \gtrsim T_2$ to $T_1 \gg T_2$. In other words, the dephasing process that leads to $S_y(t)$ and $S_z(t) \to 0$ might occur much faster than the relaxation of $S_x(t)$ to equilibrium.

Now that we have analyzed the equations of motion for the spin components from a magnetic resonance point of view, let us return to our specific application. Since the variable that is physically relevant to our truncated spin–boson problem is actually $\langle \sigma_z \rangle$, we should follow the time evolution of $P(t) \equiv S_z(t)$, which results from iterating the y and z components in (9.19) and (9.20) and reads

$$\frac{d^2 P}{dt^2} + \frac{1}{T_2}\frac{dP}{dt} + \Delta^2 P = 0, \tag{9.22}$$

the equation of a damped harmonic oscillator. This means that the coherent tunneling of a damped particle between the two degenerate states ψ_R and ψ_L of a bistable potential vanishes exponentially within a time interval $2T_2$, after which the final state of the system becomes a totally incoherent mixture of those two states. It is worth stressing here that in the spin–boson model, both T_1 and T_2 have the same origin.

The point we want to focus on now is the role played by the high-frequency oscillators we have neglected above in the dynamics of our fictitious spin. Moreover, in so doing we will at the same time show how the remaining dependence of Δ on the cutoff ω_c disappears. As we argued before, if (9.16) is to be reliable as describing the dynamics of the Brownian particle at very low temperatures, this sort of dependence must be canceled. The way to achieve this is by the implementation of the *adiabatic renormalization*, which we explain below.

9.2.2 Adiabatic renormalization

In its new form, (9.16) has most of its oscillators with frequencies much higher than the splitting frequency Δ and these can be adiabatically eliminated from our problem. In order to do this we compute the effective splitting, evaluating the overlap of the two states of the composite system corresponding to the particle and fast oscillators at $q_0/2$ with the same kind of state at $-q_0/2$. These respectively read

$$|\Psi_+^{(0)}\rangle = |+\rangle \prod_k |g_{k+}\rangle \qquad \text{and} \qquad |\Psi_-^{(0)}\rangle = |-\rangle \prod_k |g_{k-}\rangle, \qquad (9.23)$$

where $|\pm\rangle$ denote the eigenstates corresponding to the eigenvalues $\sigma_z = \pm 1$ and $|g_{k\pm}\rangle$ denote the ground state of the kth fast oscillator displaced to $\pm q_0/2$. We can write them explicitly as

$$|g_{k\pm}\rangle = \exp\left(\pm \frac{1}{2} i \hat{\Omega}_k\right)|0\rangle_k \qquad \text{where} \qquad \hat{\Omega}_k \equiv \frac{q_0 C_k}{\hbar m_k \omega_k^2} \hat{p}_k \qquad (9.24)$$

and $|0\rangle_k$ is the ground state of the kth oscillator centered at the origin.

Now we can easily evaluate the overlap between the states in (9.23) as

$$\Delta'(\omega_l) = \Delta \prod_k \langle g_{k+} | g_{k-} \rangle, \qquad (9.25)$$

where we have introduced a low-frequency cutoff ω_l meaning that only oscillators with frequencies higher than this will enter the computation of the overlap. In other words, we have chosen only those oscillators that adjust adiabatically to the slow oscillation of the particle to evaluate the effective splitting. Computing the overlap $\langle g_{k+} | g_{k-} \rangle$ and using the definition of $J(\omega)$ in (9.10), we get

$$\Delta'(\omega_l) = \Delta \exp - \int_{\omega_l}^{\infty} \frac{q_0^2}{2\pi \hbar} \frac{J(\omega)}{\omega^2} d\omega. \qquad (9.26)$$

For ohmic dissipation, it can be shown (see Leggett *et al.* (1987)) that (9.26) can be solved self-consistently for Δ' by choosing $\omega_l = p\Delta'$, where $p \gg 1$, which gives us

$$\Delta_r = \begin{cases} \Delta \, (\Delta/\omega_c)^{\frac{\alpha}{(1-\alpha)}} & \text{if} \quad \alpha < 1 \\ 0 & \text{if} \quad \alpha \geq 1, \end{cases} \qquad (9.27)$$

where α is the dimensionless damping constant defined below equation (9.14). It should be noticed that within this specific combination and because of (9.14), the dependence on the cutoff frequency ω_c disappears from the renormalized splitting Δ_r. Therefore, we should expect that all our future results on the dynamics of this system will be expressible solely in terms of this parameter.

9.2.3 Path integral approach

Once this issue has been settled, we can study the dynamics of our two-state system in an oscillator bath. In order to accomplish that we will adopt the same procedure to describe the time evolution of the reduced density operator of a system in a bath as we have done before. Suppose we have the two-state system or, for short, the "spin" held at state $|+\rangle$ (or $q_0/2$) and the environment in the state of equilibrium in the absence of the interaction with the spin. Then, at $t_0 < 0$ we allow them to interact, still holding the spin at $|+\rangle$, and compute $P(t) \equiv \langle \sigma_z(t) \rangle$ after relaxing this condition at $t = 0$. It has been argued thoroughly in Leggett *et al.* (1987) that this quantity can be evaluated with the same initial condition we used earlier.

Then, using the fact that $x' = y' = x_i$ in (4.61), we can write the reduced density operator at $x = y = x_f$ as

$$\rho(x_f, x_f, t) = \int_{x_i}^{x_f} \mathcal{D}x(\tau) \int_{x_i}^{x_f} \mathcal{D}y(\tau)\, \mathcal{A}[x(\tau)]\, \mathcal{A}^*[y(\tau)]\, \mathcal{F}[x(\tau), y(\tau)], \qquad (9.28)$$

where $\mathcal{A}[x(\tau)]$ is the amplitude for a path $x(\tau)$ ignoring the coupling to the environment and $\mathcal{F}[x(\tau), y(\tau)]$ is the well-known influence functional (6.8) which we now rewrite as

$$\mathcal{F}[x(\tau), y(\tau)] = \exp \frac{i}{\pi \hbar} \int_0^t \int_0^\tau d\tau d\sigma [x(\tau) - y(\tau)] L_1(\tau - \sigma)[x(\sigma) + y(\sigma)]$$

$$\times \exp - \frac{1}{\pi \hbar} \int_0^t \int_0^\tau d\tau d\sigma [x(\tau) - y(\tau)] L_2(\tau - \sigma)[x(\sigma) - y(\sigma)],$$

$$(9.29)$$

where we have modified our previous notation from (6.13), (6.14) to

$$L_2(\tau - \sigma) \equiv \pi \alpha_R(\tau - \sigma) = \int_0^\infty d\omega\, J(\omega) \coth \frac{\hbar \omega}{2 k_B T} \cos \omega(\tau - \sigma),$$

$$L_1(\tau - \sigma) \equiv -\pi \alpha_I(\tau - \sigma) = \int_0^\infty d\omega\, J(\omega) \sin \omega(\tau - \sigma), \qquad (9.30)$$

to allow for a more direct comparison with Leggett *et al.* (1987).

It is now time to analyze the above expressions to the spin–boson problem. In this description, the "spin paths" can only take discrete values $\pm q_0/2$ with fast hops between them. As we have a pair of paths, $[x(\tau), y(\tau)]$, we will find it more convenient to think of a single path hopping between four states. We consider the states $\{+, +\}$, $\{+, -\}$, $\{-, +\}$, and $\{-, -\}$, where $(+) = q_0/2$ and $(-) = -q_0/2$, which we label A, B, C, and D, respectively.

Next we define the new paths $\xi(\tau) \equiv q_0^{-1}[x(\tau) - y(\tau)]$, which is 0 for A and D, 1 for B, and -1 for C and $\chi(\tau) \equiv q_0^{-1}[x(\tau) + y(\tau)]$, which is 0 for B and C, 1 for A, and -1 for D, in terms of which

$$\mathcal{F}[x(\tau), y(\tau)] = \exp\frac{iq_0^2}{\pi\hbar} \int_0^t \int_0^\tau d\tau d\sigma \xi(\tau) L_1(\tau - \sigma)\chi(\sigma)$$

$$\times\exp - \frac{q_0^2}{\pi\hbar} \int_0^t \int_0^\tau d\tau d\sigma \xi(\tau) L_2(\tau - \sigma)\xi(\sigma). \quad (9.31)$$

Suppose we want to compute (9.28) for paths that start and end in A. Then, we have to consider $2n$ flips between those four states defined above. If we look at the behavior of the path $\xi(\tau)$ we see that it starts at $\xi = 0$, hops from zero to $\pm q_0$ at odd times, and hops back to zero at even times, coming finally to $\xi = 0$ again after the last flip. So, we can say that $\xi(\tau)$ stays at zero and makes n excursions to $\pm q_0$ that last $t_{2k} - t_{2k-1}$ each. Following the terminology established in Leggett *et al.* (1987), we will call these excursions *blips* and those parts for which $\xi = 0$, *sojourns*. Notice that $\chi(\tau)$ behaves in the opposite way, namely when $\xi(\tau) = 0$, $\chi(\tau) = \pm q_0$ and $\chi(\tau) = 0$ for the blips. Therefore, when building the structure of the influence functional (9.31) for these broken paths, it is convenient to assign to each of the intervals $t_{2j-1} < t < t_{2j}$ a label ζ_j, which is $+1$ (-1) if the system spends this time interval in B (C), and similarly to each of the intervals $t_{2j} < t < t_{2j+1}$ a label η_j, which is $+1$ (-1) if the system spends this interval in A (D). Notice that $\eta_0 \equiv \eta_n \equiv 1$.

Now, we must determine the factors that should enter the path integral (9.28) for each individual portion of the specific path we are studying. Let us consider the more general case of a biased system where we have chosen the zero of energy in such a way that the state $q_0/2$ ($-q_0/2$) has energy $\epsilon/2$ ($-\epsilon/2$). In dealing with (9.28) we should first consider the structure of the amplitude $\mathcal{A}$. The probability amplitude to stay at $q_0/2$ ($-q_0/2$) during dt is $\exp -i(\epsilon/2\hbar)dt$ ($\exp +i(\epsilon/2\hbar)dt$), whereas the amplitude to flip from one state to the other is, to first order in dt, $i(\Delta/2)dt$. Applying these results to the four-state path we see that the amplitude to stay in the same state is $\exp -i\epsilon\xi(t)dt$ and that to flip between states is $i\lambda(\Delta/2)dt$, where λ is 0 for $A \rightleftharpoons D$ and $B \rightleftharpoons C$, -1 for $A \rightleftharpoons B$ and $D \rightleftharpoons C$, and $+1$ for $A \rightleftharpoons C$ and $B \rightleftharpoons D$.

Putting all these together, we can write our final result as (see Leggett *et al.* (1987) for details)

$$P(t) = \sum_{n=0}^{\infty} (-1)^n \Delta^{2n} K_n(t), \tag{9.32}$$

$$K_n(t) \equiv 2^{-n} \sum_{\{\zeta_j\}} \int_0^t dt_{2n} \int_0^{t_{2n}} dt_{2n-1} \cdots \int_0^{t_2} dt_1$$
$$\times F_n(t_1, t_2, \ldots, t_{2n}; \zeta_1, \zeta_2, \ldots, \zeta_n; \epsilon), \tag{9.33}$$

$$F_n(\{t_m\}; \zeta_i; \epsilon) \equiv F_1\{t_m\} F_2\{t_m, \zeta_i\} F_3\{t_m, \zeta_i\} F_4\{t_m, \zeta_i, \epsilon\}, \tag{9.34}$$

where

$$F_1\{t_m\} = \exp\left(-\frac{q_0^2}{\pi\hbar} \sum_{j=1}^{n} S_j\right), \tag{9.35}$$

$$F_2\{t_m, \zeta_i\} = \exp\left(-\frac{q_0^2}{\pi\hbar} \sum_{k=1}^{n} \sum_{j=k+1}^{n} \zeta_j \zeta_k \Lambda_{jk}\right), \tag{9.36}$$

$$F_3\{t_m, \zeta_i\} = \prod_{k=1}^{n-1} \cos\left(\frac{q_0^2}{\pi\hbar} \sum_{j=k+1}^{n} \zeta_j X_{jk}\right), \tag{9.37}$$

$$F_4\{t_m, \zeta_i, \epsilon\} = \cos\left(\sum_{j=1}^{n} \zeta_j \left((t_{2j} - t_{2j-1})\frac{\epsilon}{\hbar} - \frac{q_0^2}{\pi\hbar} X_{j0}\right)\right), \tag{9.38}$$

with

$$S_j = Q_2(t_{2j} - t_{2j-1}), \tag{9.39}$$

$$\Lambda_{jk} = Q_2(t_{2j} - t_{2k-1}) + Q_2(t_{2j-1} - t_{2k})$$
$$- Q_2(t_{2j} - t_{2k}) - Q_2(t_{2j-1} - t_{2k-1}), \tag{9.40}$$

$$X_{jk} = Q_1(t_{2j} - t_{2k+1}) + Q_1(t_{2j-1} - t_{2k})$$
$$- Q_1(t_{2j} - t_{2k}) - Q_1(t_{2j-1} - t_{2k+1}), \tag{9.41}$$

where $Q_1(t)$ and $Q_2(t)$ are the second integrals of $L_1(t)$ and $L_2(t)$, which read

$$Q_1(t) = \int_0^{\infty} \frac{J(\omega)}{\omega^2} \sin \omega t \, d\omega, \tag{9.42}$$

$$Q_2(t) = \int_0^{\infty} \frac{J(\omega)}{\omega^2} (1 - \cos \omega t) \coth\left(\frac{\beta\hbar\omega}{2}\right) d\omega. \tag{9.43}$$

Expressions (9.32)–(9.43) represent an exact solution of our problem despite the difficulty we might have in evaluating those summations. The latter represent interactions between blips, sojourns and blips, and also interference terms between them. Nevertheless, there is a particular approximation in which analytical results are available. This is the so-called *non-interacting blip approximation* (NIBA), which has been analyzed extensively in Leggett *et al.* (1987). It consists of two main prescriptions:

(i) Set $X_{jk} = 0$ in (9.37), (9.38) if $k \neq j - 1$ and $X_{j,j-1} = Q_1(t_{2j} - t_{2j-1})$.
(ii) Set $\Lambda_{jk} = 0$ in (9.36).

The reader can follow their justification in great detail in the preceding reference. Here we quote only the final expressions.

The unbiased case. When $\epsilon = 0$ the integrand of (9.33) becomes

$$F_n(t_1 \ldots t_{2n}) = \prod_{j=1}^{n} \cos\left(\frac{q_0^2}{\pi\hbar} Q_1(t_{2j} - t_{2j-1})\right) \exp -\frac{q_0^2}{\pi\hbar} Q_2(t_{2j} - t_{2j-1}) \quad (9.44)$$

which can be taken to (9.32) yielding

$$P(t) = \sum_{n=0}^{\infty} (-1)^n \int_0^t dt_{2n} \int_0^{t_{2n}} dt_{2n-1} \ldots \int_0^{t_2} dt_1 \prod_{j=1}^{n} f(t_{2j} - t_{2j-1}), \quad (9.45)$$

where

$$f(t) = \Delta^2 \cos\left(\frac{q_0^2}{\pi\hbar} Q_1(t)\right) \exp -\frac{q_0^2}{\pi\hbar} Q_2(t). \quad (9.46)$$

Taking the Laplace transform of (9.45) we have

$$\tilde{P}(\lambda) = \sum_{n=0}^{\infty} (-1)^n \int_0^\infty dt \int_0^\infty dt_1 \ldots \int_0^\infty dt_{2n} e^{-\lambda(t_1 + t_2 + \cdots + t_{2n})} \prod_{j=1}^{n} f(t_{2j})$$

$$= \sum_{n=0}^{\infty} (-1)^n \frac{[f(\lambda)]^n}{\lambda^{n+1}} = \frac{1}{\lambda + f(\lambda)}, \quad (9.47)$$

where

$$f(\lambda) \equiv \Delta^2 \int_0^\infty dt \cos\left[\frac{q_0^2}{\pi\hbar} Q_1(t)\right] \exp -\left[\lambda t + \frac{q_0^2}{\pi\hbar} Q_2(t)\right]. \quad (9.48)$$

Notice that $f(\lambda)$ can be evaluated, in principle, for any $J(\omega)$. Then, taking its inverse Laplace transform, we have

$$P(t) = \frac{1}{2\pi i} \int_C d\lambda \, e^{\lambda t} \, \tilde{P}(\lambda)$$

$$\equiv \frac{1}{2\pi i} \int_C d\lambda \, e^{\lambda t} \, [\lambda + f(\lambda)]^{-1} \tag{9.49}$$

where the contour C is the well-known Bromwich contour.

As we are particularly interested in the case of ohmic dissipation, we must replace $J(\omega)$ by (5.22) and evaluate the integrals above to get

$$Q_1(t) = \eta \tan^{-1} \omega_c t, \tag{9.50}$$

$$Q_2(t) = \eta \ln(1 + \omega_c^2 t^2) + \eta \ln \left[\frac{\beta \hbar}{\pi t} \sinh \frac{\pi t}{\beta \hbar} \right]. \tag{9.51}$$

Carrying this procedure forward, we get a very rich dynamical behavior of the system for different regions of the parameter space (α, T) (see Fig. 9.3), as described below.

(a) For $\alpha \geq 1$ and $T = 0$ (see Fig. 9.4), the particle localizes in one of the two minima because the renormalized splitting (9.27) vanishes in this region of the parameter space. The bistable minimum gets completely blocked.

(b) For high temperatures and/or strong dissipation (see Fig. 9.5),

$$P(t) = \exp - \frac{t}{\tau}, \tag{9.52}$$

where

$$\tau^{-1} = \frac{\sqrt{\pi}}{2} \frac{\Gamma(\alpha)}{\Gamma(\alpha + \frac{1}{2})} \frac{\Delta_r^2}{k_B T/\hbar} \left[\frac{\pi k_B T}{\hbar \Delta_r} \right]^{2\alpha}, \tag{9.53}$$

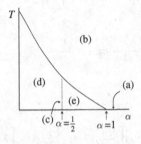

Figure 9.3 Phase diagram

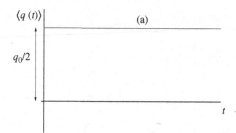

Figure 9.4 Complete blocking

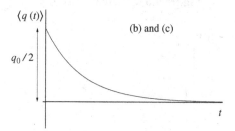

Figure 9.5 Exponential decay

which is valid for temperatures such that

$$\alpha k_B T / \hbar \gg \Delta_r, \tag{9.54}$$

and therefore the equilibrium state is reached without oscillatory behavior.

(c) $\alpha = 1/2$ (see Fig. 9.5):

$$P(t) = \exp\left[-\frac{\pi}{2} \frac{\Delta^2}{\omega_c} t \right], \tag{9.55}$$

which is an exact solution of the problem. Here too the system approaches equilibrium without oscillating. Notice that since $\Delta \sim (\omega_c/\omega_b)^\alpha$ and $\alpha = 1/2$, the dependence on the unphysical cutoff once again drops from our result.

(d) $T = 0$ and $0 \le \alpha < 1/2$ (see Fig. 9.6). Defining the dimensionless time $y \equiv \Delta_{eff} t$ where

$$\Delta_{eff} \equiv [\Gamma(1 - 2\alpha) \cos \pi \alpha]^{\frac{1}{2(1-\alpha)}} \Delta_r, \tag{9.56}$$

we have

$$P(y) = P_{coh}(y) + P_{inc}(y) \qquad \text{where} \tag{9.57}$$

$$P_{coh}(y) \equiv \frac{1}{1-\alpha} \cos\left\{ \left[\cos\left(\frac{\pi}{2} \frac{\alpha}{1-\alpha} \right) \right] y \right\} \exp - \left\{ \left[\sin\left(\frac{\pi}{2} \frac{\alpha}{1-\alpha} \right) \right] y \right\} \tag{9.58}$$

Dissipative coherent tunneling

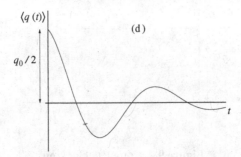

Figure 9.6 Coherent relaxation

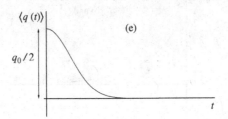

Figure 9.7 Incoherent relaxation

and

$$P_{inc}(y) \equiv -\frac{\sin 2\pi\alpha}{\pi} \int_0^\infty dz \frac{z^{2\alpha-1}e^{-zy}}{z^2 + 2z^{2\alpha}\cos 2\pi\alpha + z^{4\alpha-2}}. \qquad (9.59)$$

In this case it is possible for the two-state system to decay to its equilibrium state in an oscillatory fashion, depending on the competition between the coherent and incoherent behaviors of $P(t)$.

(e) $T = 0$ and $1/2 \leq \alpha < 1$ (see Fig. 9.7). This is the region of the parameter space over which the NIBA is least reliable. It seems that for these values of damping and temperature the system decays to equilibrium in an incoherent way.

Concluding this section we can say that we have been able to map, at least at very low temperatures and within the WKB approximation, the problem of a Brownian particle in a quartic (bistable) potential onto that of a two-state system coupled to a bath of non-interacting oscillators. The resulting model was the spin–boson Hamiltonian for which there are many methods of finding approximate dynamical solutions. We have chosen a specific one, the currently nicknamed NIBA, which provides us with very elegant analytical expressions for the unbiased ohmic case. Nevertheless, although this method allows for solutions to be obtained in closed form, they are not valid for the whole parameter space of the problem. As has been clearly stressed in Leggett *et al.* (1987), NIBA is not a very good approximation for long times when $\alpha < 1/2$ and low temperatures ($k_B T < \hbar \Delta_r$). In contrast, it can be shown to be an excellent approximation for strong damping ($\alpha > 1$) or high temperatures ($k_B T > \hbar \Delta_r$). At any rate, if we are interested in studying the intermediate-time dynamics of such systems, as is the case when we investigate the existence of macroscopic quantum effects, NIBA can be very suitable.

A more careful and complete analysis of this problem, including the finite-bias case and non-ohmic spectral functions, is provided, for example, in Leggett *et al.* (1987) and Weiss (1999).

10

Outlook

Having achieved this point we hope to have accomplished, at least partly, our main aim which was to convince the reader that once appropriate systems have been found (or built) they can present a very peculiar combination of microscopic parameters in such a way that quantum mechanics should be applied to general macroscopic variables to describe the collective effects therein. Moreover, the very nature of these macroscopic variables does not allow them to be treated in an isolated fashion. They must rather be considered coupled to uncontrollable microscopic degrees of freedom which is the ultimate origin of dissipative phenomena. The latter, at least in the great majority of cases, play a very deleterious role in the dynamics of the macroscopic variables and we hope to have introduced minimal phenomenological techniques in order to quantify this.

We have concentrated our discussions on questions originating from a few examples of superconducting or magnetic systems where quantum mechanics and dissipative effects coexist. In particular, superconducting devices which present the possibility of displaying several different quantum effects (quantum interference, decay by quantum tunneling, or coherent tunneling) are of special importance, as we will see below. Prior to development of the modern cryogenic techniques and/or the ability to build nanometric devices, it was unthinkable to imagine the existence of subtle quantum mechanical effects such as the entanglement of macroscopically distinct quantum states. What we mean by macroscopic here is actually nano or mesoscopic, since objects on this scale may behave, in appropriate conditions, as controllable giant atoms or molecules. Besides, we have seen that some of these devices have their dynamics quite well mimicked by two-state systems, and this entitles them to be good candidates for qubits. As we all know by now, it has been a great challenge for physicists to develop a system which would meet the usual requirements for these entities (see below). Microscopic systems (genuine spins, for example) preserve coherence for a long time but are very hard to access. Therefore, it would be desirable if we could develop nano or mesoscopic systems (in particular electronic circuits) obeying a two-state system dynamics because these are more easily handled by electronic technology similar to that already at work in conventional computers.

It is in this sense that superconducting devices can play a new role in physics. On the one hand, they look like good qubits and on the other hand, they can present macroscopic quantum effects. As the latter is indissociable from the destructive effects of dissipation, the preservation of quantum coherence will be inhibited and since this is the topic we have been investigating this far, we think that we are now in a position to address some of these questions (see below).

In the next three sections of this chapter we will respectively try to answer the following questions:

(i)   Are these "macroscopic quantum effects" really observable?
(ii) If so, where could they be applied?
(iii) What have we left out of our study?

10.1 Experimental results

Since the beginning of the 1980s the subject we have been calling macroscopic quantum phenomena, together with the vicious quantum dissipative effects, have been experiencing very fast development. There are so many papers, theoretical and experimental, about the subject that it would take us a very long time and several pages to refer to all of them in a proper way. Since many theoretical contributions have already been referred to in previous chapters, in this section we have chosen to present a short historical review of the main experimental developments in the field. We warn the reader that this is indeed very short and incomplete, but information about other equally important pieces of work can be traced from the references we provide.

Soon after the publication of Caldeira and Leggett (1981), the pioneering work of Voss and Webb (1981) on the tunneling of the phase across a Josephson junction was released, triggering a wave of experimental research on the subject. However, the results produced therein, and also subsequently, were not consensually accepted in the community as being evidence of the tunneling event referred to, and neither could the effects of dissipation be clearly extracted.

It was not until the seminal work of Clarke *et al.* (1988) that researchers in the area agreed that it was indeed the previously predicted tunneling of the phase difference of a Josephson junction that was being observed and, moreover, dissipative effects could be controlled in their experiment, confirming their influence in quantum tunneling as predicted in Caldeira and Leggett (1981, 1983a). At the end of the 1980s, while researchers working with superconducting devices were still facing the challenge of observing macroscopic coherent tunneling (for which more sophisticated procedures and experimental techniques were needed), some advances were being made in the observation of quantum tunneling of vortices in bulk superconductors. This was done through measurement of the magnetic relaxation due to collective creep of vortex lines (see, for example, Mota *et al.* (1991)) in an organic superconductor. The measurements performed in the latter corroborated the theory of collective quantum creep proposed by Blatter (see Blatter *et al.* (1994) and references cited therein). Observation of vortex tunneling in thin superconducting films has also been performed since then (Liu *et al.*, 1992; Sefrioui *et al.*, 2001; Tafuri *et al.*, 2006).

Almost concomitantly, there were considerable advances in the observation of coherent tunneling in magnetic particles; either in Mn_{12} clusters (Sessoli *et al.*, 1993; Friedman *et al.*, 1996; Sangregorio *et al.*, 1997), small ferromagnetic particles (Paulsen *et al.*, 1992), or magnetic proteins (Awschalom *et al.*, 1992). However, only a little later tunneling of domain walls started to be measured (Hong and Giordano, 1996; Brooke *et al.*, 2001; Shpyrko *et al.*, 2007). Although, in the 1990s, experimentalists succeeded in performing subtle measurements to detect quantum effects in magnetic particles and walls, bulk superconductors and superconducting films, coherent tunneling in superconducting devices (the cornerstone of this subject) was still untouched. It was only at the turn of the century that coherent superposition of macroscopically distinct quantum states was finally observed in SQUIDs (Friedman *et al.*, 2000; Chiorescu *et al.*, 2003) and Josephson junctions (Yu *et al.*, 2002). Coincidently, the latter publication followed Vion *et al.* (2002), in the same journal, where the authors reported the design of a quantum circuit containing a Cooper pair box, the quantronium, that could be operated as a quantum two-level atom. These achievements definitely gave a further boost to the possibility of using superconducting devices as qubits (see below).

In the brief survey of the experimental results we have provided above, only those experiments referring to the phenomena we used in the first two chapters of this book as the paradigms of macroscopic quantum phenomena were mentioned. However, there are other equally important examples of "cat-like states" that, if not necessarily macroscopic in the sense of involving a coherent superposition of a macroscopically large number of particles, still contain parameters which can be viewed as "macroscopic" in a different sense. This is the case of a laser-cooled trapped ion prepared in a superposition state of two well-localized Gaussians (Monroe *et al.*, 1996; Leibfried *et al.*, 2003), very similar to the state we presented in Section 7.2, or a superposition of two coherent states of photons in a microwave cavity (Brune *et al.*, 1996; Raimond *et al.*, 2001). Decoherence in the coherent-state representation can be treated with techniques very similar to those of Section 7.2, as shown in Castro Neto and Caldeira (1990).

10.2 Applications: Superconducting qubits

It is our goal now to analyze briefly the possibility of employing the superconducting devices we have introduced so far as qubits.

DiVincenzo (2000) established the minimal desiderata for a physical system to be a qubit through the following points:

(i) Very well-defined two-state systems, either exact or approximate. In the latter case we have to make sure that the remaining states of the system will not be accessible during any operation performed on it.

(ii) The initial state of the system has to be prepared with sufficient accuracy.

(iii) A very long phase coherent time to allow for a large number of coherent operations. This should be taken as greater than a standard value of 10^4.

(iv) Controllable "effective fields" at the qubit positions in order to implement the required logical operations through the application of unitary transformations to single or double qubits and switching the inter-qubit interactions on and off.

(v) Quantum measurement to read out the quantum information.

We see that, if not for item (iii) above, the superconducting devices we have introduced in this book fulfill those requirements and then could be considered good candidates for qubits. Therefore, all we need to do in order to implement them as such is to beat decoherence.

A possible model for a quantum processor whose qubits would take into account most of these points is given by

$$\mathcal{H} = \mathcal{H}_{qb} + \mathcal{H}_{meas} + \mathcal{H}_{env}, \tag{10.1}$$

where $\mathcal{H}_{meas}$ and $\mathcal{H}_{env}$ include their respective interactions with $\mathcal{H}_{qb}$ and

$$\mathcal{H}_{qb} = -\frac{1}{2}\sum_{i=1}^{N} \mathbf{B}^{(i)}(t)\cdot\sigma^{(i)} + \sum_{i\neq j}\sum_{a,b} J_{ij}^{ab}(t)\sigma_a^{(i)}\sigma_b^{(j)}, \tag{10.2}$$

with $a, b = x, y, z$, is the N-qubit Hamiltonian.

However, it is not our intention here to treat the general problem of the design of quantum processors. The problem we are going to address now refers only to a single two-state system immersed in an environment. Therefore, we are actually aiming only at points (i), (ii), and (iii) above. This means that we are going to consider only $\mathcal{H}_{qb}$ and $\mathcal{H}_{env}$ in (10.1) and, moreover, take $N = 1$ in $\mathcal{H}_{qb}$. The more technical problem of how quantum coherence between the qubits scales to $N \gg 1$ is still an open question, and certainly beyond the scope of this book.

10.2.1 Flux qubits

Our first example of a superconducting qubit is the SQUID ring we introduced earlier. As we have seen, these devices can operate when the external flux $\phi_x = \phi_0/2$. In this situation the potential energy stored in the ring has its lowest-energy eigenstates (almost flux eigenstates) as those of a bistable potential we investigated when we discussed the macroscopic quantum effects in SQUIDs (see Fig. 3.6). Therefore, if the external conditions are such that our Hilbert space can be restricted to this two-dimensional space, as we have seen before, the effective Hamiltonian of this system is

$$\mathcal{H}_{fqb} = -\frac{1}{2}B_z\sigma_z - \frac{1}{2}B_x\sigma_x, \tag{10.3}$$

where B_z is the term that controls the symmetry of the double well whereas B_x allows for the transition between the two eigenstates of σ_z. The former is given by

$$B_z(\phi_x) = 4\pi\sqrt{6(\beta_L - 1)}E_J\left(\frac{\phi_x}{\phi_0} - \frac{1}{2}\right), \tag{10.4}$$

where $\beta_L \equiv 2\pi L i_0/\phi_0$ (see analysis below (3.56), (3.57)) and $B_x = \hbar\Delta$ is the tunneling amplitude between the two minima, which depends exponentially on E_J (see (9.13)). For very low damping $\Delta \approx \Delta_0$, which is given by (9.3) in terms of the appropriate device parameters.

In order to establish a connection with the general form (10.2) for $N = 1$ we need to specify a method for controlling B_x also. This can be done if we replace the junction of the SQUID by a secondary SQUID ring with two Josephson junctions (see Fig. 10.1). This smaller loop threads an amount of flux $\tilde{\phi}_x$ which controls the effective Josephson coupling of the SQUID. In this way we have both B_x and B_z tunable by external parameters.

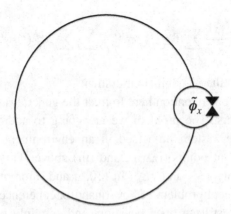

Figure 10.1 Secondary SQUID junction

Although many quantum effects have been observed in SQUIDs (Makhlin *et al.*, 2001), the existence of energy splitting near the degeneracy point was only observed for the first time in Friedman *et al.* (2000), as mentioned above.

10.2.2 Charge qubits

Now we return to the device introduced earlier as the CPB. It has $E_C \gg E_J$ and in this case charge rather than flux states are the most appropriate to describe the dynamics of the device. A charge qubit is actually a small superconducting region containing an excess of n Cooper pairs connected, by a tunnel junction, to a much larger superconductor which acts as a charge reservoir. Charge is fed into the system by a gate voltage V_g and the charging energy is then

$$E_C = \frac{e^2}{2(C_J + C_g)} \tag{10.5}$$

where $C_J \lesssim 10^{-15}$ f and C_g are, respectively, the capacitance of the junction and the gate capacitance (see Fig. 10.2), which is usually smaller than the former.

Assuming the superconducting gap to be the largest energy in the problem, so only Cooper pairs can tunnel, we can write the Hamiltonian of the system as

$$\mathcal{H}_{cqb} = 4E_C(n - n_g)^2 - E_J \cos\varphi, \tag{10.6}$$

where $n = -id/d\varphi$ is the momentum canonically conjugate to the phase $\hbar\varphi$ and $n_g \equiv C_g V_g / 2e$ the dimensionless gate charge. Since $E_C \gg E_J$, a convenient basis is formed by the charge states and we can write

$$\mathcal{H}_{cqb} = \sum_n [4E_C(n - n_g)^2 |n\rangle\langle n| - \frac{1}{2}E_J(|n\rangle\langle n+1| + |n+1\rangle\langle n|)]. \tag{10.7}$$

When $n_g = 1/2$ (see Fig. 10.3) the Josephson energy couples two adjacent states such as, for example, $|0\rangle$ and $|1\rangle$. In this case we can approximate

$$\mathcal{H}_{cqb} = -\frac{1}{2}B_z\sigma_z - \frac{1}{2}B_x\sigma_x, \tag{10.8}$$

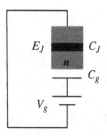

Figure 10.2 Cooper pair box

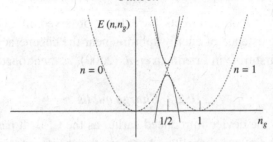

Figure 10.3 Energy eigenvalues (solid line) and energy of the charge states (dotted line) of the junction

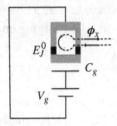

Figure 10.4 Charge qubit with tunable Josephson coupling. Dotted line represents a circuit branch furnishing flux ϕ_x to the two-junction loop

and the charge states are $| \uparrow \rangle \equiv | n = 0 \rangle$ and $| \downarrow \rangle \equiv | n = 1 \rangle$. Moreover,

$$B_z = 4E_C(1 - 2n_g) \qquad \text{and} \qquad B_x = E_J. \tag{10.9}$$

Once again we would like to write (10.8) with tunable B_x as in (10.2). This can be accomplished if we replace the junction connecting the CPB to the charge environment by two junctions in parallel (see Fig. 10.4), which forms something like a two-junction SQUID. Therefore, an external flux ϕ_x threaded by this loop can tune the effective Josephson coupling energy of this circuit, which now reads

$$B_x = E_J(\phi_x) = 2E_J^0 \cos\left(\pi \frac{\phi_x}{\phi_0}\right). \tag{10.10}$$

10.2.3 Phase qubits and transmons

Phase qubits are basically CBJJs (see Section 3.4.2) described in terms of the phase difference between the terminals of the Josephson junction with $E_J \gg E_C$, as usual. We then use the two lowest-lying energy "eigenstates" (very long-lived decaying states) in a metastable well of a tilted sinusoidal potential in the phase representation as qubits (see Fig. 10.5). The current bias can be tuned properly in such a way that only very few (probably two) of those states live in the metastable well. On top of that, we can manipulate these states with microwave pulses in

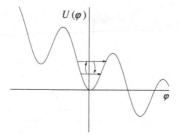

Figure 10.5 Phase qubit

resonance with the Bohr frequency $\omega_{10} \equiv (E_1 - E_0)/\hbar$ between the two lowest-lying states and also d.c. pulses in such a way that the total current through the junction is (Martinis *et al.*, 2002)

$$I(t) = I_{\mathrm{dc}} + \delta I_{\mathrm{dc}}(t) + I_{\mu\mathrm{wc}}(t)\cos\omega_{10}t + I_{\mu\mathrm{ws}}(t)\sin\omega_{10}t. \tag{10.11}$$

With the hypothesis of slow variations of $\delta I_{\mathrm{dc}}(t)$, $I_{\mu\mathrm{wc}}(t)$, and $I_{\mu\mathrm{ws}}(t)$ in comparison with $2\pi/(\omega_{10} - \omega_{21})$, the dynamics of the system is restricted to the two-dimensional Hilbert space spanned by the two lowest-lying states within the well and, in a frame rotating with ω_{10}, its Hamiltonian reads

$$H(t) = \frac{\sigma_z}{2}\delta I_{\mathrm{dc}}(t)\frac{\partial E_{10}}{\partial I_{\mathrm{dc}}} + \frac{\sigma_x}{2}\sqrt{\frac{\hbar}{2\omega_{10}C}}I_{\mu\mathrm{wc}}(t) + \frac{\sigma_y}{2}\sqrt{\frac{\hbar}{2\omega_{10}C}}I_{\mu\mathrm{ws}}(t). \tag{10.12}$$

The microwave terms above induce Rabi oscillations (Rabi, 1937) between the two states that could be measured (Martinis *et al.*, 2002; Yu *et al.*, 2002). Therefore, if logical operations can be performed in this system before the excited state decays by quantum tunneling, it can be viewed as a good qubit. Notice that since the decay rates are exponentials of the ratio between the barrier height of the nth excited state and $\hbar\omega_p$, where ω_p is the classical oscillation frequency about the local minimum of Fig. 10.5, the ratio between successive decay rates is such that $\Gamma_{n+1}/\Gamma_n \sim 1000$ (see Martinis *et al.* (2002)) and we are safe to operate only within the decay time of the excited state. However, if we try to avoid qubits whose states are not strictly stable, there is an alternative method to operate phase qubits in a different range of device parameters that culminates in the creation of a new generation of qubits, the so-called *transmons* (Koch *et al.*, 2007).

Transmons are basically phase qubits still with $E_J > E_C$ (but not $E_J \gg E_C$ as above[1]) operated as a charge qubit, namely with a gate voltage bias instead of a current bias. Therefore, the transmon charge energy bands (see Figs. 10.6 and

---

[1] Actually, both are operated for $E_J \gg E_C$ but phase qubits are such that $E_J \ggg E_C$ (see Koch *et al.*, 2007). Usually $E_J/E_C \sim 10^2$ for transmons and $\sim 10^4$ for phase qubits.

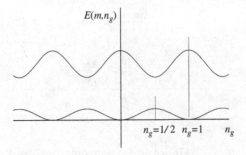

Figure 10.6 Qualitative behavior of the energy bands for $E_J \gtrsim E_C$

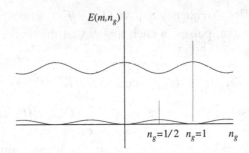

Figure 10.7 Qualitative behavior of the energy bands for $E_J \approx 5E_C$

10.7) are much narrower than those of the charge qubits, which makes them much more insensitive to charge noise than the latter.

For the sake of completeness we write here the expressions for the energy spectrum of a transmon. In the limit $E_J \gg E_C$, when the dispersion can be approximated by a cosine, we have (Koch *et al.*, 2007)

$$E_m(n_g) = E_m(n_g = 1/4) - \frac{\epsilon_m}{2} \cos(2\pi n_g), \qquad (10.13)$$

where $\epsilon_m \equiv E_m(n_g = 1/2) - E_m(n_g = 0)$ is the width of the mth charge energy band which, within the WKB approximation (again valid for $E_J \gg E_C$), is given by

$$\epsilon_m \approx (-1)^m E_C \frac{2^{4m+5}}{m!} \sqrt{\frac{2}{\pi}} \left(\frac{E_J}{2E_C}\right)^{\frac{2m+3}{4}} \exp -\sqrt{\frac{8E_J}{E_C}}. \qquad (10.14)$$

In contrast, the approximate level structure of the transmon for $E_J \gg E_C$ can be obtained by expanding (10.6) about $\varphi \approx 0$ as

$$E_m(n_g = 0) \approx -E_J + \sqrt{8E_J E_C} \left(m + \frac{1}{2}\right) - \frac{E_C}{12}(6m^2 + 6m + 3), \qquad (10.15)$$

which, together with (10.13) and (10.14), allows us to determine the parameters of (10.8) close to $n_g = 1/2$ as $B_z = 4E_C(1 - 2n_g)$ and $E_J \lesssim B_x \lesssim \sqrt{8E_J E_C}$ for $1 \lesssim E_J/E_C \lesssim 50$.

10.2.4 Decoherence

Now that we know the two-state systems to which our superconducting devices correspond, we can study the decoherence effects in their time evolution by taking into account how they couple to their respective environments.

For the single-junction device, this has already been done. That is exactly the spin–boson model (9.16). Since we are interested in the limit for which the system performs many coherent oscillations, we must look for very weakly damped systems (see (9.57) and (9.58)). Moreover, even in this limit we should keep in mind that $\hbar \Delta E \gg \alpha k_B T$.[2] This is the so-called *system-dominated* regime where the coupling to the environment can be treated perturbatively, as opposed to the *environment-dominated* regime ($\hbar \Delta E \ll \alpha k_B T$) in which the details of the coupling to the environment do matter and, for example, the exponential decay (9.52), (9.53) holds. In the system-dominated regime we can show (Makhlin *et al.*, 2001) that two decay rates are particularly important:

$$\tau_{rel}^{-1} = \pi \alpha \sin^2\lambda \ \Delta E \coth\frac{\hbar \Delta E}{2k_B T} \quad \text{and} \tag{10.16}$$

$$\tau_{\varphi}^{-1} = \frac{1}{2} \tau_{rel}^{-1} + \pi \alpha \cos^2\lambda \ \frac{2k_B T}{\hbar}, \tag{10.17}$$

where $\lambda \equiv \tan^{-1}(B_x/B_z)$ and $\Delta E = \sqrt{B_x^2 + B_z^2}$. These are respectively related to the decay rates of the average values of the matrix elements of the operators $\rho_z(t)$ and $\rho^{(\pm)}(t) \equiv [\rho_x(t) \pm i \, \rho_y(t)]/2$, where $\rho_i(t)$ ($i = x, y, z$) are Pauli matrices in the representation of the eigenstates $\{|0\rangle, |1\rangle\}$ of (10.8) whereas $\sigma_i(t)$ are represented in the usual $\{|\uparrow\rangle, |\downarrow\rangle\}$ basis. These two sets are related by

$$|0\rangle = \cos\frac{\lambda}{2}|\uparrow\rangle + \sin\frac{\lambda}{2}|\downarrow\rangle,$$

$$|1\rangle = -\sin\frac{\lambda}{2}|\uparrow\rangle + \cos\frac{\lambda}{2}|\downarrow\rangle. \tag{10.18}$$

In the particular case $\lambda = \pi/2$ (or $B_z = 0$), the rate τ_{rel}^{-1} is that for which $\langle \sigma_x(t) \rangle$ decays to equilibrium whereas τ_{φ}^{-1}, the dephasing or decoherence rate, measures how the oscillatory motion of $\langle \sigma_z(t) \rangle$ is attenuated.

Decoherence times in superconducting devices are usually measured by Ramsey interferometry (Ramsey, 1950; Vion *et al.*, 2002; Yu *et al.*, 2002) and recently very

---

[2] See definition of ΔE below (10.17). When $B_z = 0$ it simply becomes $B_x = \Delta$.

long coherence times (~ 0.1 ms) for transmons have been reported in Rigetti *et al.* (2012).

In general, the qubit is coupled to different environments through different components of the pseudo-spin by which it is being represented. Therefore, we generalize our spin–boson Hamiltonian to

$$\mathcal{H} = \mathcal{H}_{qb} + \sum_i \boldsymbol{\sigma} \cdot \mathbf{n}_i \left(\sum_a C_a^{(i)} q_a^{(i)} \right) + \sum_i \mathcal{H}_{env}^{(i)}, \tag{10.19}$$

where $\mathbf{n}_i$ is the direction through which the ith environment couples to the qubit and q_a^i is the ath "coordinate" of this environment. This new Hamiltonian generates the decay rates of relaxation and decoherence, respectively, as

$$\tau_{rel}^{-1} = \sum_i \pi \, \alpha_i \sin^2\!\lambda_i \, \Delta \coth\frac{\hbar\Delta}{2k_B T}, \tag{10.20}$$

$$\tau_{\varphi}^{-1} = \frac{1}{2}\,\tau_{rel}^{-1} + \sum_i \pi \, \alpha_i \cos^2\!\lambda_i \, \frac{2k_B T}{\hbar}, \tag{10.21}$$

where $\cos\lambda_i \equiv \mathbf{B} \cdot \mathbf{n}_i / |\mathbf{B}|$. Here we should be aware that the system-dominated regime is valid if $\hbar\Delta E \gg \sum_i \alpha_i k_B T$.

Although we have only presented the ohmic case here, the procedure is quite general. Once we identify how the relevant dynamical variable of the qubit is coupled to external noise sources, we can use the fluctuation–dissipation theorem for the corresponding environment operator (Makhlin *et al.*, 2001) in order to extract the spectral function $J(\omega)$ associated with that damping mechanism. Consequently, we determine the relevant dimensionless damping α and write generalized spin–boson Hamiltonians from which we can compute relaxation and decoherence times as in (10.20) and (10.21).

With this brief introduction to the physics of superconducting devices viewed as qubits, we hope to have convinced the reader of the possibility of practical applications of macroscopic quantum effects.

10.3 Final remarks

As our main goal from the very beginning of this work has been to give the reader an introduction to macroscopic quantum phenomena and quantum dissipation and provide him (or her) with the mathematical tools needed to make some quantitative developments on the problems presented in the two introductory chapters, there is a point at which we must content ourselves with what has been achieved. As a matter of fact, we think this is a good place to stop and let the reader choose which path to take from now on depending on his (or her) main interests. Nevertheless, even without trying to analyze in any depth other subjects that could be handled with the techniques developed here, it will be desirable to enumerate at least those we think would be natural extensions of what we have done so far. That is what follows.

Other dissipative mechanisms. In our introductory examples of magnetism and superconductivity, the classical equations of motion describing the dissipative systems of interest were quite general and the paradigmatic example of the Langevin equation turns out to be a good approximation only for some of them. In general, we may have dissipative terms which are linearly dependent on higher-order time derivatives of the dynamical variable being studied, or even present non-linear dependence in those variables. We have chosen to present three specific models that cover many possible situations but which under some hypotheses could again be described, in the classical limit, by Langevin equations. However, we have said very little (or almost nothing) about those systems whose dynamics can be well described by the minimal model with general spectral functions $J(\omega)$. In all our examples we have dealt with ohmic dissipation and entirely omitted the super-ohmic and subohmic cases. Although they are probably less ubiquitous than the ohmic case, we have only put them aside because once we have understood the concepts in the latter case, the results in other cases can be worked out without much effort (see, for example, Leggett *et al.* (1987) and Weiss (1999)). Actually, in many applications it is the ohmic case that requires more sophisticated analysis.

Other examples in the same systems. We have given examples of systems which posed questions that could be reduced to simple quantum mechanical problems of a mass particle. That was the case, for example, of tunneling (coherent or not) of flux-oids in superconducting devices. Because of these, we were led to study dissipative quantum tunneling or coherence, but did not say anything about Bloch oscillations or Zener tunneling in these systems, which can be found in, for example, Gefen *et al.* (1987) and Schön and Zaikin (1990).

Dissipation in microscopic models. The models we have been using in this book were set up to describe the dissipative dynamics of a collective macroscopic variable. However, there is nothing to prevent them describing microscopic variables instead. Actually, we have already mentioned this when introducing the non-linear and collision models. Transport problems, Kondo-like models, and dissipative quantum phase transitions are just a few examples of where the sort of models we have employed here, together with the physics coming out of them, can also be useful. More on these possibilities can be found in Leggett *et al.* (1987), Schön and Zaikin (1990) and Weiss (1999).

Quantum thermal field theory. A glimpse at our list of references is enough to show the reader how broad and general the subject addressed in this book is. We have quoted articles from very different areas of physics, ranging from the physics of the early universe to qubits. One issue that we have only briefly touched upon is field theoretical models very far from equilibrium, of which finite-temperature quantum nucleation with dissipation is an example. We refer the interested reader to Kapusta and Gale (2006) for recent developments in this area.

Other omissions and more applications. There are still other areas where the competition between quantum effects and dissipation is present and of crucial importance. In particular, those we will list in what follows are very modern topics which are currently attracting the attention of the physical community. With the machinery we have developed so far and extensions thereof, we think much can be done in particularly prepared systems in Bose–Einstein condensates (BECs), nano-electromechanical devices (NEMs), cavity quantum electrodynamics (CQED), or circuit quantum electrodynamics (cQED).

Appendix A

Path integrals, the quantum mechanical propagator, and density operators

A.1 Real-time path integrals

In this appendix, it is our intention to introduce the path (or functional) integral representation of the quantum mechanical propagator $K(x, t; x', 0)$ which, as we know, is given by

$$K(x, t; x', 0) = \langle x|e^{-i\mathcal{H}t/\hbar}|x'\rangle, \tag{A.1}$$

where we are assuming a system described by the Hamiltonian $\mathcal{H} = p^2/2M + V(q)$. If we now subdivide the time interval $[0, t]$ into small partitions such that $[0, t] = \bigcup_{i=0}^{N-1}[t_{i+1}, t_i]$, where $t_0 = 0$ and $t_N = t$, we can write

$$K(x, t; x', 0) = \langle x|e^{-i\mathcal{H}(t-t_{N-1})/\hbar}\ldots e^{-i\mathcal{H}(t_k-t_{k-1})/\hbar}\ldots e^{-i\mathcal{H}t_1/\hbar}|x'\rangle, \tag{A.2}$$

where we have the product of N exponentials. If we introduce a completeness relation $\int_{-\infty}^{+\infty} dx_k |x_k\rangle\langle x_k| = \mathbb{1}$ $(1 \leq k \leq N-1)$ between each exponential term, we have

$$K(x, t; x', 0) = \int_{-\infty}^{+\infty}\ldots\int_{-\infty}^{+\infty} dx_1\ldots dx_k\ldots dx_{N-1}\langle x|e^{-i\mathcal{H}(t-t_{N-1})/\hbar}|x_{N-1}\rangle$$

$$\times\langle x_{N-1}|\ldots\ldots|x_k\rangle\langle x_k|e^{-i\mathcal{H}(t_k-t_{k-1})/\hbar}|x_{k-1}\rangle\langle x_{k-1}|\ldots|x_1\rangle\langle x_1|e^{-i\mathcal{H}t_1/\hbar}|x'\rangle. \tag{A.3}$$

If we now make the length of each time interval $t_k - t_{k-1} \equiv \epsilon = t/N \to 0$ $(N \to \infty)$, we can write for the kth interval

$$K(x_k, t_k; x_{k-1}, t_{k-1}) \approx \langle x_k|1 - \frac{i\epsilon}{\hbar}\mathcal{H}|x_{k-1}\rangle, \tag{A.4}$$

with $x_N = x$ and $x_0 = x'$, or yet

$$K(x_k, t_k; x_{k-1}, t_{k-1}) \approx \langle x_k | x_{k-1} \rangle - \frac{i\epsilon}{\hbar} \langle x_k | \frac{p^2}{2M} | x_{k-1} \rangle - \frac{i\epsilon}{\hbar} \langle x_k | V(q) | x_{k-1} \rangle.$$
(A.5)

Next, introducing the identity $\int_{-\infty}^{+\infty} dp_k |p_k\rangle\langle p_k| = \mathbb{1}$ in the equation above we get

$$K(x_k, t_k; x_{k-1}, t_{k-1}) \approx \int_{-\infty}^{+\infty} dp_k \langle x_k | p_k \rangle \langle p_k | x_{k-1} \rangle$$

$$- \frac{i\epsilon}{\hbar} \int_{-\infty}^{+\infty} dp_k \frac{p_k^2}{2M} \langle x_k | p_k \rangle \langle p_k | x_{k-1} \rangle - \frac{i\epsilon}{\hbar} \int_{-\infty}^{+\infty} dp_k V(x_k) \langle x_k | p_k \rangle \langle p_k | x_{k-1} \rangle.$$
(A.6)

Now, as

$$\langle x_k | p_k \rangle = \frac{1}{\sqrt{2\pi\hbar}} e^{ip_k x_k/\hbar},$$
(A.7)

we can rewrite (A.6) as

$$K(x_k, t_k; x_{k-1}, t_{k-1}) \approx \frac{1}{\sqrt{2\pi\hbar}} \int_{-\infty}^{+\infty} dp_k \exp \frac{ip_k}{\hbar}(x_k - x_{k-1})$$

$$\times \left(1 - \frac{i\epsilon}{\hbar}\left(\frac{p_k^2}{2M} + V(x_k)\right)\right),$$
(A.8)

whose second part of the integrand can be exponentiated to yield

$$K(x_k, t_k; x_{k-1}, t_{k-1}) = \frac{1}{\sqrt{2\pi\hbar}} \int_{-\infty}^{+\infty} dp_k \exp \frac{ip_k}{\hbar}(x_k - x_{k-1})$$

$$\times \exp \frac{i\epsilon}{\hbar}\left(\frac{p_k^2}{2M} + V(x_k)\right).$$
(A.9)

The integral over p_k can now be evaluated easily and reads

$$K(x_k, t_k; x_{k-1}, t_{k-1}) = \sqrt{\frac{M}{2\pi i\hbar\epsilon}} \exp \frac{i\epsilon}{\hbar}\left(\frac{M}{2}\frac{(x_k - x_{k-1})^2}{\epsilon^2} - V(x_k)\right), \quad (A.10)$$

which substituted in (A.3) finally gives

$$K(x, t; x', 0) = \int\limits_{-\infty}^{+\infty} \cdots \int\limits_{-\infty}^{+\infty} dx_1 \ldots dx_k \ldots dx_{N-1} \left[\prod_{k=1}^{N} \sqrt{\frac{M}{2\pi i \hbar \epsilon}} \right]$$

$$\times \exp \sum_{k=1}^{N} \frac{i\epsilon}{\hbar} \left(\frac{M}{2} \frac{(x_k - x_{k-1})^2}{\epsilon^2} - V(x_k) \right). \tag{A.11}$$

Taking the limit $\epsilon \to 0$ and remembering that $\epsilon = \Delta t_k \equiv t_k - t_{k-1}$, we rewrite (A.11) in the following symbolic form:

$$K(x, t; x', 0) = \prod_{t'=0}^{t} \int\limits_{-\infty}^{\infty} \frac{dx(t')}{\mathcal{N}} \exp \frac{i}{\hbar} S[x(t')], \tag{A.12}$$

where

$$S[x(t')] = \int\limits_{0}^{t} dt' \left(\frac{1}{2} M \dot{x}^2 - V(x) \right) \tag{A.13}$$

is the classical action of the particle, or

$$K(x, t; x', 0) = \int\limits_{x'}^{x} \mathcal{D}x(t') \exp \frac{i}{\hbar} S[x(t')], \tag{A.14}$$

where the measure of integration $\mathcal{D}x(t')$ already contains the normalization factor $\mathcal{N}$ of (A.12). Therefore, we have succeeded in writing the propagator as a sum of exponentials of the classical actions of all geometrical paths joining x and x'. This is the famous *path integral* representation of the quantum mechanical propagator, which is due to Feynman (Feynman and Hibbs, 1965). Here it should be stressed that although our development was performed in the particular case of a conservative one-dimensional system, expression (A.14) is also valid for systems with explicit time-dependent Hamiltonians as we show even for more general representations in Appendix C.

Our next step is to provide a couple of specific examples which will prove very helpful in operating with this formalism; the free particle and the quadratic Lagrangian.[1]

---

[1] Working out these examples we will be following closely Schulman (1981).

Free particle. In this case $V(q) = 0$, which implies that

$$K(x, t; x', 0) = \int\limits_{-\infty}^{+\infty} \cdots \int\limits_{-\infty}^{+\infty} dx_1 \ldots dx_{N-1}$$

$$\times \prod_{k=1}^{N} \sqrt{\frac{M}{2\pi i \hbar \epsilon}} \exp \frac{i\epsilon}{\hbar} \left(\frac{M}{2} \frac{(x_k - x_{k-1})^2}{\epsilon^2} \right). \qquad (A.15)$$

Now, using the well-known result of Gaussian integration,

$$\int\limits_{-\infty}^{\infty} du \sqrt{\frac{a}{\pi}} \exp{-a(x - u)^2} \sqrt{\frac{b}{\pi}} \exp{-b(u - y)^2}$$

$$= \sqrt{\frac{ab}{\pi(a + b)}} \exp{-\frac{ab}{a + b}(x - y)^2}, \qquad (A.16)$$

we have, for example,

$$\frac{M}{2\pi i \hbar \epsilon} \int\limits_{-\infty}^{\infty} dx_1 \exp \frac{iM}{2\hbar\epsilon}(x_2 - x_1)^2 \exp \frac{iM}{2\hbar\epsilon}(x_1 - x')^2$$

$$= \sqrt{\frac{M}{2\pi i \hbar(2\epsilon)}} \exp \frac{iM}{2\hbar(2\epsilon)}(x_2 - x')^2. \qquad (A.17)$$

Evaluating the integral in x_3 and then successively until x_N, it is straightforward to show that

$$K(x, t; x', 0) = \sqrt{\frac{M}{2\pi i \hbar t}} \exp \frac{iM}{2\hbar} \frac{(x - x')^2}{t}. \qquad (A.18)$$

Quadratic Lagrangian. Let us now consider a system whose dynamics is governed by a Lagrangian of the form

$$L = \frac{1}{2}M\dot{x}^2 + b(t)x\dot{x} - \frac{1}{2}c(t)x^2 - e(t)x, \qquad (A.19)$$

and we wish to compute the quantum propagator of this system through the functional integral

$$K(x, t; x', 0) = \int\limits_{x'}^{x} \mathcal{D}x(t') \exp \frac{i}{\hbar} S[x(t')]. \qquad (A.20)$$

In order to do that, we first perform a change of variables

$$y(t') = x(t') - x_c(t'), \qquad (A.21)$$

where $x_c(t')$ is chosen appropriately to be the solution of the Euler–Lagrange equation of (A.19):

$$M\ddot{x}_c + (c(t) + \dot{b}(t))x_c + e = 0. \tag{A.22}$$

Then, replacing (A.21) in $S = \int_0^t L\, dt$, integrating by parts, and using (A.22) we have

$$K(x, t; x', 0) = G(t) \exp \frac{i}{\hbar} S_c(x, x', t), \tag{A.23}$$

where $S_c(x, x', t)$ is the classical action of the trajectory $x_c(t')$ and

$$G(t) = \int_0^0 \mathcal{D}y(t') \exp \frac{i}{\hbar} \int_0^t dt' \left(\frac{1}{2}M\dot{y}^2 - \frac{1}{2}\tilde{c}(t)y^2 \right), \tag{A.24}$$

with $\tilde{c} = c + \dot{b}$, is an integral over trajectories $y(t')$ such that $y(t) = y(0) = 0$. This integral can be solved in its discrete version, as we show in what follows.

Using (A.11), equation (A.24) reduces to

$$G(t) = \lim_{N \to \infty, \epsilon \to 0} \int_{-\infty}^{\infty} \cdots \int_{-\infty}^{\infty} dy_1 \ldots dy_k \ldots dy_{N-1} \left[\frac{M}{2\pi i \hbar \epsilon} \right]^{N/2}$$

$$\times \exp \frac{i}{\hbar} \sum_{k=1}^{N} \left[\frac{M}{2} \frac{(y_k - y_{k-1})^2}{\epsilon} - \frac{1}{2} \epsilon\, \tilde{c}_{k-1}\, y_{k-1}^2 \right], \tag{A.25}$$

where $\tilde{c}_k \equiv \tilde{c}(t_k)$. Then, defining a vector

$$\zeta = \begin{pmatrix} y_1 \\ \cdot \\ \cdot \\ y_{N-1} \end{pmatrix} \tag{A.26}$$

and the matrix

$$\sigma = \frac{M}{2i\hbar} \begin{pmatrix} 2 & -1 & . & 0 & 0 \\ -1 & 2 & . & 0 & 0 \\ . & . & . & . & . \\ 0 & 0 & . & 2 & -1 \\ 0 & 0 & . & -1 & 2 \end{pmatrix} + \frac{i\epsilon}{2\hbar} \begin{pmatrix} \tilde{c}_1 & 0 & . & 0 & 0 \\ 0 & \tilde{c}_2 & . & 0 & 0 \\ . & . & . & . & . \\ 0 & 0 & . & \tilde{c}_{N-2} & 0 \\ 0 & 0 & . & 0 & \tilde{c}_{N-1} \end{pmatrix}, \tag{A.27}$$

we can write (A.25) as

$$G(t) = \lim_{N \to \infty, \epsilon \to 0} \left[\frac{M}{2\pi i \hbar \epsilon} \right]^{N/2} \int d^{N-1}\zeta \exp -\zeta^T \sigma\, \zeta. \tag{A.28}$$

This matrix σ can be diagonalized formally through a similarity transformation

$$\sigma = U^\dagger \sigma_D U, \tag{A.29}$$

where U is unitary. If $\xi = U\zeta$, we have

$$\int d^{N-1}\zeta \exp{-\zeta^T \sigma \zeta} = \int d^{N-1}\xi \exp{-\xi^T \sigma_D \xi} = \prod_{\alpha=1}^{N-1} \sqrt{\frac{\pi}{\sigma_\alpha}} = \frac{\pi^{(N-1)/2}}{\sqrt{\det \sigma}} \tag{A.30}$$

whenever $\det \sigma \neq 0$.

Therefore, replacing (A.30) in (A.28) we obtain

$$G(t) = \lim_{N\to\infty,\,\epsilon\to 0} \left(\left(\frac{M}{2\pi i\hbar}\right) \frac{1}{\epsilon} \frac{1}{\left(\frac{2i\hbar\epsilon}{M}\right)^{N-1}} \frac{1}{\det\sigma} \right)^{1/2} \tag{A.31}$$

and our problem becomes the evaluation of

$$f(t) = \lim_{N\to\infty,\,\epsilon\to 0} \left(\epsilon \left(\frac{2i\hbar\epsilon}{M}\right)^{N-1} \det\sigma \right) \tag{A.32}$$

which, in turn, implies computation of the following determinant:

$$\left(\frac{2i\epsilon\hbar}{M}\right)^{N-1} \det\sigma$$

$$= \det \left[\begin{pmatrix} 2 & -1 & . & 0 & 0 \\ -1 & 2 & . & 0 & 0 \\ . & . & . & . & . \\ 0 & 0 & . & 2 & -1 \\ 0 & 0 & . & -1 & 2 \end{pmatrix} - \frac{\epsilon^2}{M} \begin{pmatrix} \tilde{c}_1 & 0 & . & 0 & 0 \\ 0 & \tilde{c}_2 & . & 0 & 0 \\ . & . & . & . & . \\ 0 & 0 & . & \tilde{c}_{N-2} & 0 \\ 0 & 0 & . & 0 & \tilde{c}_{N-1} \end{pmatrix} \right]$$

$$\equiv \det\tilde{\sigma}_{N-1} \equiv p_{N-1}. \tag{A.33}$$

Using the expansion of $\tilde{\sigma}_{j+1}$ in first minors we can show that

$$p_{j+1} = \left(2 - \frac{\epsilon^2}{M}\tilde{c}_{j+1}\right) p_j - p_{j-1}, \qquad j = 1,\ldots,N-2 \tag{A.34}$$

where $p_1 = 2 - (\epsilon^2 c_1/M)$ and $p_0 = 1$. Rewriting (A.34) as

$$\frac{p_{j+1} - 2p_j + p_{j-1}}{\epsilon^2} = \frac{\tilde{c}_{j+1}p_j}{M}, \tag{A.35}$$

and defining $\varphi(t) \equiv \epsilon p_j$, we can easily see that when $\epsilon \to 0$ this finite differences equation becomes

$$\frac{d^2\varphi(t)}{dt^2} = -\frac{\tilde{c}(t)\varphi(t)}{M}, \tag{A.36}$$

with initial conditions that can be obtained through

$$\varphi(0) = \epsilon p_0 \to 0 \tag{A.37}$$

and

$$\frac{d\varphi(t)}{dt} = \epsilon \left(\frac{p_1 - p_0}{\epsilon} \right) = 2 - \frac{\epsilon^2 c_1}{M} - 1 \to 1. \tag{A.38}$$

Consequently, the function $f(t)$ defined in (A.32) is a solution of

$$\frac{d^2 f(t)}{dt^2} + \frac{\tilde{c}(t) f(t)}{M} = 0, \tag{A.39}$$

with $f(0) = 0$ and $df(t)/dt|_{t=0} = 1$, and finally we can write the propagator (A.23) as

$$K(x, t; x', 0) = \sqrt{\frac{M}{2\pi i \hbar f(t)}} \exp \frac{i}{\hbar} S_c(x, x', t). \tag{A.40}$$

In the particular case of a harmonic oscillator, $\tilde{c}(t) = c(t) = M\omega^2$, $f(t)$ is such that

$$\frac{d^2 f(t)}{dt^2} + \omega^2 f(t) = 0, \tag{A.41}$$

with the initial conditions (A.39), which leads us to

$$f(t) = \frac{\sin(\omega t)}{\omega}. \tag{A.42}$$

Although the method we have just presented to evaluate the prefactor of the quantum mechanical propagator is quite laborious, another way to write it is presented below.

Stationary phase (or semi-classical) approximation. As we have seen before, the quantum mechanical propagator admits a path integral representation (A.14) where the action is given by (A.13). Although we have deduced (A.14) from a particular form of the Hamiltonian of the system, namely that just below (A.1), this representation is absolutely general and also applies to time-dependent Hamiltonians as we have already assumed implicitly in treating the quadratic form given by (A.19). In our present case we will assume that the action is, without loss of generality, given by

$$S[q(t')] = \int_0^t dt' \, L(q(t'), \dot{q}(t'), t'). \tag{A.43}$$

In the limit $\hbar \to 0$ the integrand of (A.14) oscillates very rapidly and the most important contribution to the integral turns out to be the stationary points (paths in this case) of the action, which is the essence of the *stationary phase approxima-tion*. Therefore, expanding the action about a path $q_c(t')$ such that $q_c(0) = x'$ and $q_c(t) = x$ up to terms quadratic in $\delta q(t') \equiv q(t') - q_c(t')$, we have

$$
S[q(t')] = S[q_c(t')] + \int_0^t dt' \left\{ \frac{\partial L}{\partial q}\bigg|_{q_c} \delta q(t') + \frac{\partial L}{\partial \dot{q}}\bigg|_{q_c} \delta \dot{q}(t') \right\}
$$

$$
+ \frac{1}{2} \int_0^t dt' \left\{ \frac{\partial^2 L}{\partial q^2}\bigg|_{q_c} \delta q(t')\, \delta q(t') + 2\frac{\partial^2 L}{\partial \dot{q}\partial q}\bigg|_{q_c} \delta \dot{q}(t')\, \delta q(t') \right.
$$

$$
\left. + \frac{\partial^2 L}{\partial \dot{q}^2}\bigg|_{q_c} \delta \dot{q}(t')\, \delta \dot{q}(t') \right\} + \dots \tag{A.44}
$$

In general, the expansion of the action up to second order in the variations reads

$$
S[q(t')] = S[q_c(t')] + \int_0^t dt'\, \delta q(t') \frac{\delta S[q(t')]}{\delta q(t')}\bigg|_{q=q_c}
$$

$$
+ \frac{1}{2} \int_0^t \int_0^t dt'dt''\, \delta q(t')\, \delta q(t'') \frac{\delta^2 S[q(t')]}{\delta q(t')\, \delta q(t'')}\bigg|_{q=q_c} + \dots \tag{A.45}
$$

but in our case, since the Lagrangian in (A.43) is instantaneous, the second func-tional derivative in the third term on the r.h.s. of (A.45) is proportional to $\delta(t' - t'')$ and, consequently, the functional expansion contains only a single time t'. Notice that the structure of the second variation of the action is such that, under discretiza-tion, it would be an N-dimensional bilinear form ($N \to \infty$) where the second functional derivative would become an $N \times N$ matrix and $\delta q(t')$ an N-dimensional vector.

Now, integrating (A.44) by parts to recover the general form (A.45) and remem-bering that the endpoint variations must vanish, $\delta q(t) = \delta q(0) = 0$, we get

$$
S[q(t')] = S[q_c(t')] + \int_0^t dt'\, \delta q(t') \left\{ \frac{\partial L}{\partial q}\bigg|_{q_c} - \frac{d}{dt'}\frac{\partial L}{\partial \dot{q}}\bigg|_{q_c} \right\}
$$

$$
+ \frac{1}{2} \int_0^t dt'\, \delta q(t') \left\{ \left[\frac{\partial^2 L}{\partial q^2}\bigg|_{q_c} - \frac{d}{dt'}\left(\frac{\partial^2 L}{\partial \dot{q}\partial q}\bigg|_{q_c} \right) \right] \right.
$$

$$
\left. - \frac{d}{dt'}\left(\frac{\partial^2 L}{\partial \dot{q}^2}\bigg|_{q_c} \right)\frac{d}{dt'} - \frac{\partial^2 L}{\partial \dot{q}^2}\frac{d^2}{dt'^2} \right\} \delta q(t'), \tag{A.46}
$$

which allows us to identify the first and second functional derivatives of the action in (A.45) directly in terms of the Lagrangian L.

The stationary path condition implies that the first variation of $S[q(t')]$ – the second term on the r.h.s. of (A.46) – vanishes at $q_c(t')$, which means that it is the solution of the Euler–Lagrange equation

$$\left.\frac{\partial L}{\partial q}\right|_{q_c} - \frac{d}{dt'}\left.\frac{\partial L}{\partial \dot{q}}\right|_{q_c} = 0 \qquad (A.47)$$

or, in other words, the classical path of the system. Therefore, the main contribution to the functional integral we are evaluating is the complex exponential of the action of the classical path $q_c(t')$ multiplied by a prefactor given by the remaining Gaussian integral to be performed over paths measuring the variations about $q_c(t')$ as in (A.23) and (A.24). However, instead of discretizing the action as before, we are going to present another way to evaluate it.

Let us consider the eigenvalue problem involving the second variation of the action, namely the last integral on the r.h.s. of (A.46). It reads

$$\left[\left.\frac{\partial^2 L}{\partial q^2}\right|_{q_c} - \frac{d}{dt'}\left(\left.\frac{\partial^2 L}{\partial \dot{q}\partial q}\right|_{q_c}\right) - \frac{d}{dt'}\left(\left.\frac{\partial^2 L}{\partial \dot{q}^2}\right|_{q_c}\frac{d}{dt'}\right)\right]\delta q(t') = \lambda\delta q(t'). \qquad (A.48)$$

In order to carry on simpler expressions, but by no means losing generality, let us consider the particular case of (A.13) that allows for a much simpler eigenvalue problem

$$-M\frac{d^2}{dt'^2}\delta q(t') - V''(q_c(t'))\delta q(t') = \lambda\delta q(t'), \qquad (A.49)$$

which is a Schrödinger-like equation of a fictitious particle in a potential $-V''(q_c(t'))$ described by a "wave function" $\delta q(t')$ with boundary conditions $\delta q(t) = \delta q(0) = 0$. Now, expanding the variation in the orthonormal basis $\{\varphi_n(t')\}$ of the eigenstates of (A.49) as $\delta q(t') = \sum_n c_n\varphi_n(t')$ and using the orthogonality relation $\int_0^t dt'\varphi_n(t')\varphi_m(t') = \delta_{mn}$, we transform the expansion of the action (A.46) into

$$S[q(t')] = S[q_c(t')] + \sum_n \frac{1}{2}\lambda_n c_n^2. \qquad (A.50)$$

Notice that the variations are now expressible in terms of variable coefficients c_n, and then

$$K(x, t; x', 0) \approx \exp \frac{i}{\hbar} S[q_c] \int \cdots \int \frac{\mathcal{J}}{\mathcal{N}} dc_0 \, dc_1 \ldots dc_n \ldots \exp \frac{i}{2\hbar} \sum_{n=0}^{\infty} \lambda_n c_n^2 \tag{A.51}$$

where $\mathcal{N}$ was introduced previously in (A.12) as a normalization constant and $\mathcal{J}$ is the Jacobian of the transformation from the element of integration $\prod_{t'=0}^{t} dq(t')$ to $\prod_{n=0}^{N} dc_n$. These remaining integrals are of the *Fresnel type* and can easily be performed to yield

$$K(x, t; x', 0) \approx \frac{1}{\mathcal{N}_R} \frac{1}{\sqrt{\det[M \partial_{t'}^2 + V''(q_c)]}} \exp \frac{i}{\hbar} S[q_c], \tag{A.52}$$

where we have defined an effective normalization constant by

$$\frac{1}{\mathcal{N}_R} \equiv \lim_{n \to \infty} \left[\frac{\mathcal{J}}{\mathcal{N}} (2\pi i \hbar)^{n/2} \right] \tag{A.53}$$

and replaced d^2/dt'^2 by $\partial_{t'}^2$, which applies to more general cases as well.

The problem is then the computation of the functional determinant in the denominator of (A.52). Although this appears to be a hard task, a theorem demonstrated by Coleman (1988, p. 340) allows us to circumvent this problem. The theorem states that

$$\det \left[\frac{-\partial_{t'}^2 + W^{(1)}(t') - \lambda}{-\partial_{t'}^2 + W^{(2)}(t') - \lambda} \right] = \frac{\psi_\lambda^{(1)}(t')}{\psi_\lambda^{(2)}(t')}, \tag{A.54}$$

where $W^{(i)}(t')$ and $\psi_\lambda^{(i)}(t')$ $(i = 1, 2)$ are such that

$$[-\partial_{t'}^2 + W^{(i)}(t')] \psi_\lambda^{(i)}(t') = \lambda \psi_\lambda^{(i)}(t') \tag{A.55}$$

with $\psi_\lambda^{(i)}(0) = 0$ and $\partial_{t'} \psi_\lambda^{(i)}(t')|_{t'=0} = 1$. Therefore, we can easily see that

$$\frac{\det[-\partial_{t'}^2 + W(t')]}{\psi_0(t)} \tag{A.56}$$

must be independent of $W(t')$ and consequently we can use it to determine the effective normalization constant $\mathcal{N}_R$. This can be done by comparing the prefactor of (A.52) with that of an already known problem such as, for example, the harmonic oscillator. Comparing (A.56) with (A.41) and (A.42), on top of realizing that the previous prefactor has been obtained by solving a differential equation such as (A.55) for $\lambda = 0$, we easily identify the effective normalization constant as $\mathcal{N}_R = \sqrt{2\pi i \hbar / M}$.

It can also be shown (Schulman, 1981) that the prefactor can be written further as

$$G(t) \equiv \sqrt{\frac{M}{2\pi i \hbar f(t)}} = \sqrt{\frac{i}{2\pi \hbar} \frac{\partial^2 S_c(x, x', t)}{\partial x \partial x'}}, \tag{A.57}$$

which can be generalized to the multidimensional case as

$$G(t) = \sqrt{\det \left(\frac{i}{2\pi \hbar} \frac{\partial^2 S_c(\mathbf{x}, \mathbf{x}', t)}{\partial x_i \partial x'_j} \right)}. \tag{A.58}$$

It should be stressed that expressions like (A.40) or (A.58) are only valid for Gaussian forms or, equivalently, within the stationary phase approximation (Schulman, 1981). For general functionals the more elaborate discretized version of the functional integral must be applied. Nevertheless, we must mention that, in practice, once the propagator has the form (A.23) the prefactor $G(t)$ can always be determined through the unitarity condition, which reads

$$\int_{-\infty}^{+\infty} dx \, K(x, t; x', 0) K^*(x, t; x'', 0) = \delta(x' - x''). \tag{A.59}$$

Once the action $S_c(x, x', t)$ in the exponent of $K(x, t; x', 0)$ is a quadratic function of x and x', the integral in (A.59) can easily be performed and $G(t)$ finally determined without recourse to more sophisticated techniques.

A.2 Imaginary-time path integrals

The representation of the matrix element $K(x, t; x', 0)$ by the functional (or path) integral (A.14) can also be used to represent the non-normalized density operator $\rho_N(x, x', \beta)$ of a system in thermodynamic equilibrium. As we know,

$$\rho_N(x, x', \beta) = \langle x | e^{-\mathcal{H}\beta/\hbar} | x' \rangle, \tag{A.60}$$

where $\beta \equiv 1/k_B T$, which allows us to face the matrix element of the density operator as the quantum mechanical propagator of the same system for complex times

$$t = -i \hbar \beta. \tag{A.61}$$

Using then equation (A.14) and performing the change of variables as in (A.61), we have

$$\rho_N(x, x', \beta) = \int_{x'}^{x} \mathcal{D}x(\tau) \exp -\frac{1}{\hbar} S_E[x(\tau)] \tag{A.62}$$

where

$$S_E[x(\tau)] = \int\limits_0^{\hbar\beta} \left(\frac{1}{2}M\dot{x}^2 + V(x(\tau))\right) d\tau \qquad (A.63)$$

is the so-called "Euclidean action" of the system. Now, in order to find the normalized density operator, all we have to do is divide $\rho_N(x, x', \beta)$ by the partition function $\mathcal{Z} = \text{tr}\rho_N = \int \rho_N(x, x, \beta)\, dx$.

This representation of the quantum mechanical density operator can also be seen as the integral of the functional

$$F[x(\tau)] = \exp -\frac{1}{\hbar} \int\limits_0^{\hbar\beta} V(x(\tau))d\tau \qquad (A.64)$$

over the *conditional Wiener measure* (Gelfand and Yaglom, 1960; Schulman, 1981)

$$dw(x, t) = dx_1 \ldots dx_{N-1} \prod_{k=1}^{N} \sqrt{\frac{M}{2\pi\hbar^2\epsilon}} \; \exp -\frac{M\epsilon}{2\hbar^2}\frac{(x_k - x_{k-1})^2}{\epsilon^2}, \qquad (A.65)$$

where $dw(x, t)$ means that we must include all paths connecting x' at $t' = 0$ to x at $t' = t$. This measure was introduced in the context of the classical theory of the Brownian motion (see Gelfand and Yaglom, 1960 and references cited therein).

Finally, we must also emphasize that both representations (A.14) and (A.62), (A.63) can be generalized to an arbitrary number of components and/or dimensions of the system. Explicit applications of the imaginary time version of the path integration are provided in Appendix D.

Appendix B

The Markovian master equation

In Section 6.1 of this book we have shown that the reduced density operator of a particle at a given time t can be written as in (4.61) with $\mathcal{J}(x, y, t; x', y', 0)$ given by (6.21), in the specific case of the minimal model. As we have mentioned before, we are interested in deducing a differential equation for $\tilde{\rho}(x, y, t)$ in the high-temperature limit, or else, when the expression for $\mathcal{J}(x, y, t; x', y', 0)$ reduces to the form (6.23). Actually, it is shown in Section 7.1 that the high-temperature (or semi-classical) limit is not the only situation in which (6.21) reduces to (6.23). Therefore, let us derive a Markovian master equation assuming a more general memoryless super-propagator where $2M\gamma k_B T \rightarrow D(T)$. The explicit form of $D(T)$ depends on the specific limit treated.

In order to proceed, we must write the evolution equation of $\tilde{\rho}(x, y, t)$ for an infinitesimal time interval ϵ,

$$\tilde{\rho}(x, y, t + \epsilon) = \int dx' \, dy' \, \mathcal{J}(x, y, t + \epsilon; x', y', t) \, \tilde{\rho}(x', y', t), \qquad (B.1)$$

where the function $J(x, y, t + \epsilon; x', y', t)$ can be approximated by

$$\mathcal{J}(x, y, t + \epsilon; x', y', t) \approx \frac{1}{A^2} \exp \frac{i}{\hbar} f(x, y, x', y')$$

$$\times \exp \frac{i}{\hbar} \left\{ \int_t^{t+\epsilon} \left(\frac{1}{2} M \dot{x}^2 - V_0(x) \right) dt' + \int_t^{t+\epsilon} \left(\frac{1}{2} M \dot{y}^2 - V_0(y) \right) dt' \right.$$

$$\left. - \int_t^{t+\epsilon} M\gamma \, (x\dot{y} - y\dot{x}) \, dt' \right\} \exp -\frac{D(T)}{\hbar^2} \int_t^{t+\epsilon} (x - y)^2 dt' \qquad (B.2)$$

and A is a normalization constant. In (B.2) we have used the fact that any regular path within an infinitesimal time interval can be approximated by a straight line and, therefore, the integrals over this short time interval are equal to the integrand multiplied by a normalization constant.

Using the fact that

$$\dot{x} \approx \frac{x - x'}{\epsilon}, \quad \dot{y} \approx \frac{y - y'}{\epsilon}, \quad \text{and} \quad \int_t^{t+\epsilon} f(x(t'))dt' \approx \epsilon f\left(\frac{x + x'}{2}\right) \quad \text{(B.3)}$$

where $x = x(t + \epsilon)$, $y = y(t + \epsilon)$, $x' = x(t)$, and $y' = y(t)$, those integrals in (B.2) will be approximated in such a way that, when $\epsilon \to 0$, equation (B.1) becomes

$$\tilde{\rho}(t + \epsilon) \approx \int_{-\infty}^{\infty} \int_{-\infty}^{\infty} \frac{d\beta_1 d\beta_2}{A^2}$$

$$\times \exp\left\{\frac{iM\beta_1^2}{2\epsilon\hbar} - \frac{i\epsilon V_0}{\hbar}\left(x - \frac{\beta_1}{2}\right) - \frac{iM\beta_2^2}{2\epsilon\hbar} + \frac{i\epsilon V_0}{\hbar}\left(y - \frac{\beta_2}{2}\right)\right\}$$

$$\times \exp\left\{-\frac{iM\gamma}{\hbar}\left(x - \frac{\beta_1}{2}\right)\beta_2 + \frac{iM\gamma}{\hbar}\left(y - \frac{\beta_2}{2}\right)\beta_1\right.$$

$$\left.-\frac{iM\gamma}{\hbar}\left(x - \frac{\beta_1}{2}\right)\beta_1 + \frac{iM\gamma}{\hbar}\left(y - \frac{\beta_2}{2}\right)\beta_2\right\}$$

$$\times \exp\left\{-\frac{D(T)\epsilon}{\hbar^2}(x - y)^2 - \frac{D(T)\epsilon}{\hbar^2}(x - y)(\beta_1 - \beta_2)\right.$$

$$\left.-\frac{D(T)}{2\hbar^2}(\beta_1 - \beta_2)^2\right\} \tilde{\rho}(x - \beta_1, y - \beta_2, t), \quad \text{(B.4)}$$

where $\beta_1 \equiv x - x'$ and $\beta_2 \equiv y - y'$.

Now we must evaluate the integral (B.4) when $\epsilon \to 0$. In this limit, we see that the first and third terms of the integrand oscillate very fast and then the main contribution to the integral comes from the regions where $\beta_1 \approx \beta_2 \approx (\epsilon\hbar/M)^{1/2}$, because in this region the phase of both exponentials will change by an amount of the order of 1. We might also worry at this point about the contribution of the region $\beta_1 \approx \beta_2 \approx \beta$, with finite β. In this case the phase of the two exponentials together would change by an amount of the order of 1 only when

$$\Delta\beta \equiv \beta_1 - \beta_2 \approx \frac{\epsilon\hbar}{M\beta},$$

and a direct inspection of (B.4) leads us to conclude that all the terms depending on $\Delta\beta$ in its integrand will be $\mathcal{O}(\epsilon^2)$, which allows us to neglect the region $\beta_1 \approx \beta_2 \approx \beta$ with finite β.

Therefore, let us define new variables $\beta_1' \equiv \beta_1 - \gamma(x - y)\epsilon$ and $\beta_2' \equiv \beta_2 + \gamma(x - y)\epsilon$ and expand $\tilde{\rho}(x - \beta_1, y - \beta_2, t)$ in (B.4) for $\beta_1 \approx \beta_2 \approx 0$ and keep all terms of the order of ϵ in the product. In terms of β_1' and β_2', it reads

$$\tilde{\rho}(x, y, t) + \epsilon \frac{\partial \tilde{\rho}}{\partial t} \approx \int_{-\infty}^{\infty} \int_{-\infty}^{\infty} \frac{d\beta_1' d\beta_2'}{A^2} \exp \frac{iM\beta_1'^2}{2\epsilon\hbar} \exp -\frac{iM\beta_2'^2}{2\epsilon\hbar}$$

$$\times \left\{ \tilde{\rho}(x, y, t) - \frac{\partial \tilde{\rho}}{\partial x} \beta_1' - \frac{\partial \tilde{\rho}}{\partial y} \beta_2' - \frac{\partial \tilde{\rho}}{\partial x}(x - y)\gamma\epsilon + \frac{\partial \tilde{\rho}}{\partial y}(x - y)\gamma\epsilon \right.$$

$$+ \frac{1}{2} \frac{\partial^2 \tilde{\rho}}{\partial x^2} \beta_1'^2 + \frac{\partial^2 \tilde{\rho}}{\partial x \partial y} \beta_1' \beta_2' + \frac{1}{2} \frac{\partial^2 \tilde{\rho}}{\partial y^2} \beta_2'^2 - \frac{i\epsilon}{\hbar} V_0(x)\tilde{\rho} + \frac{i\epsilon}{\hbar} V_0(y)\tilde{\rho}$$

$$\left. - \frac{D(T)\epsilon}{\hbar^2}(x - y)^2 \tilde{\rho} \right\}. \tag{B.5}$$

The integrals are extended from $-\infty$ to $+\infty$ because we know that their main contribution comes from those regions about $\beta_1 \approx \beta_2 \approx (\epsilon\hbar/M)^{1/2}$. Evaluating the integrals above, we have:

(a) A term independent of ϵ which gives us the normalization constant, $A^2 = 2\pi\epsilon\hbar/M$.

(b) A term proportional to ϵ which furnishes us with the desired master equation,

$$\frac{\partial \tilde{\rho}}{\partial t} = -\frac{\hbar^2}{2Mi} \frac{\partial^2 \tilde{\rho}}{\partial x^2} + \frac{\hbar^2}{2Mi} \frac{\partial^2 \tilde{\rho}}{\partial y^2} - \gamma(x - y) \left(\frac{\partial \tilde{\rho}}{\partial x} - \frac{\partial \tilde{\rho}}{\partial y} \right)$$

$$+ \frac{(V_0(x) - V_0(y))}{i\hbar} \tilde{\rho} - \frac{D(T)}{\hbar^2}(x - y)^2 \tilde{\rho}. \tag{B.6}$$

Appendix C

Coherent-state representation

In equation (6.61) we have written the Hamiltonian of a particle coupled with many bosons or fermions in one dimension in its second quantized form. In so doing we naturally make use of the creation and annihilation operators a_n and $a_n^\dagger$, which satisfy

$$[a_n, a_m]_\mp = [a_n^\dagger, a_m^\dagger]_\mp = 0 \qquad \text{and} \qquad [a_n, a_m^\dagger]_\mp = \delta_{nm} \qquad \text{(C.1)}$$

where the upper (lower) sign indicates the commutation (anti-commutation) relations between the bosonic (fermionic) operators. Notice that we have changed our notation for commutators and anti-commutators above only to describe them with the same symbol for bosons and fermions when appropriate.

Since these new operators do not carry any dependence explicitly on the coordinates and momenta of the particles, the path integral representation deduced in Appendix A is no longer useful for our needs. Therefore, we will develop an alternative representation for the time evolution of the system in the so-called *coherent-state* representation. In order to do that, let us restrict ourselves to working with a single mode just to simplify matters. The generalization to many modes is straightforward and will be presented as it arises.

A coherent state is defined as an eigenstate of the annihilation operator (see, for example, Louisell (1990), Merzbacher (1998), and Negele and Orland (1998))

$$a|\alpha\rangle = \alpha|\alpha\rangle \qquad \Rightarrow \qquad \langle\alpha|a^\dagger = \alpha^*\langle\alpha|, \qquad \text{(C.2)}$$

which applies either for bosons or fermions. Since these states are labeled by the parameter (c-number) α and the operators obey either commutation or anti-commutation rules, we must define them properly in order to obtain a representation consistent with the operator formalism. However, since the structure of the bosonic and fermionic c-numbers is quite different, as we shall see shortly, we have decided to develop these two cases separately.

The bosonic case. An equivalent way to define the bosonic coherent state is

$$|\alpha\rangle \equiv N(\alpha) \exp(\alpha a^\dagger)|0\rangle, \tag{C.3}$$

where α is an ordinary complex number and $N(\alpha)$ is a normalization constant. Expanding the exponential of the creation operator and using the bosonic commutation relations for a and $a\dagger$, we can easily show that this alternative form does indeed obey (C.2). Let us then explore some features of this state.

The first one refers to the internal product of two coherent states. Using the well-known Baker–Hausdorff formula which states that, whenever $[A, [A, B]] = [B, [A, B]] = 0$,

$$\exp A + B = \exp A \, \exp B \, \exp -\frac{1}{2}[A, B], \tag{C.4}$$

we can show that

$$\langle \alpha | \alpha' \rangle = N^*(\alpha^*) N(\alpha') \exp \alpha^* \alpha'. \tag{C.5}$$

Now, let us choose the normalization constant such that $\langle \alpha | \alpha \rangle = 1$, which implies that

$$|\alpha\rangle = \exp -\frac{|\alpha|^2}{2} \exp(\alpha a^\dagger)|0\rangle, \tag{C.6}$$

and consequently

$$\langle \alpha | \alpha' \rangle = \exp -\frac{|\alpha|^2}{2} \exp -\frac{|\alpha'|^2}{2} \exp \alpha^* \alpha' \tag{C.7}$$

or

$$|\langle \alpha | \alpha' \rangle|^2 = \exp -|\alpha - \alpha'|^2, \tag{C.8}$$

which means that the coherent states are only approximately orthogonal.

Another useful expression is the expansion of the coherent state in the number eigenstate basis $|n\rangle$, which reads

$$|\alpha\rangle = \sum_n |n\rangle \langle n | \alpha\rangle = \sum_n |n\rangle \langle n| \exp -\frac{|\alpha|^2}{2} \exp(\alpha a^\dagger)|0\rangle$$

$$= \exp -\frac{|\alpha|^2}{2} \sum_n |n\rangle \langle n| \sum_k \frac{(\alpha a^\dagger)^k}{k!} |0\rangle = \exp -\frac{|\alpha|^2}{2} \sum_n \frac{\alpha^n}{\sqrt{n!}} |n\rangle, \tag{C.9}$$

with which we can deduce the probability of finding the component of the coherent state $|\alpha\rangle$ along $|n\rangle$ as the *Poisson distribution*

$$P_n(\alpha) = \frac{|\alpha|^{2n}}{n!} \exp -|\alpha|^2, \tag{C.10}$$

where $|\alpha|^2 = \langle \alpha|a^\dagger a|\alpha \rangle \equiv \bar{n}$ is the average of the number operator in $|\alpha\rangle$. Moreover, this expansion can be used to deduce the *overcompleteness relation*

$$\frac{1}{\pi} \int d^2\alpha \, |\alpha\rangle \langle \alpha| = \frac{1}{\pi} \int d\,\mathrm{Re}\,\alpha \, d\,\mathrm{Im}\,\alpha \, |\alpha\rangle \langle \alpha| = \frac{1}{2\pi i} \int d\alpha \, d\alpha^* \, |\alpha\rangle \langle \alpha| = 1,$$

(C.11)

which can be proved using (C.9) with $\alpha = \rho \exp i\theta$, integrating over the surface element $\rho \, d\rho \, d\theta$ and finally using the completeness relation $\sum_n |n\rangle \langle n| = 1$. We can also make use of (C.11) to define the integration measure previously introduced in (6.63) as

$$d\mu(\alpha) \equiv \frac{d^2\alpha}{\pi} = \frac{d\,\mathrm{Re}\,\alpha \, d\,\mathrm{Im}\,\alpha}{\pi} = \frac{d\alpha \, d\alpha^*}{2\pi i}.$$

(C.12)

One last step toward the development of a new representation for the quantum mechanical propagator is the representation of a general *normally ordered* operator $\hat{A}$ in terms of coherent states. This ordering procedure means that all creation operators must be placed on the left of the annihilation operators. In this way

$$\hat{A} = \sum_{nm} A_{nm} a^{\dagger m} a^n,$$

(C.13)

which is already a normally ordered operator and has its matrix elements in the coherent state representation given by

$$\langle \alpha|\hat{A}|\alpha'\rangle \equiv A(\alpha^*, \alpha') = \sum_{nm} A_{nm} \alpha^{*m} \alpha'^n.$$

(C.14)

Notice that all the above introduced concepts can easily be generalized to multimode states. All we do is extend our previous definitions concerning the single-mode coherent state to the multimode case whenever the commutation relations established in (C.1) allow us to do so. For example, we can start by generalizing the expression of the normalized coherent state to the multimode case as

$$|\alpha\rangle \equiv |\alpha_1, \ldots \alpha_n \ldots\rangle = \exp - \sum_n \frac{|\alpha_n|^2}{2} \exp\left(\sum_n \alpha_n a_n^\dagger\right)|0\rangle,$$

(C.15)

and consequently

$$\langle \alpha|\alpha'\rangle = \exp - \sum_n \frac{|\alpha_n|^2}{2} \exp - \sum_n \frac{|\alpha_n'|^2}{2} \exp \sum_n \alpha_n^* \alpha_n'.$$

(C.16)

The integration measure can also be extended straightforwardly to

$$d\mu(\boldsymbol{\alpha}) \equiv \prod_n \frac{d^2\alpha_n}{\pi} = \prod_n \frac{d\,\mathrm{Re}\,\alpha_n\,d\,\mathrm{Im}\,\alpha_n}{\pi} = \prod_n \frac{d\alpha_n\,d\alpha_n^*}{2\pi i}. \qquad \text{(C.17)}$$

Now we can apply this representation to any time evolution operator whenever it is described in terms of bosonic operators. Let us start with a proper generalization of (A.1) in order to deal with explicitly time-dependent Hamiltonians

$$K(\alpha^*, t; \alpha', 0) = \langle \alpha | \mathcal{T} \exp -\frac{i}{\hbar} \int_0^t dt'\, \mathcal{H}(t') | \alpha' \rangle, \qquad \text{(C.18)}$$

where $\mathcal{T}$ stands for the well-known *time-ordering operator* (see, for example, Fetter and Walecka (2003)) which, applied to the exponential operator, means that each term of its expansion must be properly ordered.

Proceeding analogously to what we have done in (A.2) and (A.3), we can split the time interval $[0, t]$ into infinitesimal ones of length ϵ and write

$$K(\alpha^*, t; \alpha', 0) = \langle \alpha_N | \prod_{j=0}^{N-1} \mathcal{T} \exp -\frac{i}{\hbar} \int_{j\epsilon}^{(j+1)\epsilon} dt'\, \mathcal{H}(t') | \alpha_0 \rangle, \qquad \text{(C.19)}$$

where $\alpha = \alpha_N$ and $\alpha' = \alpha_0$ are the fixed endpoints of the integrand. In the limit $\epsilon \to \infty$ we can write further

$$K(\alpha^*, t; \alpha', 0) \approx \langle \alpha_N | \prod_{j=0}^{N-1} \exp -\frac{i}{\hbar} \epsilon \mathcal{H}(t_j) | \alpha_0 \rangle, \qquad \text{(C.20)}$$

where inserting the identity operator (C.11) with appropriately chosen integration variables α_k ($k = 1, \ldots, N - 1$) between any two infinitesimal time evolution operators we end up with

$$K(\alpha^*, t; \alpha', 0) = \left\{ \prod_{j=1}^{N-1} \int \frac{d^2\alpha_j}{\pi} \right\} \prod_{j=0}^{N-1} \langle \alpha_{j+1} | \exp -\frac{i}{\hbar} \epsilon \mathcal{H}(t_j) | \alpha_j \rangle. \qquad \text{(C.21)}$$

Notice that discretizing the time and associating a bosonic operator with each instant has naturally led us to deal with a multidimensional coherent state representation of bosonic operators.

Now, we expand the complex exponentials above up to first order in ϵ to get

$$K(\alpha^*, t; \alpha', 0) \approx \left\{ \prod_{j=1}^{N-1} \int \frac{d^2\alpha_j}{\pi} \right\} \prod_{j=0}^{N-1} \langle \alpha_{j+1} | \alpha_j \rangle \left(1 - \frac{i\epsilon}{\hbar} \frac{\langle \alpha_{j+1} | \mathcal{H}(t_j) | \alpha_j \rangle}{\langle \alpha_{j+1} | \alpha_j \rangle} \right),$$

$$\text{(C.22)}$$

whose term in parentheses can be re-exponentiated to give

$$K(\alpha^*, t\,; \alpha', 0) \approx \left\{ \prod_{j=1}^{N-1} \int \frac{d^2\alpha_j}{\pi} \right\} \prod_{j=0}^{N-1} \langle \alpha_{j+1} | \alpha_j \rangle \exp - \sum_{j=0}^{N-1} \frac{i}{\hbar} \epsilon \mathcal{H}_{j+1,j}, \quad \text{(C.23)}$$

where

$$\mathcal{H}_{j+1,j} \equiv \mathcal{H}(\alpha^*_{j+1}, \alpha_j) = \frac{\langle \alpha_{j+1} | \mathcal{H}(t_j) | \alpha_j \rangle}{\langle \alpha_{j+1} | \alpha_j \rangle}$$

is the matrix element of the Hamiltonian operator in the coherent state representation. Expression (C.23) can also be rewritten in a more convenient way if we use (C.7) for $\langle \alpha_{j+1} | \alpha_j \rangle$, which leads us to

$$K(\alpha^*, t\,; \alpha', 0) \approx \left\{ \prod_{j=1}^{N-1} \int \frac{d^2\alpha_j}{\pi} \right\} \exp \frac{i}{\hbar} \sum_{j=0}^{N-1} \epsilon \left\{ \frac{i\hbar}{2} \alpha^*_{j+1} \frac{(\alpha_{j+1} - \alpha_j)}{\epsilon} \right.$$
$$\left. - \frac{i\hbar}{2} \frac{(\alpha^*_{j+1} - \alpha^*_j)}{\epsilon} \alpha_j - \mathcal{H}_{j+1,j} \right\}. \quad \text{(C.24)}$$

Once we have come this far we can perform the limit $\epsilon \to 0$ ($N \to \infty$) in (C.24) and write

$$K(\alpha^*, t\,; \alpha', 0) = \int \mathcal{D}\mu(\alpha(t')) \exp \frac{i}{\hbar} S[\alpha(t'), \alpha^*(t')], \quad \text{(C.25)}$$

where

$$\mathcal{D}\mu(\alpha(t')) \equiv \lim_{N \to \infty} \left\{ \prod_{j=1}^{N-1} \int \frac{d^2\alpha_j}{\pi} \right\} \quad \text{(C.26)}$$

is the bosonic path integration measure and

$$S[\alpha(t'), \alpha^*(t')] \equiv \int_0^t dt' \left\{ \frac{i\hbar}{2} \left[\alpha^*(t') \dot{\alpha}(t') - \alpha(t') \dot{\alpha}^*(t') \right] - \mathcal{H}(\alpha^*, \alpha) \right\} \quad \text{(C.27)}$$

is the bosonic action.

Similarly to what we have done in Appendix A for the conventional representation of path integrals in the configuration space, there are two ways of solving it. The first one is to return to the discretized version (C.24) and solve the resulting multidimensional integral by whatever method proves to be appropriate for the maneuver. This procedure would be equivalent to those employed to solve the free particle and quadratic Lagrangians in Appendix A.

Another alternative has also been presented in Appendix A and refers to the stationary phase approximation. In the latter we have to expand the action functional

(C.27) about its stationary complex "paths" $\alpha_c(t')$ which means that, if we write $\alpha(t') = \alpha_c(t') + \delta\alpha(t')$,

$$S[\alpha(t'), \alpha^*(t')] = S[\alpha_c(t'), \alpha_c^*(t')] + \int_0^t dt' \left[\delta\alpha(t') \frac{\delta S}{\delta\alpha} \bigg|_{\alpha_c} + \delta\alpha^*(t') \frac{\delta S}{\delta\alpha^*} \bigg|_{\alpha_c} \right]$$

$$+ \frac{1}{2} \int_0^t dt' \left[\delta^2\alpha(t') \frac{\delta^2 S}{\delta\alpha^2(t')} \bigg|_{\alpha_c} + \delta^2\alpha(t')\delta^2\alpha^*(t') \frac{2\delta^2 S}{\delta\alpha^2(t')\delta\alpha^{*2}(t')} \bigg|_{\alpha_c} \right.$$

$$\left. + \delta^2\alpha^*(t')^2 \frac{\delta^2 S}{\delta\alpha^{*2}(t')} \bigg|_{\alpha_c} \right], \tag{C.28}$$

where the first functional derivatives

$$\frac{\delta S}{\delta\alpha^*} \bigg|_{\alpha_c} = i\hbar \frac{\partial\alpha_c}{\partial t} - \frac{\partial\mathcal{H}}{\partial\alpha^*} \bigg|_{\alpha_c} = 0 \qquad \text{and}$$

$$\frac{\delta S}{\delta\alpha} \bigg|_{\alpha_c} = -i\hbar \frac{\partial\alpha_c^*}{\partial t} - \frac{\partial\mathcal{H}}{\partial\alpha} \bigg|_{\alpha_c} = 0 \tag{C.29}$$

must obey the boundary conditions $\alpha_c(0) = \alpha'$ and $\alpha_c^*(t) = \alpha^*$. Notice that although α and α^* are complex conjugates of one another, we have to be very careful in treating the two equations of motion in (C.29). Since they are first order in time, establishing the conditions for $\alpha_c(t')$ at $t' = 0$ ($\alpha_c^*(t')$ at $t' = t$) determines its value uniquely at $t' = t$ ($t' = 0$). Therefore, the imposition that one equation must be the complex conjugate of the other would lead us to an overdetermined problem (see, for example, Baranger *et al.* (2001)). The way out of this dilemma is to replace $\alpha_c(t') \to u(t')$ and $\alpha_c^*(t') \to v(t')$, where $u(t')$ and $v(t')$ are independent complex variables such that

$$u(0) = u' = \alpha' \qquad \text{and} \qquad v(t) = v = \alpha^*. \tag{C.30}$$

In other words, we are saying that we should regard $\alpha^*(t')$ as a new independent complex variable $\bar{\alpha}(t')$ which is not necessarily the complex conjugate of $\alpha(t')$ at any t'.

The solutions of the equations of motion (C.29) for $u(t')$ and $v(t')$ with the boundary conditions specified above allow us to calculate $u(t) = u \neq \alpha$ and $v(0) = v' \neq \alpha'^*$, respectively. These solutions are then used to evaluate the classical action $S_c \equiv S[\alpha_c(t'), \alpha_c^*(t')]$ in (C.28). However, some care must be taken in so doing (Baranger *et al.*, 2001). If we return to the discretized form of (C.27) in (C.24) we easily see that the endpoints showing up there at $t' = t$ and $t' = 0$ are respectively $|\alpha|^2$ and $|\alpha'|^2$, instead of $u\alpha^*$ and $v'\alpha'$, which would be the only possible endpoint dependence coming from the evaluation of S_c with the classical solutions $u(t')$ and $v(t')$. In order to correct this mistake, we must subtract from S_c the wrong endpoint contribution and replace it with the correct one. Therefore, we should make the following replacement in (C.28):

$$S[u(t'), v(t')] \to S[u(t'), v(t')] - \frac{i\hbar}{2}(u\alpha^* + v'\alpha') + \frac{i\hbar}{2}(|\alpha|^2 + |\alpha'|^2), \quad (C.31)$$

which accounts for the exponential contribution to (C.25).

The remaining contribution to the latter comes from the second and third lines on the r.h.s. of (C.28) and refers to the second functional derivative of the action at $\alpha_c(t')$, $\delta^2 S$, which in terms of these newly defined functions can be cast into the form

$$\delta^2 S = \frac{1}{2} \int_0^t dt' \left(\bar{z}(t'), \; z(t') \right) \begin{pmatrix} i\hbar\frac{\partial}{\partial t'} - \frac{\partial^2 \mathcal{H}}{\partial \alpha \partial \bar{\alpha}} & -\frac{\partial^2 \mathcal{H}}{\partial \bar{\alpha}^2} \\ -\frac{\partial^2 \mathcal{H}}{\partial \alpha^2} & -i\hbar\frac{\partial}{\partial t'} - \frac{\partial^2 \mathcal{H}}{\partial \alpha \partial \bar{\alpha}} \end{pmatrix} \begin{pmatrix} z(t') \\ \bar{z}(t') \end{pmatrix},$$
$$(C.32)$$

where all the path-dependent partial derivatives must be taken at $\alpha(t') = u(t')$ and $\bar{\alpha}(t') = v(t')$, and $z(t') \equiv \delta\alpha(t') = \alpha(t') - u(t')$ and $\bar{z}(t') \equiv \delta\bar{\alpha}(t') = \bar{\alpha}(t') - v(t')$ must be treated as independent variations which obey $z(0) = \bar{z}(t) = 0$.

Notice that (C.32) is once again a bilinear form in the variations and as such its functional integration can be performed by either method presented in Appendix A, resulting in a time-dependent function $G_B(t)$. On top of that, the remaining integrals we have to evaluate in (6.67) are all multidimensional complex Gaussian integrals over the endpoints $\alpha, \alpha', \alpha^*$, and γ'^*, and therefore we think it is time to briefly sketch the strategy for evaluating Gaussian integrals for coherent states. For a very thorough treatment of the latter in a slightly different context, we refer the reader to Baranger *et al.* (2001).

Our starting point is the well-known result of complex Gaussian integrals (Fresnel integrals)

$$\int_{-\infty}^{+\infty} \frac{dx}{\sqrt{2\pi i}} \exp\left(\frac{i}{2}xAx\right) = A^{-1/2} \quad (C.33)$$

for $A > 0$, which can readily be generalized to the multidimensional case as

$$\prod_n \int_{-\infty}^{+\infty} \frac{dx_n}{\sqrt{2\pi i}} \exp\left(\frac{i}{2}\sum_n x_n A_n x_n\right) = \left(\prod_n A_n\right)^{-1/2}. \quad (C.34)$$

If the exponent is given by a real bilinear form $i/2 \sum_{nm} x_n A_{nm} x_m$, where A_{nm} is a real symmetric positive matrix, there is an orthogonal transformation which brings the exponent to the form (C.34), with A_n being the eigenvalues of A, and also leaves the integration measure unchanged. Then,

$$\prod_n \int_{-\infty}^{+\infty} \frac{dx_n}{\sqrt{2\pi i}} \exp\left(\frac{i}{2} \sum_{nm} x_n A_{nm} x_m\right) = (\det A)^{-1/2}. \qquad (C.35)$$

The latter result can be generalized further to the complex case. If we replace the previous exponent by $i/2 \sum_{nm} z_n^* A_{nm} z_m$ where z_n is a complex number and A is now a Hermitian and positive matrix, there is a unitary transformation that diagonalizes A and the former result holds for both the real and imaginary parts of z_n, yielding

$$\prod_n \int_{-\infty}^{+\infty} \frac{dz_n}{\sqrt{2\pi i}} \frac{dz_n^*}{\sqrt{2\pi i}} \exp\left(\frac{i}{2} \sum_{nm} z_n^* A_{nm} z_m\right) = (\det A)^{-1}. \qquad (C.36)$$

From (C.36) we easily recognize the integration measure $d\mu(\mathbf{z})$ of our coherent-state integrals. Therefore, we must first bring the integral to this bilinear form and then evaluate the resulting determinant as, for example, in (6.71).

Note that in evaluating the previously defined prefactor $G_B(t)$ by this method, we still have to deal with the continuum limit of the final expression as we have done in (A.35) and (A.36). Nevertheless, its lengthy explicit evaluation can be avoided if we remember that it can always be computed with the help of the unitarity condition of the time evolution operator, which, with the help of (C.7) and (C.11), reads

$$\langle \alpha' | \alpha'' \rangle = \int d\mu(\alpha) K(\alpha^*, t; \alpha'', 0) K^*(\alpha, t; \alpha'^*, 0)$$

$$= \exp -\frac{|\alpha'|^2}{2} \exp -\frac{|\alpha''|^2}{2} \exp \alpha'^* \alpha'', \qquad (C.37)$$

in analogy with (A.59). Notice that $K^*(\alpha, t; \alpha'^*, 0) = [K(\alpha^*, t; \alpha', 0)]^*$, as usual.

The fermionic case. Once we have developed the concept of the coherent-state representation for bosonic operators, we will try to create a "c-number" representation for fermionic operators as well. However, due to the anti-commutation relations obeyed by different fermionic operators in (C.1), there is no way to reproduce the correct properties of fermions using a representation in terms of ordinary complex numbers (Negele and Orland, 1998). Instead, what we should do is work with a set of anti-commuting variables (actually generators of an algebra) which are called *Grassmann variables* and are defined in the following way.

Given a set of $2N$ Grassmann variables $\{\xi_i^*, \xi_i\}$ $(i = 1, \ldots, N)$, where $(\xi_i)^* = \xi_i^*$ is the variable conjugate to ξ_i, they, by definition, obey

$$\{\xi_i, \xi_j\} = \{\xi_i^*, \xi_j^*\} = \{\xi_i, \xi_j^*\} = \xi_i^2 = \xi_i^{*2} = 0. \tag{C.38}$$

Moreover, these anti-commutation relations also hold between any Grassmann variable of the set $\{\xi_i^*, \xi_i\}$ and the fermionic operators c_n and $c_n^\dagger$, which means

$$\{\xi_i, c_j\} = \{\xi_i^*, c_j^\dagger\} = \{\xi_i, c_j^\dagger\} = \{\xi_i^*, c_j\} = 0. \tag{C.39}$$

Other important properties of the Grassmann variables are that

$$(\xi_n^*)^* = \xi_n; \qquad (\lambda\xi_n)^* = \lambda^*\xi_n^* \text{ if } \lambda \in \mathbb{C} \qquad \text{and}$$
$$(\xi_1\ldots\xi_n\ldots)^* = \ldots\xi_n^*\ldots\xi_1^*. \tag{C.40}$$

Equipped with these definitions and properties we can now define our single-mode fermionic coherent state as

$$|\xi\rangle = \exp-\frac{\xi^*\xi}{2} \exp-(\xi c^\dagger)|0\rangle, \tag{C.41}$$

which after having its exponential terms expanded reads

$$|\xi\rangle = \left(1 - \frac{\xi^*\xi}{2}\right)(1 - \xi c^\dagger)|0\rangle = \left(1 - \xi c^\dagger - \frac{\xi^*\xi}{2}\right)|0\rangle. \tag{C.42}$$

Using again the Grassmann variables properties we can easily show that $c|\xi\rangle = \xi|\xi\rangle$, as desired, and also that

$$\langle\xi|\xi'\rangle = \exp-\frac{\xi^*\xi}{2} \exp-\frac{\xi'^*\xi'}{2} \exp\xi^*\xi', \tag{C.43}$$

which has the same form as the same relation for bosonic coherent states (C.7). The only difference is that we are explicitly keeping $\xi^*\xi$ in our notation instead of $|\xi|^2$, because the latter does not make sense for Grassmann variables since $\xi^*\xi = -\xi\xi^*$.

The generalization of (C.41) to the multimode case is also straightforward since $[\xi_i c_i, \xi_j^* c_j^\dagger] = 0$ and reads

$$|\boldsymbol{\xi}\rangle = \exp-\sum_i \frac{\xi_i^*\xi_i}{2} \exp-\sum_i (\xi_i c_i^\dagger)|0\rangle, \tag{C.44}$$

which allows us to deduce

$$\langle\boldsymbol{\xi}|\boldsymbol{\xi'}\rangle = \exp-\sum_i \frac{\xi_i^*\xi_i}{2} \exp-\sum_i \frac{\xi_i'^*\xi_i'}{2} \exp\sum_i \xi_i^*\xi_i'. \tag{C.45}$$

In order to follow the same steps to obtain the functional integral representation for the bosonic propagator, we need some other definitions concerning the Grassmann variables.

Let us start with the definition of an analytic function of a Grassmann variable ξ. A function $f(\xi)$ is said to be analytic in ξ if it admits a power series expansion

$$f(\xi) = f_0 + f_1\xi + f_2\xi^2 + \dots \tag{C.46}$$

where $f_n \in \mathbb{C}$. But, by the properties of the Grassmann variables, we know that $\xi^n = 0$ if $n > 1$, which leaves us with the linear function

$$f(\xi) = f_0 + f_1\xi. \tag{C.47}$$

We can then take the complex conjugation of this expression to write the expression of an analytic function of ξ^* as

$$[f(\xi)]^* = f^*(\xi^*) = f_0^* + f_1^*\xi^*. \tag{C.48}$$

Now, we extend these definitions to functions of ξ and ξ^*, which we frequently have to deal with, and write

$$A(\xi^*, \xi) = a_0 + a_1\xi + a_1^*\xi + a_{12}\xi^*\xi. \tag{C.49}$$

Armed with these definitions we can easily establish the properties of the derivatives of functions of ξ and ξ^* as

$$\partial_\xi f(\xi) = f_1,$$
$$\partial_{\xi^*} f^*(\xi^*) = f_1^*,$$
$$\partial_\xi (\xi^*\xi) = -\partial_\xi (\xi\xi^*) = -\xi^*, \tag{C.50}$$

and consequently

$$\partial_\xi A(\xi^*, \xi) = a_1 - a_{12}\xi^*,$$
$$\partial_{\xi^*} A(\xi^*, \xi) = a_1^* + a_{12}\xi,$$
$$\partial_{\xi^*}\partial_\xi A(\xi^*, \xi) = -\partial_\xi \partial_{\xi^*} A(\xi^*, \xi) = -a_{12}, \tag{C.51}$$

where the latter equation implies that $\partial_\xi \partial_{\xi^*} = -\partial_{\xi^*}\partial_\xi$ and $\partial_\xi^2 = \partial_{\xi^*}^2 = 0$.

Let us turn to the most pertinent definitions for our present purposes, namely the "definite" integral over Grassmann variables. These read (see the reasoning for the forthcoming definitions in Negele and Orland (1998))

$$\int d\xi\, 1 = \int d\xi\, \partial_\xi \xi = 0 \quad \text{and}$$

$$\int d\xi\, \xi = 1. \tag{C.52}$$

If we compare the definitions of the derivatives with those of the integrals, it becomes clear that for Grassmann variables

$$\int d\xi \Leftrightarrow \partial_\xi. \tag{C.53}$$

Other easily derivable results using (C.52) are

$$\int d\xi f(\xi) = f_1,$$

$$\int d\xi A(\xi^*, \xi) = a_1 - a_{12}\xi^*,$$

$$\int d\xi^* A(\xi^*, \xi) = a_1^* + a_{12}\xi,$$

$$\int d\xi^* d\xi A(\xi^*, \xi) = -\int d\xi d\xi^* A(\xi^*, \xi) = -a_{12}. \tag{C.54}$$

These results, together with our definition of multimode fermionic coherent states, finally make it possible to show the overcompleteness fermionic relation

$$\left\{ \prod_n \int d\xi_n^* d\xi_n \right\} |\xi\rangle\langle\xi| = \mathbb{1}, \tag{C.55}$$

which also defines the integration measure as

$$d\mu(\xi) \equiv \left\{ \prod_n d\xi_n^* d\xi_n \right\}. \tag{C.56}$$

Given all that, we can follow step-by-step the development of the bosonic functional representation of the evolution operator from (C.19) to (C.25) only taking care of the ordering of the Grassmann variables. The only difference between the resulting fermionic functional integral and the bosonic one is the measure of integration and the commutation relation obeyed by the variables of integration at different times. Once again, we can solve this functional integral by either reducing it back to its discretized version and evaluating a multidimensional Gaussian integral over Grassmann variables or employing the stationary phase approximation which leads us to the already familiar product of a prefactor, $G_F(t)$, with the complex exponential of the classical action. The fermionic prefactor, $G_F(t)$, is the functional integral of the complex exponential of the second variation of the action (C.32), where now $z(t')$ and $\bar{z}(t')$ are Grassmann variables. Therefore, we proceed exactly as before with the only difference that the Gaussian integrals we have to evaluate are over Grassmann variables. Actually, the most general form we usually end up with is

$$\prod_n \int d\xi_n d\xi_n^* \exp\left(\frac{i}{2} \sum_{nm} \xi_n^* A_{nm} \xi_m\right). \tag{C.57}$$

Performing a similarity transformation in the exponent of (C.57) with the help of a unitary operator and applying the rules of integration of Grassmann variables to the resulting integral, we have

$$\prod_n \int d\zeta_n d\zeta_n^* \exp\left(\frac{i}{2}\sum_n \zeta_n^* A_n \zeta_m\right) = \prod_n A_n = \det A \qquad \text{(C.58)}$$

where A_n are the eigenvalues of A. Therefore, we see that the only difference between the functional integrals in the bosonic and fermionic cases lies in the final form of the prefactors through the power of the determinant of the operator that appears in the second variation of the action. Here we should stress that our results are only valid for a $2N$-dimensional set of Grassmann variables. For an N-dimensional set the term $\sqrt{\det A}$ appears instead.

Appendix D

Euclidean methods

In this appendix, it is our intention to employ the path integral method to study the low-energy sector of some physical systems. We shall be particularly interested in the dynamics of a single particle in a general potential.

Our starting point is the well-known path integral representation of the non-normalized equilibrium density operator of the system at temperature $T = (k_B \beta)^{-1}$ which reads (see Section A.2)

$$\rho_N(x, y, \beta) = \langle x | e^{-\beta \mathcal{H}} | y \rangle = \int_y^x \mathcal{D}q(\tau)\, e^{-S_E[q(\tau)]/\hbar}, \qquad (D.1)$$

where

$$S_E[q(\tau)] = \int_0^{\hbar \beta} d\tau\, L_E(q, \dot{q}) \qquad \text{and} \qquad L_E(q, \dot{q}) = \frac{1}{2} M \dot{q}^2 + V(q). \quad (D.2)$$

This is called the Euclidean version of Feynman's theory. It should be noticed that the Euclidean Lagrangian is that of a particle in a potential $-V(q)$. Now, inserting the completeness relation $\sum_n |\psi_n\rangle\langle\psi_n| = \mathbb{1}$ of the eigenstates $|\psi_n\rangle$ of $\mathcal{H}$ on each side of the exponential in the middle of (D.1), we have

$$\rho_N(x, y, \beta) = \langle x | e^{-\beta \mathcal{H}} | y \rangle = \sum_{n=0}^{\infty} \psi_n(x)\, \psi_n^*(y)\, e^{-\beta E_n}. \qquad (D.3)$$

The leading term of this expansion when $\beta \to \infty$ is given by $\psi_0(x)\, \psi_0^*(y) e^{-\beta E_0}$, where E_0 denotes the ground-state energy of the system. Thus, with the help of (D.1) we can write

$$\lim_{\beta \to \infty} \int_y^x \mathcal{D}q(\tau)\, e^{-S_E[q(\tau)]/\hbar} \approx \psi_0(x)\, \psi_0^*(y)\, e^{-\beta E_0}. \qquad (D.4)$$

It may also happen that we need to consider other E_n when we deal with many degenerate minima of the potential $V(q)$ (see below).

From now on we are going to evaluate the path integral in (D.4) in order to compute the lowest-energy levels of the system. However, we all know that only very few examples of functional integrals can be solved exactly, and then we shall employ a particular approximation method to achieve our goal. In our specific case, the most appropriate one turns out to be the stationary phase method, which we have already treated in Appendix A for the real-time case.

Let us then repeat the main steps of that development and start by expanding the Euclidean action S_E about the solution of the equation

$$\left.\frac{\delta S_E}{\delta q}\right|_{q_c} = -M\ddot{q}_c + V'(q_c) = 0, \tag{D.5}$$

where $q_c(0) = y$ and $q_c(\hbar\beta) = x$ when $\hbar\beta \to \infty$. Then we have

$$S_E[q(\tau)] \simeq S_E[q_c(\tau)] + \frac{1}{2}\int_0^{\hbar\beta}\int_0^{\hbar\beta} d\tau' \, d\tau'' \, \delta q(\tau') \, \delta q(\tau'')\left.\frac{\delta^2 S_E}{\delta q(\tau') \, \delta q(\tau'')}\right|_{q_c}, \tag{D.6}$$

where

$$\left.\frac{\delta^2 S_E}{\delta q(\tau') \, \delta q(\tau'')}\right|_{q_c} = -M\frac{d^2}{d\tau'^2}\delta(\tau' - \tau'') + V''(q_c)\delta(\tau' - \tau'') \tag{D.7}$$

and $\delta q(\tau') = q(\tau') - q_c(\tau')$, which implies that $\delta q(\hbar\beta) = \delta q(0) = 0$. With this boundary condition we can expand the variations $\delta q(\tau')$ in a basis of orthogonal functions that vanish at the same points. Therefore,

$$\delta q(\tau') = \sum_{n=0}^{\infty} c_n q_n(\tau') \qquad \text{with} \qquad q_n(\hbar\beta) = q_n(0) = 0 \qquad \text{and}$$

$$\int_0^{\hbar\beta} d\tau' \, q_n(\tau') \, q_m(\tau') = \delta_{mn}. \tag{D.8}$$

Using (D.8) and (D.9) in (D.6) we have

$$S_E[q(\tau')] \to S_E(c_0, \ldots c_n, \ldots) = S_E[q_c] + \frac{1}{2}\sum_{n=0}^{\infty}\lambda_n c_n^2. \tag{D.9}$$

Thus, we have been able to reduce our functional integral (D.4) to an integral over the discrete set $\{c_n\}$ as

$$\int_y^x \mathcal{D}q(\tau) \, e^{-S_E[q(\tau)]/\hbar}$$

$$\approx e^{-S_E[q_c]/\hbar} \frac{\mathcal{J}}{\mathcal{N}} \int \ldots \int dc_0 \, dc_1 \ldots dc_n \ldots \exp - \sum_n \frac{\lambda_n c_n^2}{2\hbar}, \quad (D.10)$$

where the variations $\delta q(\tau')$ of the paths $q(\tau')$ have been replaced by the variations of $\{c_n\}$. In (D.10), $\mathcal{J}$ is the Jacobian of the transformation $\delta q(\tau') \to \{c_n\}$ as in the real-time case and $\mathcal{N}$ is the already known normalization factor for functional integrals. Performing this integration, we have

$$\int_y^x \mathcal{D}q(\tau) \, e^{-S_E[q(\tau)]/\hbar} \approx e^{-S_E[q_c]/\hbar} \frac{\mathcal{J}}{\mathcal{N}} \prod_n \left(\frac{2\pi\hbar}{\lambda_n} \right)^{1/2}$$

$$\equiv \frac{1}{\mathcal{N}_R} \left(\frac{1}{\det[-M\partial_\tau^2 + V''(q_c)]} \right)^{1/2} e^{-S_E[q_c]/\hbar}, \quad (D.11)$$

where $\mathcal{N}_R \equiv \mathcal{N}/(\mathcal{J} \prod_n \sqrt{2\pi\hbar})$ and $\prod_n \lambda_n$ is obviously the determinant of the operator $\delta^2 S_E / \delta q(\tau') \, \delta q(\tau'')|_{q_c}$. Despite its elegant and compact form, (D.11) must be analyzed carefully for each specific case in which we are interested.

D.1 Harmonic approximation

Let us now specialize (D.11) to the cases where the potential $V(q)$ has only one isolated non-degenerate minimum, say at $q = 0$, and is bounded below. Moreover, suppose we want to compute only the matrix element $\rho(0, 0, \beta)$ by evaluating that functional integral. Then, the only possible solution of (D.5) with the appropriate boundary condition is a constant solution $q_c(\tau) = 0$. The value assumed by the classical Euclidean action is $S_E[q_c] = 0$. We are then left with the computation of the prefactor of (D.11). This can be done by computing the determinant of the operator $-M\partial_\tau^2 + V''(0)$ and the normalization factor $\mathcal{N}_R$ directly, or by explicitly using the fact that, when $V''(0) \equiv M\omega^2 \neq 0$, (D.1) is the equilibrium density operator of a harmonic oscillator of frequency ω (see (6.7)) at $x = y = 0$. In the limit of large $\hbar\beta$ this reads

$$\frac{1}{\mathcal{N}_R} \left(\frac{1}{\det[-M\partial_\tau^2 + M\omega^2]} \right)^{1/2} \approx \left(\frac{M\omega}{\pi\hbar} \right)^{1/2} \exp - \frac{\hbar\omega\beta}{2}, \quad (D.12)$$

which inserted in (D.4) allows us to identify $|\psi_n(0)|^2 = (M\omega/\pi\hbar)^{1/2}$ and $E_0 = \hbar\omega/2$ as expected.

D.2 Bistable potential

Now we want to treat the problem of a particle in a bistable potential such as the one in Fig. 9.1. In order to simplify our notation we shall write the positions of the minima as $\pm q_0/2 = \pm a$.

In this case it is very easy to see that there is a solution of (D.5) that interpolates between $-a$ and a (instantons) or between a and $-a$ (anti-instantons) within the imaginary time interval $\hbar\beta \to \infty$. These are solutions of zero Euclidean energy in the inverted potential of Fig. 9.2 and turn out to have finite action, which will be important in the evaluation of the path integral (D.1) if we wish to compute the matrix elements $\rho(a, a) = \rho(-a, -a)$ and $\rho(a, -a) = \rho(-a, a)$ of the density operator in that expression. The instanton (and also the anti-instanton) solution is easily obtained by integration of the equation

$$\frac{1}{2}M\dot{q}_c^2 - V(q_c) = 0, \tag{D.13}$$

where $q_c(-\hbar\beta/2) = -a$ and $q_c(\hbar\beta/2) = a$.

This solution rests at $-a$ for a long time and about a given instant (the center of the instanton) makes a quick excursion to a, where it rests again for another long period. The anti-instanton does the same, only exchanging a for $-a$. Since these solutions contain quick excursions from one maximum (of the inverted potential) to the other, we can imagine that a dilute gas of instantons followed by anti-instantons would also be a solution of (D.5) and, therefore, in order to evaluate matrix elements such as $\rho(a, a) = \rho(-a, -a)$ we would need an even number of objects (excursions) whereas for $\rho(a, -a) = \rho(-a, a)$ we need an odd number.

So, if we want to evaluate (D.11) we need to consider three points.

(a) If the action due to a single object is B, the action due to N of them will be NB. The computation of B is quite straightforward. As the instanton (or anti-instanton) is a solution of zero Euclidean energy (see (D.13)), it allows us to write

$$S_E[q_c] = \int_{-\infty}^{\infty} d\tau' \left\{ \frac{1}{2}M\dot{q}_c^2 + V(q_c) \right\} = \int_{-\infty}^{\infty} d\tau' \, M\dot{q}_c^2 \equiv B \tag{D.14}$$

and then $NB/\hbar$ will be the exponent of (D.11).

(b) The term involving the second functional derivative of $S_E[q_c]$ can also be evaluated in a simple way for N objects. Since the solution differs from the constant value at either maxima only for those rapid excursions from one to the other, let us assume the prefactor in (D.11) can be written as

$$\frac{1}{\mathcal{N}_R}\left(\frac{1}{\det[-M\partial_\tau^2 + V''(q_c)]}\right)^{1/2} = \frac{K^N}{\mathcal{N}_R}\left(\frac{1}{\det[-M\partial_\tau^2 + M\omega^2]}\right)^{1/2}$$

$$= K^N\left(\frac{M\omega}{\pi\hbar}\right)^{1/2}\exp{-\frac{\hbar\omega\beta}{2}}, \quad \text{(D.15)}$$

where the term involving the determinant on the r.h.s. is due to trivial solutions $q_c = \pm a$ and $M\omega^2 \equiv V''(\pm a)$. Notice that we have used our previous result (D.12).

(c) Finally, as there is no restriction on the N instants $\bar{\tau}_i$ when the ith excursion takes place, we must integrate over all of them, or

$$\lim_{\beta\to\infty}\int_{-\hbar\beta/2}^{\hbar\beta/2} d\bar{\tau}_1 \int_{-\hbar\beta/2}^{\bar{\tau}_1} d\bar{\tau}_2 \cdots \int_{-\hbar\beta/2}^{\bar{\tau}_{N-1}} d\bar{\tau}_N = \lim_{\beta\to\infty}\frac{(\hbar\beta)^N}{N!}. \quad \text{(D.16)}$$

Now, collecting all these points we get, from (D.1),

$$\rho(-a, -a, \beta) = \rho(a, a, \beta) = \left(\frac{M\omega}{\pi\hbar}\right)^{1/2}\exp{-\frac{\hbar\omega\beta}{2}}\sum_{\text{even }N}\frac{(Ke^{-B/\hbar}\hbar\beta)^N}{N!}$$

$$\text{(D.17)}$$

whereas $\rho(a, -a, \beta) = \rho(-a, a, \beta)$ is given by the same expression summed over odd N. Performing the summations we get

$$\rho(\pm a, -a, \beta) = \left(\frac{M\omega}{\pi\hbar}\right)^{1/2}\exp{-\frac{\hbar\omega\beta}{2}}$$

$$\times\frac{1}{2}\left[\exp(Ke^{-B/\hbar}\hbar\beta) \mp \exp(-Ke^{-B/\hbar}\hbar\beta)\right]. \quad \text{(D.18)}$$

Comparing these expressions to (D.3) for the two lowest-energy eigenstates of H, $|\psi_E\rangle \equiv |+\rangle$ and $|\psi_O\rangle \equiv |-\rangle$, we get

$$E_\pm = \frac{\hbar\omega}{2} \mp \hbar Ke^{-B/\hbar} \quad \text{(D.19)}$$

where $E_\pm$ denote the energy eigenvalues corresponding to the above-defined eigenstates. We can also identify

$$|\langle+|\pm a\rangle|^2 = |\langle-|\pm a\rangle|^2 = \langle+|-a\rangle\langle a|+\rangle = -\langle-|-a\rangle\langle a|-\rangle = \frac{1}{2}\left(\frac{M\omega}{\pi\hbar}\right)^{1/2}.$$

$$\text{(D.20)}$$

The only thing left to be done is the determination of K. For this, we are going to study (D.11) for a single instanton. To start, let us notice that $\dot{q}_c(\tau)$ is an eigenstate of $-M\partial_\tau^2 + V''(q_c)$ with zero eigenvalue. This is easily seen if we derive (D.5) with respect to τ, which yields

$$- M\partial_\tau \ddot{q}_c + V''(q_c)\dot{q}_c = -M\partial_\tau^2 \dot{q}_c + V''(q_c)\dot{q}_c = 0, \qquad (D.21)$$

confirming what we have just said. As the determinant of this operator appears in the denominator of (D.11), we apparently have a problem with the existence of this eigenvalue. We will show next that this is not the case.

Let us take the function q_0 in (D.8) as the one with zero eigenvalue. Notice that it also happens to be the ground state of this operator, since it is a nodeless function. In this case, due to the normalization of the q'_ns, we have

$$q_0(\tau) = \frac{\dot{q}_c(\tau)}{||\dot{q}_c||} = \left(\frac{B}{M}\right)^{1/2} \dot{q}_c(\tau), \qquad (D.22)$$

where we have used (D.14). The variation of any function $q(\tau')$ in the direction $q_0(\tau')$ can be written as

$$\delta q(\tau') = dc_0\, q_0(\tau') = \frac{dq_c(\tau')}{d\tau'}\, d\bar{\tau}. \qquad (D.23)$$

Inserting (D.22) into (D.23) we conclude that

$$\frac{dc_0}{\sqrt{2\pi\hbar}} = \left(\frac{B}{2\pi\hbar M}\right)^{1/2} d\bar{\tau}, \qquad (D.24)$$

and therefore the integration over $\delta c_0/\sqrt{2\pi\hbar}$ is nothing but the integration over the center of a single instanton. Fortunately, this integration has been performed explicitly in (D.16) for $N = 1$. Thus, from the definition of K in (D.15) and the one instanton contribution for (D.17), we can suppress the zero eigenvalue from the evaluation of the determinant and multiply it by $(B/2\pi\hbar M)^{1/2}$. The expression for K finally reads

$$K = \left(\frac{B}{2\pi M\hbar}\right)^{1/2} \left(\frac{\det(-M\partial_\tau^2 + M\omega^2)}{\det'(-M\partial_\tau^2 + V''(q_c))}\right)^{1/2}, \qquad (D.25)$$

where det′ means that the zero eigenvalue must be omitted from the evaluation of the ratio of determinants.

D.3 Metastable potential

Now we want to study the problem of the decay of a particle initially placed at the minimum of a cubic potential such as the one in Fig. 8.1 via quantum mechanical tunneling.

In order to compute the tunneling rate of a particle in this potential, we need to establish a method that allows us to compute the imaginary part of the energy of an approximate eigenstate within the metastable well. In this way

$$\psi(q, t) \propto e^{-i(E_R + i E_I)t/\hbar}, \tag{D.26}$$

which implies that (for $E_I < 0$)

$$\psi^*(q, t)\psi(q, t) \propto e^{-2|E_I|t/\hbar}, \tag{D.27}$$

and then

$$\Gamma = \frac{2|E_I|}{\hbar} \tag{D.28}$$

is the tunneling rate we want to compute. A more refined argument for this hypothesis can be found in Langer (1967).

Our starting point to obtain E_I is again (D.1), from which we have been addressing successfully the low-energy sector of our systems. However, a word of caution is now necessary. Expressions such as (D.3) and (D.4) only make sense for eigenstates and energy eigenvalues of potentials bounded below, which is not the case now. Therefore, we must regard them as expressions depending on a given control parameter ϵ that may change the shape of the potential. Suppose, for instance, that for $\epsilon > 0$ the potential is well-behaved and bounded below. Therefore, we can safely implement our path integral approach and interpret the results as before. However, as we change ϵ to negative values, suppose the global minimum of the potential becomes only local and the function is no longer bounded below. Usually, the resulting expressions from (D.3) or (D.4) do not make sense unless we perform an appropriate analytical extension of them for negative values of ϵ. That is exactly what we do in the following.

The procedure to solve the present problem is very similar to what we have done for the bistable potential. On top of having the trivial constant solution $q_c(\tau) = 0$ to (D.5), there is another one that stays at the local maximum $q = 0$ (in the inverted potential in Fig. 8.2) for a very long imaginary time, makes a rapid excursion to q_0 at $\bar{\tau}$, and returns to the original configuration. The whole motion lasts $\hbar\beta \to \infty$. This is the so-called bounce solution. These are also solutions of zero Euclidean energy in the inverted potential (Fig. 8.2), and turn out to have finite action. Therefore, we can compute their contribution to the path integral (D.1) through the evaluation of the matrix elements $\rho(0, 0)$ of the density operator in that expression. The bounce is easily obtained by integration of equation (D.13) with the boundary condition $q_c(\pm\hbar\beta/2) = 0$.

Contrary to what we have seen for the bistable potential, where there were instantons and anti-instantons, the bounce is symmetric about its center $\bar{\tau}$ and it is the only non-trivial solution there is for (D.11). Actually, since the excursions to

q_0 are very rapid, N widely spaced bounces will also be a solution of (D.11) with the appropriate boundary conditions.

So, in order to evaluate (D.11) for $x = y = 0$, we need to consider again the three essential points of the previous section. These still hold here and therefore expressions such as (D.14), (D.15), and (D.16) can be used in our present case with the only difference that instantons and anti-instantons are now replaced by bounces due to the new boundary conditions.

Using our conclusions from items (a), (b), and (c) of the previous section we can compute

$$\rho(0, 0, \beta) = \left(\frac{M\omega}{\pi\hbar}\right)^{1/2} \exp -\frac{\hbar\omega\beta}{2} \sum_N \frac{(Ke^{-B/\hbar}\hbar\beta)^N}{N!}$$

$$= \left(\frac{M\omega}{\pi\hbar}\right)^{1/2} \exp -\frac{\hbar\omega\beta}{2} \exp(Ke^{-B/\hbar}\hbar\beta), \qquad (D.29)$$

where now $M\omega^2 \equiv V''(0)$.

Comparing this expression with (D.4) for $x = y = 0$ we get

$$E_0 = \frac{\hbar\omega}{2} - \hbar Ke^{-B/\hbar}, \qquad (D.30)$$

from which we realize that the only effect of the bounce solution is to introduce an exponentially small correction to the metastable state in the potential well otherwise treated within the harmonic approximation. However, as we will see below, it is exactly this correction that will account for the decay by quantum mechanical tunneling.

Following our steps of the previous section, let us proceed with the determination of K. Once again we are going to study (D.11), but this time for a single bounce. The fact that $\dot{q}_c(\tau)$ is an eigenstate of $-M\partial_\tau^2 + V''(q_c)$ with zero eigenvalue is demonstrated as in (D.21). The difference now is that $\dot{q}_c(\tau)$ has one node since $q_c(\tau)$ is symmetric with respect to $\bar{\tau}$. Therefore, there must be a nodeless eigenfunction in (D.8) (the ground state of the second variation operator) with negative eigenvalue which renders (D.10) divergent. So, we now have two dangerous directions in function space corresponding to a zero and a negative eigenvalue of the second variation operator.

Let us take the functions $q_1(\tau)$ and $q_0(\tau)$ in (D.8) as the eigenfunctions with zero and negative eigenvalue, respectively. The treatment of the translationally invariant mode (zero eigenvalue function) follows step-by-step that of our previous section with the replacement of the former $q_0(\tau)$ by the new $q_1(\tau)$ and the final conclusion of omitting the zero eigenvalue from the determinant in (D.8) also applies here.

Then we are left with the problem of the negative eigenvalue and, as we anticipated at the beginning of this section, we shall appeal for analytical extension techniques in order to overcome it.

As the only direction in the function space still causing us problems is $q_0(\tau)$, let us parametrize a path in this space, $f_z(\tau)$, in such a way that $f_0(\tau) = 0$, $f_1(\tau) = q_c(\tau)$, and $\partial f_z/\partial z|_{z=1} = q_0(\tau)$ and study the integral

$$I \equiv \int \frac{dz}{\sqrt{2\pi\hbar}} \exp -\frac{S_E(z)}{\hbar} \tag{D.31}$$

for potentials like the one in Fig. 8.1. For any value $z > z_0$ ($S_E(z_0) = 0$) we have $S_E(z) < 0$. Besides, we know that $S_E(1) = B$ is a maximum of the Euclidean action because $d^2 S_E(1)/dz^2 < 0$. Therefore, in order to evaluate (D.31), we need to deform the contour of integration following the steepest descent of $S_E(z)$ in the complex z plane. As we are only interested in the imaginary part of I, we can apply the saddle point method to get

$$\text{Im}\, I = \int_1^{1+i\infty} \frac{dz}{\sqrt{2\pi\hbar}} \exp -\frac{S_E(1)}{\hbar} \exp -\frac{S_E''(1)}{2\hbar}(z-1)^2$$

$$= \frac{|S_E''(1)|^{-1/2}}{2} \exp -\frac{S_E(1)}{\hbar}. \tag{D.32}$$

The factor $1/2$ comes from the integration of only half a Gaussian peak. An important fact about (D.32) is that $\text{Im}\, I$ depends only on the action and its second derivative at $z = 1$ or, in other words, at the bounce. Extending this result to the function space we have

$$\text{Im} \int \mathcal{D}q(\tau) \exp -\frac{S_E[q(\tau)]}{\hbar}$$

$$= \frac{1}{2\mathcal{N}_R} \left(\frac{B}{2\pi M\hbar}\right)^{1/2} \times \hbar\beta \left|\frac{1}{\det'[-M\partial_\tau^2 + V''(q_c)]}\right|^{1/2} \exp -\frac{B}{\hbar}. \tag{D.33}$$

As (D.33) was computed for a single bounce, we can use (D.15) to write

$$\text{Im} \int \mathcal{D}q(\tau) \exp -\frac{S_E[q(\tau)]}{\hbar} = \frac{\hbar\beta}{\mathcal{N}_R} \text{Im}\, K \left|\frac{1}{\det[-M\partial_\tau^2 + M\omega^2)]}\right|^{1/2} \exp -\frac{B}{\hbar}, \tag{D.34}$$

and consequently

$$\mathrm{Im} K = \frac{1}{2}\left(\frac{B}{2\pi M\hbar}\right)^{1/2}\left|\frac{\det[-M\partial_\tau^2 + M\omega^2]}{\det'[-M\partial_\tau^2 + V''(q_c)]}\right|^{1/2}. \qquad (D.35)$$

Finally, substituting (D.35) in (D.30) and using (D.28), we have

$$\Gamma = \left(\frac{B}{2\pi M\hbar}\right)^{1/2}\left|\frac{\det[-M\partial_\tau^2 + M\omega^2]}{\det'[-M\partial_\tau^2 + V''(q_c)]}\right|^{1/2}\exp-\frac{B}{\hbar}. \qquad (D.36)$$

References

Affleck, I. 1981. Quantum-statistical metastability. *Phys. Rev. Lett.*, **46**, 388–391.

Akhiezer, A. I., Bar'yaktar, V. G., and Peletminskii, S. V. 1968. *Spin Waves*. Amsterdam: North-Holland.

Ambegaokar, V. 1991. Quantum Brownian motion and its classical limit. *Ber. Bunsenges. Phys. Chem.*, **95**(3), 400–404.

Anderson, P. W. 1963. Plasmons, gauge invariance, and mass. *Phys. Rev.*, **130**, 439–442.

Anderson, P. W. 1966. Considerations on the flow of superfluid helium. *Rev. Mod. Phys.*, **38**, 298–310.

Ankerhold, J. 2007. *Quantum Tunneling in Complex Systems: The Semiclassical Approach*. Springer Tracts in Modern Physics. Berlin: Springer-Verlag.

Ashcroft, N. W. and Mermin, N. D. 1976. *Solid State Physics*. Austin, TX: Holt, Rinehart and Winston.

Awschalom, D. D., Smyth, J. F., Grinstein, G., DiVincenzo, D. P., and Loss, D. 1992. Macroscopic quantum tunneling in magnetic proteins. *Phys. Rev. Lett.*, **68**, 3092–3095.

Baranger, M., de Aguiar, M. A. M., Keck, F., Korsch, H. J., and Schellhaa, B. 2001. Semiclassical approximations in phase space with coherent states. *J. Phys. A*, **34**, 7227–7286.

Bardeen, J. and Stephen, M. J. 1965. Theory of the motion of vortices in superconductors. *Phys. Rev.*, **140**, A1197–A1207.

Barone, P. M. V. B. and Caldeira, A. O. 1991. Quantum mechanics of the radiation damping. *Phys. Rev. A*, **43**, 57–63.

Bertotti, G. 1998. *Hysteresis in Magnetism: For Physicists, Materials Scientists, and Engineers*. Electromagnetism Series. New York: Academic Press.

Bjorken, J. D. and Drell, S. D. 1964. *Relativistic Quantum Mechanics (Relativistic Quantum Fields)*. International Series in Pure and Applied Physics. New York: McGraw-Hill.

Blatter, G., Feigel'man, M. V., Geshkenbein, V. B., Larkin, A. I., and Vinokur, V. M. 1994. Vortices in high-temperature superconductors. *Rev. Mod. Phys.*, **66**, 1125–1388.

Braun, H.-B., Kyriakidis, J., and Loss, D. 1997. Macroscopic quantum tunneling of ferromagnetic domain walls. *Phys. Rev. B*, **56**, 8129–8137.

Brazovskii, S. and Nattermann, T. 2004. Pinning and sliding of driven elastic systems: From domain walls to charge density waves. *Adv. Phys.*, **53**(2), 177–252.

Breuer, H. P. and Petruccione, F. 2002. *The Theory of Open Quantum Systems*. New York: Oxford University Press.

Brooke, J., Rosenbaum, T. F., and Aeppli, G. 2001. Tunable quantum tunnelling of magnetic domain walls. *Nature*, **413**, 610–613.

Brune, M., Hagley, E., Dreyer, J., Maître, X., Maali, A., Wunderlich, C., Raimond, J. M., and Haroche, S. 1996. Observing the progressive decoherence of the "meter" in a quantum measurement. *Phys. Rev. Lett.*, **77**, 4887–4890.

Caldeira, A. O. and Castro Neto, A. H. 1995. Motion of heavy particles coupled to fermionic and bosonic environments in one dimension. *Phys. Rev. B*, **52**, 4198–4208.

Caldeira, A. O. and Leggett, A. J. 1981. Influence of dissipation on quantum tunneling in macroscopic systems. *Phys. Rev. Lett.*, **46**, 211–214.

Caldeira, A. O. and Leggett, A. J. 1983a. Quantum tunnelling in a dissipative system. *Ann. Phys.*, **149**, 374–456.

Caldeira, A. O. and Leggett, A. J. 1983b. Path integral approach to quantum Brownian motion. *Physica A*, **121**(3), 587–616.

Caldeira, A. O. and Leggett, A. J. 1985. Influence of damping on quantum interference: An exactly soluble model. *Phys. Rev. A*, **31**, 1059–1066.

Caldeira, A. O., Cerdeira, H. A., and Ramaswamy, R. 1989. Limits of weak damping of a quantum harmonic oscillator. *Phys. Rev. A*, **40**, 3438–3440.

Callan, C. G. and Coleman, S. 1977. Fate of the false vacuum. II. First quantum corrections. *Phys. Rev. D*, **16**, 1762–1768.

Castella, H. and Zotos, X. 1993. Exact calculation of spectral properties of a particle interacting with a one-dimensional fermionic system. *Phys. Rev. B*, **47**, 16186–16193.

Castro Neto, A. H. and Caldeira, A. O. 1990. Quantum dynamics of an electromagnetic mode in a cavity. *Phys. Rev. A*, **42**, 6884–6893.

Castro Neto, A. H. and Caldeira, A. O. 1992. Alternative approach to the dynamics of polarons in one dimension. *Phys. Rev. B*, **46**, 8858–8876.

Castro Neto, A. H. and Caldeira, A. O. 1993. Transport properties of solitons. *Phys. Rev. E*, **48**, 4037–4043.

Castro Neto, A. H. and Fisher, M. P. A. 1996. Dynamics of a heavy particle in a Luttinger liquid. *Phys. Rev. B*, **53**, 9713–9718.

Chang, L.-D. and Chakravarty, S. 1984. Quantum decay in a dissipative system. *Phys. Rev. B*, **29**, 130–137.

Chiorescu, I., Nakamura, Y., Harmans, C. J. P. M., and Mooij, J. E. 2003. Coherent quantum dynamics of a superconducting flux qubit. *Science*, **299**(5614), 1869–1871.

Chudnovsky, E. M. and Gunther, L. 1988a. Quantum theory of nucleation in ferromagnets. *Phys. Rev. B*, **37**, 9455–9459.

Chudnovsky, E. M. and Gunther, L. 1988b. Quantum tunneling of magnetization in small ferromagnetic particles. *Phys. Rev. Lett.*, **60**, 661–664.

Chudnovsky, E. M., Iglesias, O., and Stamp, P. C. E. 1992. Quantum tunneling of domain walls in ferromagnets. *Phys. Rev. B*, **46**, 5392–5404.

Clarke, J., Cleland, A. N., Devoret, M. H., Esteve, D., and Martinis, J. M. 1988. Quantum mechanics of a macroscopic variable: The phase difference of a Josephson junction. *Science*, **239**, 992–997.

Coffey, W., Kalmykov, Y. P., and Waldron, J. T. 1996. *The Langevin Equation: With Applications in Physics, Chemistry, and Electrical Engineering*. World Scientific Series in Contemporary Chemical Physics. Singapore: World Scientific.

Coleman, S. 1977. Fate of the false vacuum: Semiclassical theory. *Phys. Rev. D*, **15**, 2929–2936.

Coleman, S. 1988. *Aspects of Symmetry: Selected Erice Lectures*. Cambridge: Cambridge University Press.

d'Agliano, E. G., Kumar, P., Schaich, W., and Suhl, H. 1975. Brownian motion model of the interactions between chemical species and metallic electrons: Bootstrap derivation and parameter evaluation. *Phys. Rev. B*, **11**, 2122–2143.

De Gennes, P. G. 1999. *Superconductivity of Metals and Alloys*. Advanced Book Classics. Advanced Book Program. Cambridge: Perseus Books.

DiVincenzo, D. P. 2000. The physical implementation of quantum computation. *Forts. Phys.*, **48**(9–11), 771–783.

Dorsey, A. T., Fisher, M. P. A., and Wartak, M. S. 1986. Truncation scheme for double-well systems with ohmic dissipation. *Phys. Rev. A*, **33**, 1117–1121.

Duarte, O. S. and Caldeira, A. O. 2006. Effective coupling between two Brownian particles. *Phys. Rev. Lett.*, **97**, 250601.

Duarte, O. S. and Caldeira, A. O. 2009. Effective quantum dynamics of two Brownian particles. *Phys. Rev. A*, **80**, 032110.

Eschenfelder, A. H. 1980. *Magnetic Bubble Technology*. Springer Series in Solid-State Sciences. Berlin: Springer-Verlag.

Fetter, A. L. and Walecka, J. D. 2003. *Quantum Theory of Many-Particle Systems*. Dover Books on Physics. Mineola, NY: Dover Publications.

Feynman, R. P. 1998. *Statistical Mechanics: A Set of Lectures*. Advanced Book Classics. Boulder, CO: Westview Press.

Feynman, R. P. and Hibbs, A. R. 1965. *Quantum Mechanics and Path Integrals*. International Series in Pure and Applied Physics. New York: McGraw-Hill.

Feynman, R. P. and Vernon Jr., F. L. 1963. The theory of a general quantum system interacting with a linear dissipative system. *Ann. Phys.*, **24**, 118–173.

Forster, D. 1990. *Hydrodynamic Fluctuations, Broken Symmetry, and Correlation Functions*. Cambridge: Perseus Books.

Frampton, P. H. 1976. Vacuum instability and Higgs scalar mass. *Phys. Rev. Lett.*, **37**, 1378–1380.

Freidkin, E., Riseborough, P., and Hanggi, P. 1986. Decay of a metastable state: A variational approach. *Phys. Rev. B*, **34**, 1952–1955.

Friedman, J. R., Sarachik, M. P., Tejada, J., and Ziolo, R. 1996. Macroscopic measurement of resonant magnetization tunneling in high-spin molecules. *Phys. Rev. Lett.*, **76**, 3830–3833.

Friedman, J. R., Patel, V., Chen, W., Tolpygo, S. K., and Lukens, J. E. 2000. Quantum superposition of distinct macroscopic states. *Nature*, **406**, 43–46.

Furuya, K. and Caldeira, A. O. 1989. Quantum tunnelling in a double sine-Gordon system near the instability. *Physica A*, **154**(2), 289–306.

Gefen, Y., Ben-Jacob, E., and Caldeira, A. O. 1987. Zener transitions in dissipative driven systems. *Phys. Rev. B*, **36**, 2770–2782.

Gelfand, I. M. and Yaglom, A. M. 1960. Integration in functional spaces and its applications in quantum physics. *J. Math. Phys.*, **1**, 48–69.

Gilbert, T. L. 2004. A phenomenological theory of damping in ferromagnetic materials. *IEEE Trans. Mag.*, **40**, 3443–3449.

Gottfried, K. 1966. *Quantum Mechanics*. Advanced Book Classics. New York: Addison-Wesley.

Grabert, H. and Weiss, U. 1984. Thermal enhancement of the quantum decay rate in a dissipative system. *Z. Phys. B*, **56**, 171–183.

Grabert, H., Olschowski, P., and Weiss, U. 1987. Quantum decay rates for dissipative systems at finite temperatures. *Phys. Rev. B*, **36**, 1931–1951.

Greenberger, D. M., Horne, M., and Zeilinger, A. 1989. *Quantum Theory and Conceptions of the Universe*. Dordrecht: Kluwer.

Gross, D. H. E. 1975. Theory of nuclear friction. *Nucl. Phys. A*, **240**(3), 472–484.

Guinea, F. 1984. Friction and particle–hole pairs. *Phys. Rev. Lett.*, **53**, 1268–1271.

Guth, A. H. 1981. Inflationary universe: A possible solution to the horizon and flatness problems. *Phys. Rev. D*, **23**, 347–356.

Hakim, V. and Ambegaokar, V. 1985. Quantum theory of a free particle interacting with a linearly dissipative environment. *Phys. Rev. A*, **32**, 423–434.

Harris, R. A. and Silbey, R. 1983. On the stabilization of optical isomers through tunneling friction. *J. Chem. Phys.*, **78**(12), 7330–7333.

Hedegård, P. and Caldeira, A. O. 1987. Quantum dynamics of a particle in a fermionic environment. *Phys. Scripta*, **35**(5), 609–622.

Hida, K. and Eckern, U. 1984. Quantum dynamics of the sine-Gordon model in the presence of dissipation. *Phys. Rev. B*, **30**, 4096–4098.

Hong, K. and Giordano, N. 1996. Evidence for domain wall tunnelling in a quasi-one dimensional ferromagnet. *J. Phys.: Condens. Matter*, **8**(19), L301.

Huang, K. 1987. *Statistical Mechanics*. New York: Wiley.

Jaynes, E. T. and Cummings, F. W. 1963. Comparison of quantum and semiclassical radiation theories with application to the beam maser. *Proc. IEEE*, **51**(1), 89–109.

Josephson, B. D. 1962. Possible new effects in superconductive tunnelling. *Phys. Lett.*, **1**(7), 251–253.

Kapusta, J. I. and Gale, C. 2006. *Finite-Temperature Field Theory: Principles and Applications*. Cambridge: Cambridge University Press.

Kittel, C. 1987. *Quantum Theory of Solids*. New York: Wiley.

Kittel, C. 2004. *Introduction to Solid State Physics*. New York: Wiley.

Koch, J., Yu, T. M., Gambetta, J., Houck, A. A., Schuster, D. I., Majer, J., *et al.* 2007. Charge-insensitive qubit design derived from the Cooper pair box. *Phys. Rev. A*, **76**, 042319.

Kramers, H. A. 1940. Brownian motion in a field of force and the diffusion model of chemical reactions. *Physica*, **7**, 284–304.

Kurkijärvi, J. 1972. Intrinsic fluctuations in a superconducting ring closed with a Josephson junction. *Phys. Rev. B*, **6**, 832–835.

Landau, L. D. and Lifshitz, E. M. 1935. On the theory of the dispersion of magnetic permeability in ferromagnetic bodies. *Phys. Z. Sowjet*, **8**, 153–169.

Landau, L. D. and Lifshitz, E. M. 1974. *Course of Theoretical Physics: Statistical Physics*. New York: Addison-Wesley.

Langer, J. S. 1967. Theory of the condensation point. *Ann. Phys.*, **41**, 108–157.

Larkin, A. I. 1970. Effect of inhomogeneities on the structure of the mixed state of superconductor. *Sov. Phys. JETP*, **31**(4), 1466–1470.

Larkin, A. I. and Ovchinnikov, Yu. N. 1983. Quantum tunneling with dissipation. *JETP Lett.*, **37**, 382–385.

Larkin, A. I. and Ovchinnikov, Yu. N. 1984. Quantum-mechanical tunneling with dissipation. The pre-exponential factor. *JETP*, **59**, 420–424.

Leggett, A. J. 1980. Macroscopic quantum systems and the quantum theory of measurement. *Progr. Theor. Phys. Suppl.*, **69**, 80–100.

Leggett, A. J. 1984. Quantum tunneling in the presence of an arbitrary linear dissipation mechanism. *Phys. Rev. B*, **30**, 1208–1218.

Leggett, A. J. 2006. *Quantum Liquids: Bose Condensation and Cooper Pairing in Condensed-matter Systems*. Oxford Graduate Texts in Mathematics. Oxford: Oxford University Press.

Leggett, A. J. and Sols, F. 1991. On the concept of spontaneously broken gauge-symmetry in condensed matter physics. *Found. Phys.*, **21**(3), 353–364.

Leggett, A. J., Chakravarty, S., Dorsey, A. T., Fisher, M. P. A., Garg, A., and Zwerger, W. 1987. Dynamics of the dissipative two-state system. *Rev. Mod. Phys.*, **59**, 1–85.

Leibfried, D., Blatt, R., Monroe, C., and Wineland, D. 2003. Quantum dynamics of single trapped ions. *Rev. Mod. Phys.*, **75**, 281–324.

Likharev, K. K. 1986. *Dynamics of Josephson Junctions and Circuits*. Oxford: Taylor & Francis.

Lindblad, G. 1975. Completely positive maps and entropy inequalities. *Commun. Math. Phys.*, **40**, 147–151.

Liu, Y., Haviland, D. B., Glazman, L. I., and Goldman, A. M. 1992. Resistive transitions in ultrathin superconducting films: Possible evidence for quantum tunneling of vortices. *Phys. Rev. Lett.*, **68**, 2224–2227.

London, F. 1961. *Superfluids: Macroscopic Theory of Superconductivity*. Superfluids. Mineola, NY: Dover Publications.

Louisell, W. H. 1990. *Quantum Statistical Properties of Radiation*. Wiley Classics Library. New York: Wiley.

Makhlin, Y., Schön, G., and Shnirman, A. 2001. Quantum-state engineering with Josephson-junction devices. *Rev. Mod. Phys.*, **73**, 357–400.

Martinis, J. M., Nam, S., Aumentado, J., and Urbina, C. 2002. Rabi oscillations in a large Josephson-junction qubit. *Phys. Rev. Lett.*, **89**, 117901.

Mattis, D. C. 1988. *The Theory of Magnetism I: Statics and Dynamics*. Berlin: Springer-Verlag.

Meissner, W. and Ochsenfeld, R. 1933. Ein neuer Effekt bei Eintritt der Supraleitfhigkeit. *Naturwiss.*, **21**, 787.

Merzbacher, E. 1998. *Quantum Mechanics*. New York: Wiley.

Monroe, C., Meekhof, D. M., King, B. E., and Wineland, D. J. 1996. A Schrödinger cat superposition state of an atom. *Science*, **272**(5265), 1131–1136.

Mota, A. C., Juri, G., Visani, P., Pollini, A., Teruzzi, T., Aupke, K., and Hilti, B. 1991. Flux motion by quantum tunneling. *Physica C*, **185–189**, 343–348.

Munro, W. J. and Gardiner, C. W. 1996. Non-rotating-wave master equation. *Phys. Rev. A*, **53**, 2633–2640.

Murray, I. S., Scully, M. O., and Lamb Jr., W. E. 1978. *Laser Physics*. Advanced Book Program. Boulder, CO: Westview Press.

Negele, J. W. and Orland, H. 1998. *Quantum Many-particle Systems*. Advanced Book Classics. Boulder, CO: Westview Press.

Onnes, H. K. 1911. *Leiden Comm. 120b, 122b, 124c*. Technical reports.

Paulsen, C., Sampaio, L. C., Barbara, B., Tucoulou-Tachoueres, R., Fruchart, D., Marchand, A., Tholence, J. L., and Uehara, M. 1992. Macroscopic quantum tunnelling effects of Bloch walls in small ferromagnetic particles. *Europhys. Lett*, **19**(7), 643.

Pechukas, P. 1994. Reduced dynamics need not be completely positive. *Phys. Rev. Lett.*, **73**, 1060–1062.

Rabi, I. I. 1937. Space quantization in a gyrating magnetic field. *Phys. Rev.*, **51**, 652–654.

Raimond, J. M., Brune, M., and Haroche, S. 2001. Manipulating quantum entanglement with atoms and photons in a cavity. *Rev. Mod. Phys.*, **73**, 565–582.

Rajaraman, R. 1987. *Solitons and Instantons: An Introduction to Solitons and Instantons in Quantum Field Theory*. North-Holland Personal Library. Amsterdam: North-Holland.

Ramsey, N. F. 1950. A molecular beam resonance method with separated oscillating fields. *Phys. Rev.*, **78**, 695–699.

Reichl, L. E. 2009. *A Modern Course in Statistical Physics*. Physics Textbook. New York: Wiley.

Reif, F. 1965. *Statistical and Thermal Physics*. New York: McGraw-Hill.

Rigetti, C., Gambetta, J. M., Poletto, S., Plourde, B. L. T., Chow, J. M., Córcoles, A. D., *et al.* 2012. Superconducting qubit in a waveguide cavity with a coherence time approaching 0.1 ms. *Phys. Rev. B*, **86**, 100506.

Rosenau da Costa, M., Caldeira, A. O., Dutra, S. M., and Westfahl, H. 2000. Exact diagonalization of two quantum models for the damped harmonic oscillator. *Phys. Rev. A*, **61**, 022107.

Sangregorio, C., Ohm, T., Paulsen, C., Sessoli, R., and Gatteschi, D. 1997. Quantum tunneling of the magnetization in an iron cluster nanomagnet. *Phys. Rev. Lett.*, **78**, 4645–4648.

Santos, A. C. 2013. Decoerência de pacote gaussiano num oscilador harmônico. MSc thesis, Universidade Estadual de Campinas (in Portuguese).

Schön, G. and Zaikin, A. D. 1990. Quantum coherent effects, phase transitions, and the dissipative dynamics of ultra small tunnel junctions. *Phys. Rep.*, **198**, 237–412.

Schrödinger, E. 1935. Die gegenwärtige Situation in der Quantenmechanik. *Naturwiss.*, **23**, 807–812; 823–828; 844–849.

Schulman, L. S. 1981. *Techniques and Applications of Path Integration*. New York: Wiley-Interscience.

Sefrioui, Z., Arias, D., Morales, F., Varela, M., León, C., Escudero, R., and Santamaria, J. 2001. Evidence for vortex tunnel dissipation in deoxygenated $YBa_2Cu_3O_{6.4}$ thin films. *Phys. Rev. B*, **63**, 054509-1–4.

Sessoli, R., Gatteschi, D., Caneschi, A., and Novak, M. A. 1993. Magnetic bistability in a metal–ion cluster. *Nature*, **365**, 141–143.

Shpyrko, O. G., Isaacs, E. D., Logan, J. M., Feng, Y., Aeppli, G., Jaramillo, R., *et al.* 2007. Direct measurement of antiferromagnetic domain fluctuations. *Nature*, **447**, 68–71.

Slichter, C. P. 1996. *Principles of Magnetic Resonance*. Springer Series in Solid-State Sciences. Berlin: Springer-Verlag.

Smith, C. M., and Caldeira, A. O. 1990. Application of the generalized Feynman–Vernon approach to a simple system: The damped harmonic oscillator. *Phys. Rev. A*, **41**, 3103–3115.

Smith, C. M., Caldeira, A. O., and Blatter, G. 1996. Creep of vortices from columnar defects. *Phys. Rev. B*, **54**, R784–R787.

Smith, C. M., Caldeira, A. O., and Blatter, G. 1997. Relaxation rates of flux-lines and flux-pancakes in the presence of correlated disorder. *Physica C: Supercond.*, **276**(1–2), 42–56.

Stamp, P. C. E., Chudnovski, E. M., and Barbara, B. 1992. Quantum tunnelling of magnetization in solids. *Int. J. Mod. Phys. B*, **6**(9), 1355–1473.

Stone, M. 1976. Lifetime and decay of "excited vacuum" states of a field theory associated with nonabsolute minima of its effective potential. *Phys. Rev. D*, **14**, 3568–3573.

Tafuri, F., Kirtley, J. R., Born, D., Stornaiuolo, D., Medaglia, P. G., Orgiani, P., *et al.* 2006. Dissipation in ultra-thin current-carrying superconducting bridges; evidence for quantum tunneling of Pearl vortices. *Europhys. Lett.*, **73**(6), 948.

Takagi, S. 2002. *Macroscopic Quantum Tunneling*. Cambridge: Cambridge University Press.

Thouless, D. J. 1972. *The Quantum Mechanics of Many-body Systems*. Pure and Applied Physics. New York: Academic Press.

Tinkham, M. 2004. *Introduction to Superconductivity*, 2nd edn. Dover Books on Physics and Chemistry. Mineola, NY: Dover Publications.

Ueda, M. 1996. Transmission spectrum of a tunneling particle interacting with dynamical fields: Real-time functional-integral approach. *Phys. Rev. B*, **54**, 8676–8687.

Valente, D. M. and Caldeira, A. O. 2010. Thermal equilibrium of two quantum Brownian particles. *Phys. Rev. A*, **81**, 012117–1–7.

Venugopalan, A. 1994. Preferred basis in a measurement process. *Phys. Rev. A*, **50**, 2742–2745.

Vion, D., Aassime, A., Cottet, A., Joyez, P., Pothier, H., Urbina, C., *et al.* 2002. Manipulating the quantum state of an electrical circuit. *Science*, **296**(5569), 886–889.

Voss, R. F. and Webb, R. A. 1981. Macroscopic quantum tunneling in 1-μm Nb Josephson junctions. *Phys. Rev. Lett.*, **47**, 265–268.

Wax, N. 2003. *Selected Papers on Noise and Stochastic Processes*. Dover Phoenix Editions. Mineola, NY: Dover Publications.

Waxman, D. 1985. Macroscopic quantum coherence and tunnelling in a double well potential. *J. Phys. C: Solid State Phys.*, **18**(15), L421.

Waxman, D. and Leggett, A. J. 1985. Dissipative quantum tunneling at finite temperatures. *Phys. Rev. B*, **32**, 4450–4468.

Weiss, U. 1999. *Quantum Dissipative Systems*. Series in Modern Condensed Matter Physics. Singapore: World Scientific.

White, R. M. 2007. *Quantum Theory of Magnetism*. Berlin: Springer-Verlag.

Yang, C. N. 1962. Concept of off-diagonal long-range order and the quantum phases of liquid He and of superconductors. *Rev. Mod. Phys.*, **34**, 694–704.

Yu, Y., Han, S., Chu, X., Chu, S.-I., and Wang, Z. 2002. Coherent temporal oscillations of macroscopic quantum states in a Josephson junction. *Science*, **296**(5569), 889–892.

Zurek, W. H. 1991. Decoherence and the transition from quantum to classical. *Physics Today*, **44**, 36–44.

Zurek, W. H. 2003. Decoherence, einselection, and the quantum origins of the classical. *Rev. Mod. Phys.*, **75**, 715–775.

Index

Printed in the United States
By Bookmasters